Maximum Entropy and Bayesian Methods

Fundamental Theories of Physics

An International Book Series on The Fundamental Theories of Physics: Their Clarification, Development and Application

Volume 98

Maximum Entropy and Bayesian Methods

Boise, Idaho, U.S.A., 1997

Proceedings of the 17th International Workshop on Maximum Entropy and Bayesian Methods of Statistical Analysis

edited by

Gary J. Erickson

Joshua T. Rychert

Department of Electronical Engineering,
Boise State University,
Boise, Idaho, U.S.A.

and

C. Ray Smith

Fayetteville, Tennessee, U.S.A.

SPRINGER-SCIENCE+BUSINESS MEDIA, B.V.

A C.I.P. Catalogue record for this book is available from the Library of Congress.

ISBN 978-94-010-6111-7 ISBN 978-94-011-5028-6 (eBook)
DOI 10.1007/978-94-011-5028-6

Printed on acid-free paper

CONTENTS

IN MEMORY OF EDWIN T. JAYNES

With the passing of Edwin Thompson Jaynes on April 30, 1998, his many friends in the MAXENT community and beyond must say good-bye to a very special person.

His openness and unselfishness, his independent and original thought, and his uncompromising high standards have made an indelible impact. His written work was so lucid that it was in and of itself a pleasure to read; his speaking style was every bit as penetrating and intelligible as his writing.

Beyond his prodigious scientific contributions and wisdom, much more could be said about those personal qualities which made Ed's friendship over the years a rare privilege. But as anyone who knew him understands, Ed believed that such matters are by their nature private, and would have been uncomfortable with public profession of the grief which naturally accompanies this loss.

He will be keenly missed.

PREFACE

This volume has its origin in the Seventeenth International Workshop on Maximum Entropy and Bayesian Methods, MAXENT 97. The workshop was held at Boise State University in Boise, Idaho, on August 4 - 8, 1997. As in the past, the purpose of the workshop was to bring together researchers in different fields to present papers on applications of Bayesian methods (these include maximum entropy) in science, engineering, medicine, economics, and many other disciplines.

Thanks to significant theoretical advances and the personal computer, much progress has been made since our first Workshop in 1981. As indicated by several papers in these proceedings, the subject has matured to a stage in which computational algorithms are the objects of interest, the thrust being on feasibility, efficiency and innovation. Though applications are proliferating at a staggering rate, some in areas that hardly existed a decade ago, it is pleasing that due attention is still being paid to foundations of the subject. The following list of descriptors, applicable to papers in this volume, gives a sense of its contents: deconvolution, inverse problems, instrument (point-spread) function, model comparison, multisensor data fusion, image processing, tomography, reconstruction, deformable models, pattern recognition, classification and group analysis, segmentation/edge detection, brain shape, marginalization, algorithms, complexity, Ockham's razor as an inference tool, foundations of probability theory, symmetry, history of probability theory and computability.

MAXENT 97 and these proceedings could not have been brought to final form without the support and help of a number of people. In particular, SCP Global Technologies helped with the realization of MAXENT 97 . The editors, Gary Erickson, Josh Rychert, and Ray Smith, express their gratitude to all the speakers and are appreciative for the additional time and effort authors expended in producing a finished manuscript.

This preface must end on a sad note: Professor E. T. Jaynes died on April 30, 1998, in St. Louis, Missouri.

MASSIVE INFERENCE AND MAXIMUM ENTROPY

JOHN SKILLING
Department of Applied Mathematics and Theoretical Physics
University of Cambridge
England CB3 9EW

Abstract. In data analysis, maximum entropy (MaxEnt) has been used to reconstruct measures *i.e.* positive, additive distributions) from limited data. The MaxEnt prior was originally derived from the "monkey model" in which quanta of uniform intensity could appear randomly in the field of view. To avoid undue digitisation, the quanta had to be small, and this led to difficulties with the Law of Large Numbers, and to unavoidable approximations in computing the posterior. A better way of avoiding digitisation is to give the quanta variable intensity with an exponential prior, that being the natural MaxEnt assignment. We call this technique "Massive Inference" (MassInf). Although the entropy formula no longer appears in the prior, MassInf results show improved quality. MassInf is also capable of assigning a simple prior for polarized images.

Key words: Maximum entropy, infinitely divisible, polarization, regularization

1. History of maximum entropy

As presented by Jaynes (1957), the Principle of Maximum Entropy (PME) is a rule for assigning probability distributions: in making inferences on the basis of partial infomation we are to use that probability distribution which has maximum entropy subject to whatever ensemble-average constraints are known. Given some mean values

$$\sum_i R_{ki} p_i = D_k, \quad k = 1, 2, \ldots \tag{1}$$

that include normalisation $\sum p_i = 1$, the probability distribution p is to be assigned by maximising its entropy

$$S = -\sum p_i \log p_i \tag{2}$$

(Shannon, 1948) subject to the given constraints. In statistical mechanics, the entropy S is derivable from the combinatoric number of ways

$$\Omega = N!/\prod n_i! \approx \exp(S), \qquad S(n) = -\sum n_i \log(n_i/N) \tag{3}$$

of dividing an ensemble of N systems into cells with occupation n_i having mean Np_i. Alternatively, if individual mean values m_i are assigned, each occupation number can be given a Poisson distribution with correct mean, leading to

$$\Pr(n) = \prod e^{-m_i} m_i^{n_i} / n_i! \approx \exp(S), \qquad S(n) = \sum (n_i - m_i - n_i \log(n_i/m_i)) \quad (4)$$

The PME amounts to the entirely reasonable prescription of giving equal intrinsic weight to each individual state. When the constraints D include a defining set of physical variables such as energy and volume, the maximum entropy distribution gives accurate predictions of other average quantities with different R. Fluctuations are generally $\mathcal{O}(N^{-1/2})$ small, and deviations larger than this indicate an unacknowledged extra constraint.

It is tempting and productive to apply this successful formalism to data analysis, and this has been done both directly and indirectly. Indirect applications use PME to assign the posterior probability distribution $\Pr(f|D)$ of the quantity f being sought. It is supposed that the data are ensemble-average constraints $D_k = \int R_k(f) \Pr(f) df$, so that the PME becomes applicable. The commonest such application is the derivation of a power spectrum from autocorrelation coefficients D of a time-series f. Actually, data tend to be observations of the particular object being investigated at the time, and the proper analysis is Bayesian. A prior probability $\Pr(f)$ needs to be assigned (by PME or other insight), but it is then not determined by the data, whose rôle is to modulate the prior through the usual likelihood function $\Pr(D|f)$.

For a direct application of MaxEnt, we suppose that we seek the distribution f of some positive, additive quantity such as the intensity of light in an image, or the flux of energy along a spectrum. Mathematicians call such an object a "measure". We then proceed with the "monkey model" (Gull and Daniell, 1978) in which f is identified with a number n of quanta of some unknown but presumed small strength q. With m similarly rescaled, Stirling's approximation for large n yields the Quantified Maximum Entropy (QME) prior

$$\Pr(f) \propto \exp(\alpha S) / \prod f_i^{1/2}, \qquad S(f) = \sum (f_i - m_i - f_i \log(f_i/m_i)) \qquad (5)$$

where m is a set of weights that models any *a priori* non-equivalence of the cells i, and $\alpha = 1/q$ is an unknown but apparently large hyper-parameter. When applying this, α cannot in fact be particularly large, lest the prior dominate the likelihood. Fortunately, the same entropy form can be derived from symmetry arguments which avoid combinatoric modelling (Shore and Johnson, 1980). Related arguments (Skilling, 1989) suggest the QME prior, implicitly assuming that the desirable entropy maximum should yield a useful selection from the posterior distribution.

Were the data observations $D_k = \sum R_{ki} f_i$ to be exact, the PME could be used to assign a corresponding f, just as in statistical mechanics. Even if the data are noisy and thus subsumed into a likelihood function $\Pr(D|f)$, the PME can still be used to assign a particular f from among those with some acceptable fit $\Pr(f) \geq P_0$ to the data. Visually, distributions assigned by maximum entropy are often of high quality and utility. Entropy is a good regulariser.

Yet the reliability of the inferred f cannot be deduced from any single assignment, however distinguished its provenance. To deal with uncertainty, we must use the probability distribution in (5), and not just regularise. In each particular application, α is to be estimated probabilistically (Gull, 1989). The "best" individual result $\widehat{f}$ is then taken to be that which maximises the product $\exp(\alpha S)\,\Pr(D|f)$, so it remains a PME selection with the corresponding visual appeal. Around it lies a probability distribution from which the uncertainty of any inference can be calculated. This methodology has been used with considerable success, but limitations have become apparent.

2. MaxEnt Polarization

Suppose the object f being inferred is the intensity pattern $I(x,y)$ of light across an image. Light can be polarized, most simply into potentially independent linear channels $X(x,y)$ and $Y(x,y)$ that sum to $I = X + Y$. Each individual polarization represents an observable spatial pattern to which the symmetries of QME plausibly apply, so that it is natural to assign QME priors to each of X and Y:

$$\Pr(X) \propto e^{\alpha(S(X))}, \quad \Pr(Y) \propto e^{\alpha(S(Y))} \tag{6}$$

The prior on the total I is then determined through

$$\Pr(I) = \int\!\!\int \delta(I - X - Y)\Pr(X)\Pr(Y)\,dX\,dY \tag{7}$$

This does **not** evaluate to a QME form for I (even if the value of α differs), even though it would have been just as natural to insist on the latter assignment. Worse, if the prior on I is assigned by QME, it becomes impossible to find **any** subsidiary probability distribution that can be assigned to X and Y to yield the combination I. A subsidiary distribution can be found numerically, but parts of it are negative, which is impossible.

3. MaxEnt Pixellation

The QME prior is defined over some mesh of M cells that must be fixed in advance. It can happen that no choice of M allows a sensible reconstruction. For example, suppose we wish to infer a distribution $f(x)$ from inexact data on the lowest-order moments

$$\mu_0 = \int_0^1 f(x)\,dx, \quad \mu_1 = \int_0^1 x f(x)\,dx \tag{8}$$

The range $x \in [0,1]$ is to be divided into M equal cells, within each of which f will be approximated by a single central value. Specifically, let the likelihood function be

$$\Pr(\mathrm{Data}\,|\,f) = e^{-64\mu_1}(e^{-\mu_0} - e^{-2\mu_0})^8 \tag{9}$$

In detail, this form is chosen to enable a M-dimensional probability integral to be computed without approximation as the binomially-expanded sum of 9 integrals each of which is M-fold separable. In broad outline, the likelihood factors indicate a measurement of mean location

$$\langle x \rangle = \mu_1/\mu_0 = 0.022 \pm 0.025 \tag{10}$$

along with normalisation $\mu_0 \approx 1$. Clearly, at least 40 or so cells are needed in order to make such a small mean x visible.

Unfortunately, the law of large numbers comes into play as the number of cells increases, and the prior forces $\langle x \rangle$ ever closer to $\frac{1}{2}$. Specifically, the prior variance is

$$\text{var}(x) \leq \frac{1}{6M} \quad \forall \alpha \tag{11}$$

Consequently, when M is 40, the "measurement" of $\langle x \rangle$ has already become an improbable 7σ outlier, according to the prior. The posterior fails to cover it well and the prior predictive "evidence" value becomes small. In fact, the favoured value of M is 4, which is much too coarse to fit the data well (see Fig. 1). In our analysis, we should seek a method that behaves sensibly as the division into cells becomes finer. QME does not have this property.

This was not noticed at first because practical computations approximated the probability summit with a Gaussian, and Gaussian distributions do give error bars with sensible limits as the number of cells increases. Unfortunately, the approximation becomes indefinitely bad in this limit because it is being used out to $\mathcal{O}(f^{1/2})$ whereas f cannot decrease by more than itself without going negative, and typical f values are becoming small.

4. Massive Inference (MassInf)

When we seek a measure F, distributed with density f over a domain α of x, both the observed data and any subsequent inferences take the form of integrals

$$F(\alpha) = \int_{x \in \alpha} R(x) f(x)\, dx \tag{12}$$

or, at worst, functions of such integrals. Indeed, the main aim of this branch of data analysis is to use imperfectly observed values of some integrals in order to predict others. An integral over a finite x-domain can be defined as the limit of the sum over its division into cells, as all these M cells become arbitrarily small — it being assumed that the limit exists. Hence the probability distribution for such an integral must be the limit of the convolution of the individual probability distributions $\text{pr}(F_i)$ over the cells: a consistent family of probability distributions over sub-domains is called a "process", and a process that allows an arbitrary degree of sub-division is called "infinitely divisible".

$$F(\alpha) = \lim \sum_{i=1}^{M} F_i, \text{Pr}(F|\alpha) = \lim \int \cdots \int dF_1 \cdots dF_M\, \text{pr}(F_1) \cdots \text{pr}(F_M)\, \delta(F_1 + \cdots + F_M - F) \tag{13}$$

In writing this without extra conditionalities, we have already assumed spatial independence. It is convenient to assume spatial invariance also, so that the prior over any finite domain can be built from a single common prior assigned to an arbitrary microscopic cell. Analytically, the required convolution is conveniently performed by multiplying the Laplace transforms:

$$\widetilde{\Pr}(s|\alpha) = \widetilde{\mathrm{pr}}_1(s)\cdots\widetilde{\mathrm{pr}}_M(s) = \left(\widetilde{\mathrm{pr}}(s)\right)^M \tag{14}$$

In macroscopic units, the mean flux associated with a small cell must be correspondingly small, $\mathcal{O}(M^{-1})$. So must its variance, which also divides equally among the cells so that it becomes $\mathcal{O}(M^{-1})$ small, and the standard deviation considerably exceeds the mean. This indicates that there will be a small probability of having a finite $\mathcal{O}(1)$ flux in a small cell, rather than a substantial $\mathcal{O}(1)$ probability of a small flux: this plausibility argument can be extended into a proof that most of the flux will be tightly localised. Accordingly, the flux in a small cell dx might be defined from a density $\lambda(F)$ through

$$\mathrm{pr}(F|dx) = (1 - \Lambda dx)\delta(F) + \lambda(F)dx + \mathcal{O}(dx)^2 \tag{15}$$

where $\Lambda = \int_0^\infty \lambda(F)dF$ ensures normalisation. Laplace convolution up to a range α induces the macroscopic transform

$$\widetilde{\Pr}(s|\alpha) = \exp\Big(-\int_0^\infty F^{-1}L(F)(1 - e^{-sF})dF\Big) \tag{16}$$

known as the Lévy-Khinchin representation, $L(F) = \alpha F\lambda(F)$ being the Lévy measure. Actually, the latter integral can converge meaningfully to define an infinitely divisible process even if λ is non-integrable: a process can exist at all finite resolutions without the infinitesimal limit necessarily having the form (15). Even so, (16) is restrictive. Many prior distributions, including QME, cannot be written in Lévy-Khinchin form, and consequently cannot be resolved onto arbitrarily small cells. A standard professional reference for this material is Feller (1971).

Table 1 lists the forms that have been found with a reasonably convenient algebraic form, with the available rescaling options in the top line. Form 1 is the trivial special case of a distribution F of rigidly constant density. Form 2 is a basic Poisson distribution in which "atoms" of fixed (unit) flux are placed randomly in x. Form 3 is a Poisson variant in which the atoms have an exponential distribution of flux. Form 4 generalises the Poisson variant 3 to a model in which atoms have $m \geq 1$ exponentially-distributed components (${}_0\mathrm{F}_m$ is a hypergeometric function). Form 5 is a Gamma distribution, particularly popular among statisticians. Form 6 arises in the study of random walks on an integer lattice. Form 7 is known as the Lévy distribution. L being a simple power law, form 7 has the "stable" property (Lukacs, 1970) that self-convolution recovers the same shape. Form 8 is another stable distribution, apparently overlooked previously. Perhaps understandably, form 9 also appears to be absent from the literature ($\mathcal{D}$ is the parabolic cylinder function). Forms 10 to 13 are combinations of the above that happen to have closed form, 11 being a special case of 10. Of these, only the Poisson forms 2, 3 and 4 have the infinitesimal interpretation (15) with integrable λ.

The basic Poisson form 2 is, in fact, the original "monkey model" without any large-n approximation. However, it restricts all fluxes to integer multiples of some

Table 1
Analytic Lévy-Khinchin Representations

	$L(F)$	$\Pr(F\|\alpha) = P(F)$
	$L(\beta F)e^{-\gamma F}$	$P(\beta F)\, e^{-\gamma F} \Big/ \int_0^\infty P(\beta F)\, e^{-\gamma F} dF$
1	$\alpha\delta(F)$	$\delta(F-\alpha)$
2	$\alpha\delta(F-1)$	$e^{-\alpha}\sum_{n=0}^{\infty}\frac{\alpha^n}{n!}\,\delta(F-n)$
3	$\alpha F e^{-F}$	$e^{-\alpha}\left(\delta(F) + e^{-F}\sqrt{\frac{\alpha}{F}}\, I_1\left(2\sqrt{\alpha F}\right)\right)$
4	$\frac{\alpha F^m e^{-F}}{(m-1)!}$	$e^{-\alpha}\left(\delta(F) + \frac{\alpha F^{m-1} e^{-F}}{(m-1)!}\, {}_0\mathrm{F}_m\left(\frac{m+1}{m}, \frac{m+2}{m}, \cdots, 2; \frac{\alpha F^m}{m^m}\right)\right)$
5	αe^{-F}	$\frac{F^{-1+\alpha} e^{-F}}{\Gamma(\alpha)}$
6	$\alpha I_0(F) e^{-F}$	$\frac{\alpha}{F} I_\alpha(F)\, e^{-F}$
7	$\frac{\alpha}{2\sqrt{\pi}} F^{-1/2}$	$\frac{\alpha}{2\sqrt{\pi F^3}}\, e^{-\alpha^2/4F}$
8	$\frac{\alpha\Gamma(1/3)}{2\sqrt{3}\pi} F^{-1/3}$	$\frac{\alpha^{3/2}}{3\pi F^{3/2}}\, K_{\frac{1}{3}}\left(2\sqrt{\alpha^3/27F}\right)$
9	$\alpha e^F \operatorname{erfc}(\sqrt{F})$	$\sqrt{\frac{8}{\pi}}\,\alpha\,(2F)^{\alpha-1}\, e^{F/2}\, \mathcal{D}_{-2\alpha-1}\left(\sqrt{2F}\right)$
10	$\alpha e^{-aF} + \beta e^{-bF}$	$\frac{a^\alpha b^\beta}{\Gamma(\alpha+\beta)} F^{-1+\alpha+\beta}\, e^{-bF}\, M\left(\alpha; \alpha+\beta; (b-a)F\right)$
11	$\alpha e^{-aF} + \alpha e^{-bF}$	$\frac{\sqrt{\pi}(ab)^\alpha}{\Gamma(\alpha)}\left(\frac{F}{b-a}\right)^{\alpha-1/2} e^{-(a+b)F/2}\, I_{\alpha-\frac{1}{2}}\left(\frac{b-a}{2}F\right)$
12	$\alpha e^{-F} + \beta F e^{-F}$	$\left(\frac{F}{\beta}\right)^{(\alpha-1)/2} e^{-\beta-F}\, I_{\alpha-1}(2\sqrt{\beta F})$
13	$\alpha e^{-F} + \beta e^{-F} \Big/ \sqrt{\pi F}$	$\frac{2^\alpha}{\sqrt{2\pi}}\, F^{-1+\alpha}\, e^{2\beta-F-\beta^2/2F}\, \mathcal{D}_{1-2\alpha}\left(\beta\sqrt{2/F}\right)$

unit quantum, whereas we usually require flux to be on a continuous scale. To allow individual fluxes to be continuous, the natural prior distribution for a single atom is exponential

$$\lambda(F) = q^{-1} e^{-F/q} \tag{17}$$

because that is the PME assignment appropriate to a constraint q on the expected flux. This yields the Poisson variant form 3, now seen to be the natural prior distribution for a measure F. It has two hyper-parameters, q for the quantum size, and the Poisson mean number of atoms per unit range x (which could be a function of x). The associated macroscopic distribution over a range on which α

atoms are expected is

$$\Pr(F|\alpha, q) = e^{-\alpha}\Big(\delta(F) + e^{-F/q}\sqrt{\frac{\alpha}{qF}}\, I_1\big(2\sqrt{\alpha F/q}\big)\Big) \tag{18}$$

Interestingly, the delta function ensures that the posterior mode follows the prior mode at $F = 0$: unless the data utterly prohibit it, the most probable individual measure F will be null. This destroys the supposition of QME that the maximum might be a useful selection from the posterior distribution. Instead, the mean of the posterior is used for display purposes. Of itself, the mean lacks the symmetry properties that underlie a MaxEnt selection, but it is a sufficient statistic for reading off the mean value of any integral property of f. To determine deeper information such as uncertainty, the posterior distribution must be recorded more fully, usually as a set of random samples.

Because the flux in any sample is located as a set of "point mass" delta functions, we call (17)/(18) the "Massive Inference" (MassInf) prior. Computationally, the MassInf prior is convenient because each sample of F is fully defined by a finite number of atoms, whilst other forms have flux (mostly very small) everywhere. Also, the exponential prior on each atomic flux is conjugate to the Gaussian likelihood function that is commonly appropriate, so it can be folded in without un-necessary algorithmic complexity.

5. MassInf Pixellation

By construction, MassInf has no difficulty with pixellation. Any finite resolution with a restricted number of cells may damage the results, but inferences will have a well-defined continuum limit as the resolution increases. This is confirmed with the example (9) above, for which the upper curve in Fig. 1 shows the prior predictive "evidence" values for various numbers of cells. As it ought, this increases gently towards the continuum limit, as opposed to the awkward behaviour of QME. Towards this limit, the posterior behaves sensibly.

6. MassInf Polarization

Even if an image has the extra complication of polarization, we can still model it with a Poisson distribution of point atoms, in the spirit of the original monkey model. Instead of simply assigning an exponential prior on a single intensity, though, an atom now has polarized structure on which a prior is needed.

Among the oscillating electric fields E_x and E_y comprising a beam of radiation along z, there can be up to four different correlation coefficients among the in-phase and out-of-phase components. These define four Stokes' parameters I, Q, U, V (Stone, 1963). Of these, I is the (non-negative) total intensity. The other three are smaller and obey

$$P \equiv \sqrt{Q^2 + U^2 + V^2} \leq I \tag{19}$$

Geometrically, they lie within the "Poincaré sphere" of radius I. They can be rotated into each other by harmless coordinate rotation and time offsets, so we

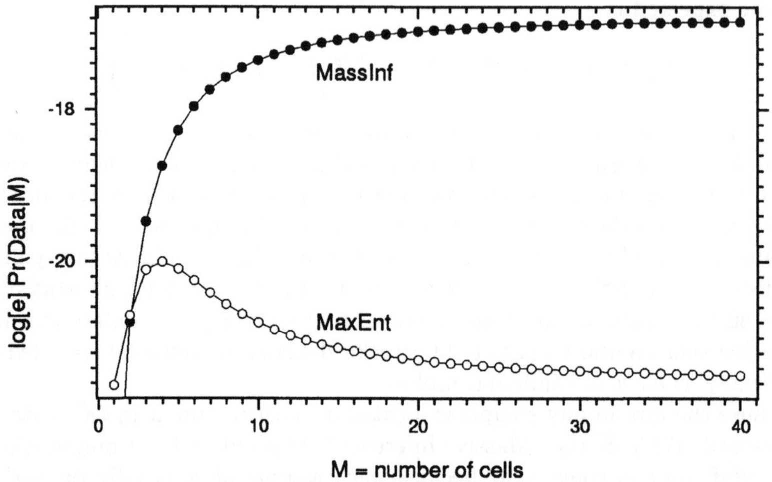

Figure 1. Test comparison of MaxEnt and MassInf "evidence" values (hyper-parameters being given their optimal values).

should assign the prior uniformly over the sphere.

$$\Pr(I, Q, U, V \mid 1 \text{ atom}) = \theta(I, P) \tag{20}$$

Let us restrict our attention to the special case $U = V = 0$. The polarization state now has two channels, linear x with intensity $X = \frac{1}{2}(I + Q)$, and linear y with intensity $Y = \frac{1}{2}(I - Q)$. These states being orthogonal, it is natural to assign exponential priors $\Pr(X) = e^{-X}$ and $\Pr(Y) = e^{-Y}$ to each as in (17), scaling to $q = 1$ for convenience. In terms of I and Q, this sets

$$\Pr(I, Q \mid U = V = 0,\ 1 \text{ atom}) = e^{-I}/2, \qquad |Q| \leq I \tag{21}$$

This implies that $\theta(I, P) \propto e^{-I}$. After normalising to unit total, we reach

$$\Pr(I, Q, U, V \mid 1 \text{ atom}) = \left\{ \begin{array}{cc} e^{-I}/8\pi, & P \leq I \\ 0, & \text{otherwise} \end{array} \right\} \tag{22}$$

With the general prior in hand, we can now investigate special cases.

1. Fixed polarization state: Q/I, U/I, V/I all known.

$$\Pr(I \mid \text{fixed polarization, } 1 \text{ atom}) = e^{-I} \tag{23}$$

Lacking any sub-structure, the atom has a simple exponential prior.

2. Observe linear polarizations from arbitrarily polarized radiation: seek X and Y after marginalizing over U and V.

$$\Pr(X \mid 1 \text{ atom}) = Xe^{-X} \qquad \text{and independently} \qquad \Pr(Y \mid 1 \text{ atom}) = Ye^{-Y} \tag{24}$$

Each of X and Y show the effect of internal structure, as if each was the convolution of two independent exponentially-distributed fluxes. The corresponding macroscopic prior can be found by setting $m = 2$ in form 4 of Table 1.

3. Observe the total intensity from arbitrarily polarized radiation: seek I after marginalizing over Q, U and V.

$$\Pr(I \mid 1 \text{ atom}) = I^3 e^{-I}/6 \tag{25}$$

The intensity appears as if it were the convolution of four independent exponentially-distributed fluxes, though these cannot be identified individually among the four polarization parameters. The corresponding macroscopic prior can be found by setting $m = 4$ in form 4 of Table 1.

7. Examples

We present a simulation and an application to practical data. The simulation is due to Bretthorst (1990) and represents data

$$D_t = 100e^{-0.03t} + 50e^{-0.05t} + e_t, \qquad t = 1, 2, \cdots, 100 \tag{26}$$

where e is unit normally distributed noise. These data (Fig. 2) are to be analyzed to recover the spectrum $f(x)$ of decay rates.

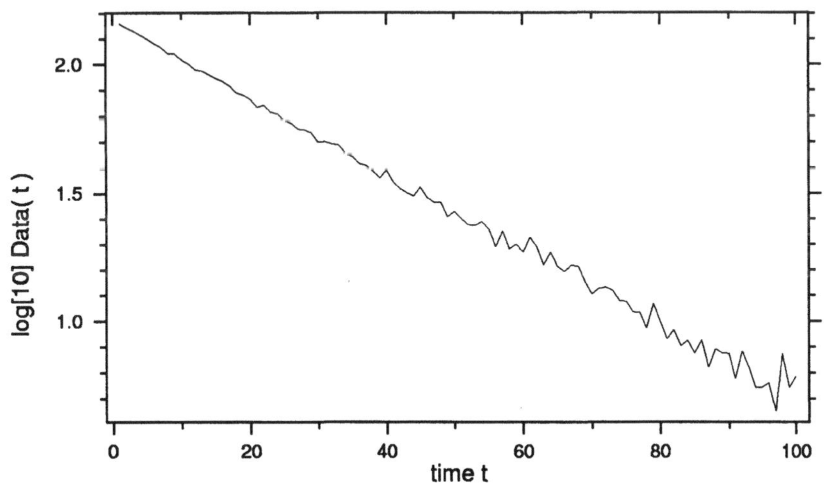

Figure 2. Bretthorst's decaying exponential data.

Bretthorst's original analysis used a Gaussian prior, which loses power because the context supposes that f is a measure that cannot be negative.

The MaxEnt selection (Skilling, 1991) is shown in Fig. 3, where it may be compared with the simulation. Only a single hump is recovered, though the rightward tail might optimistically suggest rightward structure. The mean MassInf result,

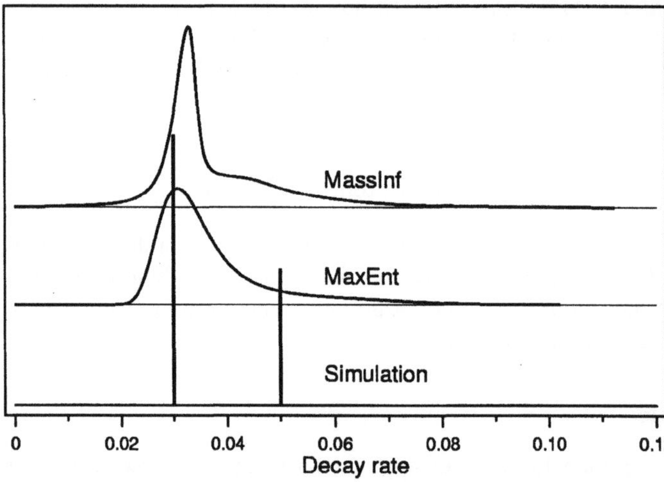

Figure 3. Comparison of MaxEnt with mean MassInf reconstructions from Bretthorst's data.

though, is sharper: the rightward extension almost gives a second local maximum. Fig. 4 shows a selection of the random samples from which the mean MassInf result was accumulated. Although each of these is necessarily sharp and positive, the mean is smooth, and localised only to the degree imposed by the data. In as many samples as not, there are two important components, though sometimes only one dominates. One can easily accumulate the probabilities of having 0, 1, 2, 3,... components.

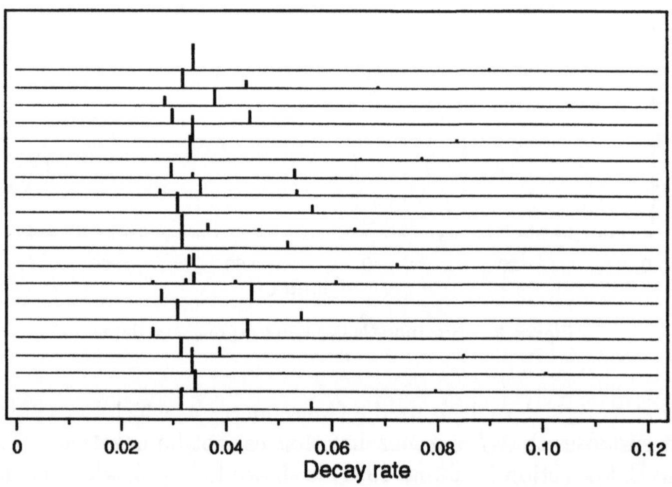

Figure 4. Figure 4: 20 typical MassInf samples with Bretthorst's data.

A reason why the MassInf result is sharper than MaxEnt can be seen in Fig. 5, which shows a selection of random samples from the MaxEnt probability distribution.

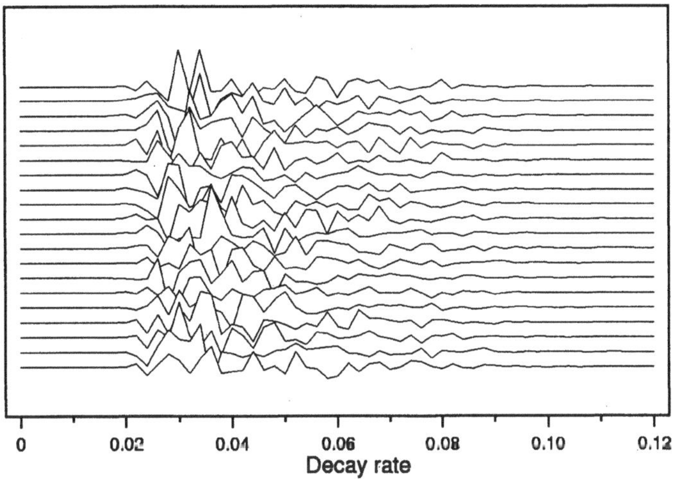

Figure 5. 20 typical QME samples (in Gaussian approximation) with Bretthorst's data.

Technically, these samples are drawn from the Gaussian approximation to the posterior, otherwise the pixellation into finite cells would have induced its bias towards uniformity. These approximated samples are **not** properly required to be positive, so that an important constraint is slackened. In effect, part of the prior information in MaxEnt has been relaxed in order to disguise the problem with pixellation, and this leads to a looser reconstruction. A practical dataset from NMR spectroscopy is shown in Fig. 6. The underlying signals have been blurred by a point-spread-function about 40 units wide (full-width-half-maximum), and there is also noise. Fig. 7 compares the mean MassInf deconvolution with that from MaxEnt. The MassInf result is again sharper and thus higher, and has a cleaner, flatter background.

A reason for this extra clarity can be seen by comparing the samples (Fig. 8 for MassInf and Fig. 9 for MaxEnt). MaxEnt samples can be locally negative, and this has allowed some low-level breakthrough at several locations across the MaxEnt reconstruction. MassInf prohibits such negatives, giving tighter control.

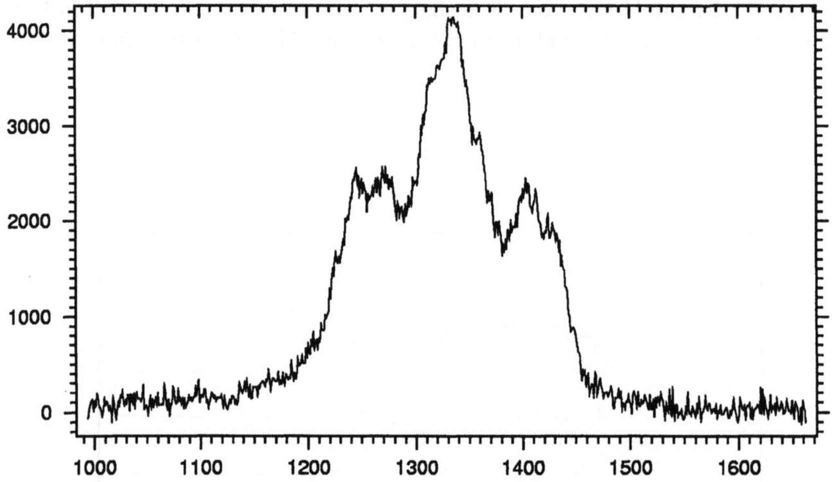

Figure 6. NMR spectroscopy data (courtesy Wellcome Research Laboratories).

Table 2 Quantified MassInf results from NMR data

Peak	Prob(>0)	Mean Position	Cumulant(%)
1	10%	1067.4±41.6	0.0±0.1
2	100%	1243.6± 0.7	12.1±0.5
3	100%	1269.8± 0.8	11.6±0.4
4	100%	1311.9± 1.0	11.3±0.7
5	100%	1335.7± 0.9	26.3±0.9
6	100%	1359.9± 0.9	11.7±0.6
7	100%	1403.0± 0.9	13.5±0.4
8	100%	1430.5± 0.7	13.0±0.5
9	70%	1517.3± 8.2	0.4±0.3
10	60%	1629.4±11.4	0.2±0.2

The MassInf analysis can be taken further by separating the mean reconstruction into constituent peaks (lying between adjacent minima). Accumulating the samples in each domain gives the usual mean and standard deviation uncertainty, but augmented with the probability that there is a signal present at all (see Table 2). This is the proportion of samples in which at least one atom is present. The seven obvious components are all definitely present, with 100% reliability. Their intensities, moreover, accord with the 1:1:1:2:1:1:1 ratio expected from the chemistry, to within 1 or 2 standard deviations. However, there are also a couple of weaker suggestions to the right, arising from a barely visible excess in the data. MassInf gives probabilities of 70% and 60% of these lines being present. Such analysis directly answers the basic question "Is structure there?" — a helpful preliminary to the usual question "How much is there?".

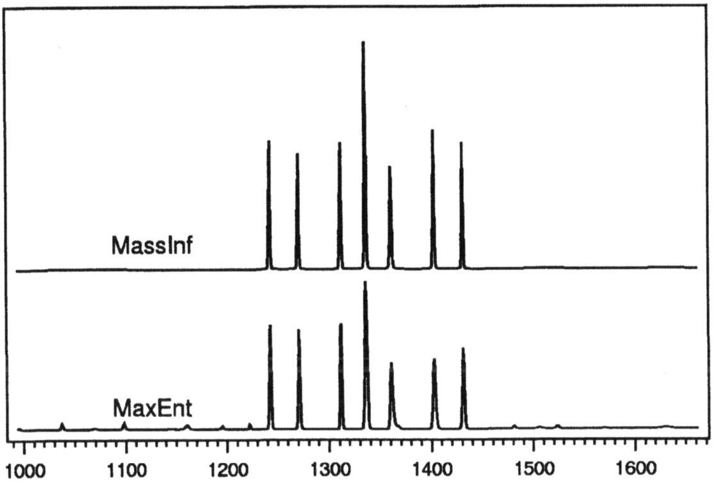

Figure 7. Comparison of MaxEnt with mean MassInf reconstructions from NMR data.

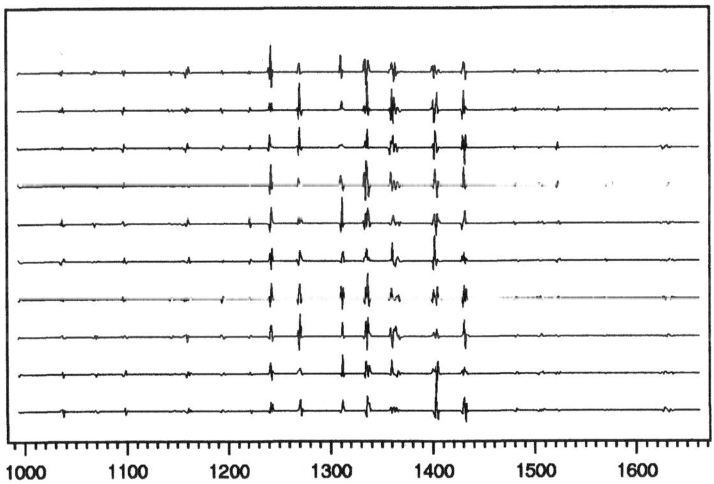

Figure 8. 10 typical MassInf samples with NMR data.

Acknowledgments

This work was supported by MaxEnt Solutions Ltd. Theoretical work on Massive Inference was carried out with S. Sibisi, and the ideas on polarization were developed with S.F. Gull. My thanks to both for long-standing collaboration.

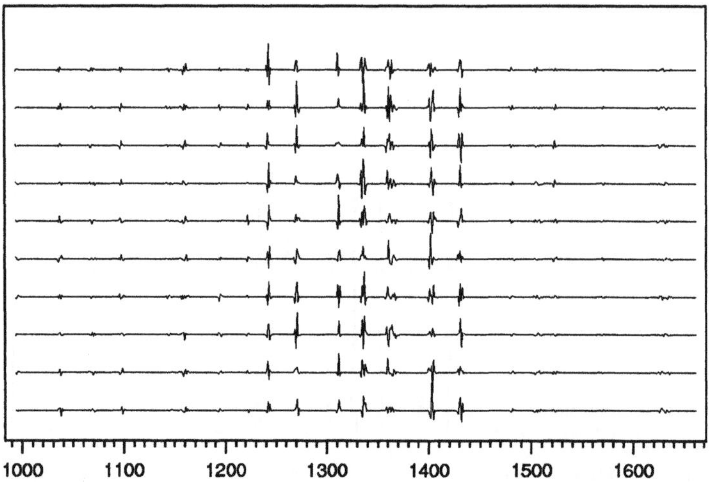

Figure 9. Figure 9: 10 typical QME samples (in Gaussian approximation) with NMR data.

References

Bretthorst, G.L.: 1990, 'Bayesian Spectrum Analysis and Parameter Estimation' in *Lecture Notes in Statistics* **48**, Springer-Verlag, New York.

Feller, W.: 1971, *An Introduction to Probability Theory and its Applications, Vol. II*, Wiley, New York.

Gull, S.F. and Daniell, G.J.: 1978, 'Image reconstruction from incomplete and noisy data', *Nature* **272**, 686-690.

Gull, S.F.: 1989, 'Developments in Maximum Entropy data analysis' in *Maximum Entropy and Bayesian Methods*, J. Skilling (ed.) Kluwer Academic Publishers, Dordrecht, 53-71.

Jaynes, E.T.: 1957, 'Information Theory and Statistical Mechanics, I', *Phys. Rev.* **106**, 620-630.

Lukacs, E.: 1970, *Characteristic Functions*, 2nd ed., Griffin, London

Shannon, C.E.: 1948, 'A Mathematical Theory of Communication', *Bell Systems Tech. J.* **27**, 379, 623.

Shore, J.E. and Johnson, R.W.: 1980, 'Axiomatic derivation of the principle of maximum entropy and the principle of minimum cross-entropy', *IEEE Trans. Info. Theory* **IT-26**, 26-39 and **IT-29**, 942-943.

Skilling, J.: 1989, 'Classic Maximum Entropy' in *Maximum Entropy and Bayesian Methods*, J. Skilling (ed.) Kluwer Academic Publishers, Dordrecht, 45-52.

Skilling, J.: 1991, 'On Parameter Estimation and Quantified MaxEnt' in *Maximum Entropy and Bayesian Methods*, W.T. Grandy, Jr. and L.H. Schick (ed.) Kluwer Academic Publishers, Dordrecht, 267-273.

Stone, J.M.: 1963, *Radiation and Optics*, McGraw-Hill, New York, 313.

CV-NP BAYESIANISM BY MCMC

Cross Validated Non Parametric Bayesianism by Markov Chain Monte Carlo

CARLOS C. RODRIGUEZ
Department of Mathematics and Statistics
University at Albany, SUNY
Albany NY 12222, USA†

Abstract. Completely automatic and adaptive non-parametric inference is a pie in the sky. The frequentist approach, best exemplified by the kernel estimators, has excellent asymptotic characteristics but it is very sensitive to the choice of smoothness parameters. On the other hand the Bayesian approach, best exemplified by the mixture of gaussians models, is optimal given the observed data but it is very sensitive to the choice of prior. In 1984 the author proposed to use the Cross-Validated gaussian kernel as the likelihood for the smoothness scale parameter h, and obtained a closed formula for the posterior mean of h based on Jeffreys's rule as the prior. The practical operational characteristics of this bayes' rule for the smoothness parameter remained unknown for all these years due to the combinatorial complexity of the formula. It is shown in this paper that a version of the metropolis algorithm can be used to approximate the value of h producing *remarkably* good completely automatic and adaptive kernel estimators. A close study of the form of the cross validated likelihood suggests a modification and a new approach to Bayesian Non-parametrics in general.

Key words: Cross validation, Density estimation, Bayes method, Kernel estimator

1. Introduction

A basic problem of statistical inference is to estimate the probability distribution that generated the observed data. In order to allow the data to *speak by themselves*, it is desirable to solve this problem with a minimum of a priori assumptions on the class of possible solutions. For this reason, the last thirty years have been burgeoning with interest in nonparametric estimation. The main positive results are almost exclusively from non-Bayesian quarters, but due to recent advances in Monte Carlo methods and computer hardware, Bayesian nonparametric estimators are now becoming more competitive.

---

†Email: carlos@math.albany.edu

This paper aims to extend the idea in [1] to a new general Bayesian approach to nonparametric density estimation. As an illustration one of the techniques is applied to compute the estimator in [1]. In particular it is shown how to approximate the bayes estimator (for Jeffreys' prior and quadratic loss) for the bandwidth parameter of a kernel using a version of the metropolis algorithm. The computer experiments with simulated data clearly show that the bayes estimator with Jeffreys' prior outperforms the standard estimator produced by plain Cross-validation.

2. Density Estimation by Summing Over Paths

Any non-Bayesian nonparametric density estimator can be turned into a Bayesian one by the following three steps:

1. Transform the estimator into a likelihood for the smoothness parameters via the sum over paths technique introduced below.
2. Put a reference prior on the smoothness parameters.
3. Use the predictive distribution as the Bayesian estimator obtainable by Markov Chain Monte Carlo.

The summing over paths method, springs from the simple idea of interchanging sums and products in the expression for the cross-validated kernel likelihood (see [1]). Without further reference to its humble origins, let us postulate the following sequence of functions,

$$\Phi_n = \Phi\left(x_0, x_1, x_2, \ldots, x_n \middle| h\right) \propto \sum_{\text{all paths}} \prod_{j=0}^{n} K_h\left(x_j - x_{i_j}\right) \quad (1)$$

where $h > 0$, $K_h(x) = \frac{1}{h}K(x/h)$ with K a density symmetric about 0. We call a vector of indices, $(i_0, \ldots, i_n)$ with the property that $i_j \in \{0, \ldots, n\} \setminus \{j\}$, a path (more specifically a general unrestricted path, see below). The sum above runs over all the n^{n+1} possible paths. The functions, Φ_n are defined up to a proportionality constant independent of the x_j's.

Notice that by flipping the sum and the product we get

$$\frac{1}{n^{n+1}} \sum_{\text{all paths}} \prod_{j=0}^{n} K_h\left(x_j - x_{i_j}\right) = f_{-0,n}(x_0) f_{-1,n}(x_1) \cdots f_{-n,n}(x_n) \quad (2)$$

where,

$$f_{-j,n}(x_j) = \frac{1}{n} \sum_{\substack{i_j=0 \\ i_j \neq j}}^{n} K_h\left(x_j - x_{i_j}\right). \quad (3)$$

Thus, $f_{-j,n}(x_j)$ is nothing but the kernel density estimator of $f(x_j)$ based on all the data except the jth observation x_j. Under mild regularity conditions the kernel estimator is known to converge (in every conceivable way) provided that $h = h_n$ is taken as a function of n such that, $h_n \to 0$ and $nh_n \to \infty$ as $n \to \infty$.

The Φ_n's can be used as a ***universal*** method for attaching a class of exchangeable one parameter models to any set of observations. The positive scalar parameter h is the only free parameter, and different models are obtained by changing the kernel function K.

These empirical parametric models are invariant under relabeling of the observations (i.e. they are exchangeable) but they do not model the observations as independent variables. Rather, these models introduce a pattern of correlations for which there is a priori no justification. This suggest that there might be improvements in performance if the sum is restricted to special subsets of the set of all n^{n+1} paths. Three of these modifications are mentioned in the following section.

Notice also that the ability of the Φ_n to adapt comes at the expense of regularity. These models are always non-regular. If the kernel has unbounded support then Φ_n integrates to infinity but the conditional distribution of x_0 given $x_1, \ldots, x_n$ and h is proper. When the kernel has compact support the Φ_n are proper but still non-regular since their support now depends on h.

The above recipe would have been a capricious combination of symbols if it were not for the fact that, under mild regularity conditions, these models ***adapt*** to the form of the true likelihood as n increases.

As a function of $x_0 = x$, the Φ_n have the following asymptotic property:

Theorem 1 ***If $x_1, x_2, \ldots, x_n$ are iid observations from an unknown pdf f which is continuous a.s. and $h = h_n$ is taken as a function of n such that, $h_n \to 0$ and $nh_n \to \infty$ as $n \to \infty$, then,***

$$\frac{\Phi_n}{\int \Phi_n dx_0} = f_{-0,n}(x) + o(1) = f(x) + o(1). \tag{4}$$

where the little o is taken in probability as $n \to \infty$

Proof (sketch only)
Just flip the sum and the product to get again,

$$\frac{1}{n^{n+1}} \sum_{\text{all paths}} \prod_{j=0}^{n} K_h \left(x_j - x_{i_j}\right) = f_{-0,n}(x) f_{-1,n}(x_1) \cdots f_{-n,n}(x_n) \tag{5}$$

Under the simple regularity conditions of the theorem, the kernel estimator is known to converge in probability as $n \to \infty$. However, even though x_0 appears in all of the $n+1$ factors, and their number goes to infinity, still all the factors are converging to the value of the true density at the given point. Therefore the theorem follows.□

It is worth noticing that the above theorem is only one of a large number of results that are readily available from the thirty years of literature on density estimation. In fact under appropriate regularity conditions the convergence can be strengthened to be pointwise a.s., uniformly a.s., or globally a.s. in L_1, or L_2.

3. Paths, Graphs and Loops

Each of the n^{n+1} paths $(i_0, \ldots, i_n)$ can be represented by a graph with nodes $x_0, \ldots, x_n$ and edges from x_j to x_k if and only if $i_j = k$. Here are some graphs

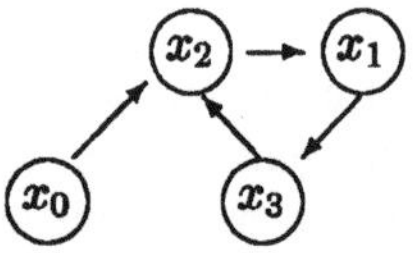

Figure 1. The graph of $(2,3,1,2)$

for paths with $n = 3$. For example, the path $(2,3,1,2)$ is given a probability proportional to

$$K_h(x_0 - x_2)K_h(x_1 - x_3)K_h(x_2 - x_1)K_h(x_3 - x_2) \tag{6}$$

and represented by the graph in figure [1]. Let's call it a 1-3-loop. The path $(1,2,3,0)$ is the single ordered loop of size four (a 4-loop), $(3,0,1,2)$ is the same loop backwards (also a 4-loop), $(2,3,0,1)$ are two disconnected loops (a 2-2-loop) and $(1,0,0,0)$ is connected and contains a loop of size two with x_0 and x_1 in it (a 1-1-2-loop). Draw the pictures!

The classification of paths in terms of number and size of loops appears naturally when trying to understand how Φ_n distributes probability mass among the different paths. To be able to provide simple explicit formulas let us take K in the definition of Φ_n to be the standard gaussian density, i.e. from now on we take,

$$K(x) = \frac{1}{\sqrt{2\pi}} \exp\left(\frac{-x^2}{2}\right). \tag{7}$$

The gaussian kernel has unbounded support and that makes the total integral of each path to diverge. Thus, the partition function

$$Z = \int \Phi_n dx_0 \ldots dx_n \tag{8}$$

$$= \sum_{\text{all paths}} \int \prod_{j=0}^{n} K_h\left(x_j - x_{i_j}\right) dx_0 \ldots dx_n \tag{9}$$

is the sum of infinities and it also diverges. Recall that this anomaly is the price we need to pay for using a model with a finite number of free parameters (only one in this case) and hoping to still adapt to the form of the true likelihood as $n \to \infty$. Even though the value of Z is in fact ∞ we can still write (formally) a useful decomposition that will help explain how the Φ_n's adapt and how to modify the set of paths to improve the convergence. We first need the following simple property of gaussians,

$$\int K_a(x-y)K_b(y-z)dy = K_{\sqrt{a^2+b^2}}(x-z). \tag{10}$$

This can be shown by straight forward integration after completing the square. Now notice that whatever the value of the integrals appearing in equation (9) that value only depends on the type of loop that is being integrated. For this reason we omit the integrand and simply denote the value of the integral with the integral sign and the type of loop. With this notation we have,

Theorem 2

$$\int_{m_1-m_2-\ldots-m_k-loop} = \left(\int_{m_1-loop}\right)\left(\int_{m_2-loop}\right)\cdots\left(\int_{m_k-loop}\right) \tag{11}$$

Moreover, $\int_{1-loop} = 1$ *and for* $m > 1$,

$$\int_{m-loop} = \frac{1}{\sqrt{2\pi}}\frac{h^{-1}}{\sqrt{m}}L \tag{12}$$

where we write formally $L = \int dx$.

Proof

Equation (11) follows from Fubini's theorem. To get (12) use Fubini's theorem and apply (10) each time to obtain,

$$\begin{aligned}
\int_{m\text{-loop}} &= \int K_h(x_0-x_1)K_h(x_1-x_2)\ldots K_h(x_{m-1}-x_0)dx_0dx_1\ldots dx_{m-1} \\
&= \int K_{\sqrt{2}h}(x_0-x_2)\ldots K_h(x_{m-1}-x_0)dx_0dx_2\ldots dx_{m-1} \\
&\ldots \\
&= \int K_{\sqrt{m-1}h}(x_0-x_{m-1})K_h(x_{m-1}-x_0)dx_0dx_{m-1} \\
&= \int K_{\sqrt{m}h}(0)dx_0 = \frac{1}{\sqrt{2\pi}}\frac{h^{-1}}{\sqrt{m}}L
\end{aligned}$$

□

Hence, by splitting the sum over all paths into

$$\sum_{\text{all paths}} = \sum_{\text{2-2...-2-loops}} + \sum_{\text{1-3-2...-2-loops}} + \ldots + \sum_{(n+1)\text{-loops}}$$

and applying the previous theorem we obtain,

$$Z = N_{2-2\ldots-2}\left(\frac{1}{\sqrt{2\pi}}\frac{h^{-1}}{\sqrt{2}}L\right)^{(n+1)/2} + \ldots + N_{n+1}\left(\frac{1}{\sqrt{2\pi}}\frac{h^{-1}}{\sqrt{n+1}}L\right)^{1} \tag{13}$$

where for simplicity we have assumed that n is odd and we denote by $N_{m_1-\ldots-m_k}$ the total number of $m_1 - \ldots - m_k$ − loops. Using simple combinatorial arguments it is possible to write explicit formulas for the number of loops of each kind. The important conclusion from the decomposition (13) is that even though the Φ_n appear to be adding equally over all paths, in reality they end up allocating almost all the probability mass on paths with maximally disconnected graphs. This is not surprising. This is the reason why there is consistency when assuming iid observations. There is a built in bias towards independence. The bias can be imposed explicitly on the Φ_n by restricting the paths to be considered in the sum. Here are three examples:

loops:
Only paths $(i_0, \ldots, i_n)$ that form a permutation of the integers $\{0, 1, \ldots, n\}$ are considered.

$2-2-\ldots-2$–loops:
Only maximally disconnected paths are considered.

QM:
Paths as above but use $|\Phi_n|^2$ instead of Φ_n as the joint likelihoods.

Preliminary simulation experiments seem to indicate that only maximally disconnected paths are not enough and that all the loops are too many. The QM method has all the maximally disconnected paths but not all the loops (e.g. with $n = 5$ the 3-3-loops can not be reached by squaring the sum of 2-2-2-loops) so it looks like the most promising among the three. What is more interesting about the QM method is the possibility of using kernels that can go negative or even complex valued. More research is needed since very little is known about the performance of these estimators.

4. Estimation by MCMC

We show in this section how to approximate the predictive distribution and the bayes rule for the smoothness parameter by using Markov Chain Monte Carlo techniques.

4.1. POSTERIOR MEAN OF THE BANDWIDTH

Apply bayes' theorem to (1) to obtain the posterior distribution of h,

$$\pi(h|x, x_1, x_2, \ldots, x_n) = \frac{\Phi(x, x_1, x_2, \ldots, x_n|h)\pi(h)}{\int_0^\infty \Phi(x, x_1, x_2, \ldots, x_n|\tau)\pi(\tau)d\tau} \tag{14}$$

where π is a function of h. It is worth noticing that π is **not** the prior on h. It is only the part of the prior on h that we can manipulate. Recall that Φ_n integrates to the function of h given by (13) so effectively the prior that is producing (14) is,

$$\Pi(h) \propto \frac{\pi(h)}{h^{(n+1)/2}} \tag{15}$$

The posterior mean is then given by,

$$E(h|x, x_1, x_2, \ldots, x_n) = \hat{h}_x = \frac{\int_0^\infty h\Phi(x, x_1, x_2, \ldots, x_n|h)\pi(h)dh}{\int_0^\infty \Phi(x, x_1, x_2, \ldots, x_n|h)\pi(h)dh} \tag{16}$$

Equation (16) provides a different estimator for each value of x. To obtain a single global estimate for h just erase the x's from (16) and change n to $n-1$ in the formulas below. When K is the univariate gaussian kernel and $\pi(h) = h^{-\delta}$ equation (16) simplifies to:

$$\hat{h}_x = C_{n,\delta} \frac{\sum_{\text{all paths}} \alpha(\underline{i}) s(\underline{i})}{\sum_{\text{all paths}} \alpha(\underline{i})} \tag{17}$$

where

$$C_{n,\delta} = \frac{1}{\sqrt{2}} \frac{\Gamma\left(\frac{n+\delta-1}{2}\right)}{\Gamma\left(\frac{n+\delta}{2}\right)} \tag{18}$$

$\underline{i} = (i_0, \ldots, i_n)$ is a path, $\alpha = s^{-(n+\delta)}$ and

$$s^2(\underline{i}) = \sum_{j=0}^{n} (x_j - x_{i_j})^2. \tag{19}$$

Equation (17) follows from two applications of the formula,

$$\int_0^\infty h^{-(\beta+1)} \exp\left\{-\frac{s^2}{2h^2}\right\} dh = \frac{1}{2} 2^{\beta/2} \Gamma(\beta/2) s^{-\beta} \tag{20}$$

4.1.1. *Bandwidth by Metropolis*
To approximate equation (17) we use the fact that the ratio of the two sums is the expected value of a random variable that takes the value $s(\underline{i})$ on the path $\underline{i}$ which is generated with probability proportional to $\alpha(\underline{i})$. The following version of the Metropolis algorithm produces a sequence of averages that converge to the expected value,

Algorithm

0) Start from somewhere
$\underline{i} \leftarrow (1, 2, \ldots, n, 0)$
$s2 \leftarrow \sum_{j=0}^{n} (x_j - x_{i_j})^2$
$\alpha \leftarrow (s2)^{-(n+\delta)/2}$
$N \leftarrow 0$, sum $\leftarrow 0$, ave $\leftarrow 0$

1) Sweep along the components of $\underline{i}$
for k from 0 to n do

{
$i'_k \leftarrow$ Uniform on $\{0, \ldots, n\} \setminus \{k, i_k\}$
$\Delta_k \leftarrow (x_{i'_k} - x_{i_k})(x_{i'_k} + x_{i_k} - 2x_k)$
$s2' \leftarrow s2 + \Delta_k$
$\alpha' \leftarrow (s2')^{-(n+\delta)/2}$
$R \leftarrow \alpha'/\alpha$

if $R > 1$ or Unif[0,1] $< R$ then $\{i_k \leftarrow i'_k, s2 \leftarrow s2', \alpha \leftarrow \alpha'\}$
}

2) Update the estimate for the average,
sum $\leftarrow$ sum $+\sqrt{s2}$
$N \leftarrow N+1$
ave $\leftarrow$ sum$/N$
goto 1)

4.2. THE PREDICTIVE DISTRIBUTION BY GIBBS

To sample from the predictive distribution, $f(x|x_1, x_2, \ldots, x_n)$ we use Gibbs to sample from the joint distribution, $f(x, h|x_1, x_2, \ldots, x_n)$. Hence, we only need to know how to sample from the two conditionals, a) $f(x|h, x_1, x_2, \ldots, x_n)$ and b) $\pi(h|x, x_1, x_2, \ldots, x_n)$. To sample from a) we use the fact that this is (almost) the classical kernel so all we need is to generate from an equiprobable mixture of gaussians. To get samples from b) just replace the gaussian kernel in the numerator of equation (14) to obtain, for $\pi(h) \propto h^{-\delta}$,

$$\pi(h|x, x_1, x_2, \ldots, x_n) \propto \sum_{\text{all paths}} h^{-(n+\delta+1)} \exp\left\{-\frac{s^2}{2h^2}\right\}. \tag{21}$$

The integral with respect to h of each of the terms being added in (21) is proportional to $s^{-(n+\delta)}$ (see (20)). Thus, by multiplying and dividing by this integral each term, we can write,

$$\pi(h|x, x_1, x_2, \ldots, x_n) \propto \sum_{\text{all paths}} \alpha(\underline{i})\pi_{s(\underline{i})}(h) \tag{22}$$

where $\alpha(\underline{i}) = (s(\underline{i}))^{-(n+\delta)}$ as before and,

$$\pi_s(h) \propto h^{-(n+\delta+1)} \exp\left\{-\frac{s^2}{2h^2}\right\} \tag{23}$$

From the change of variables theorem it follows that if y is Gamma$(\frac{n+\delta}{2}, 1)$ then $h = \sqrt{\frac{s^2}{2y}}$ follows the distribution (23). This shows that the posterior distribution of h is a mixture of transformed gamma distributions. This mixture can be generated by a simple modification to the algorithm used to get the posterior mean of h.

5. Experiments on simulated and real data

I have coded (essentially) the algorithm described in the previous section in MAPLE and tested it dozens of times on simulated data for computing a global value for the smoothness parameter h. All the experiments were carried out with $\delta = 1$

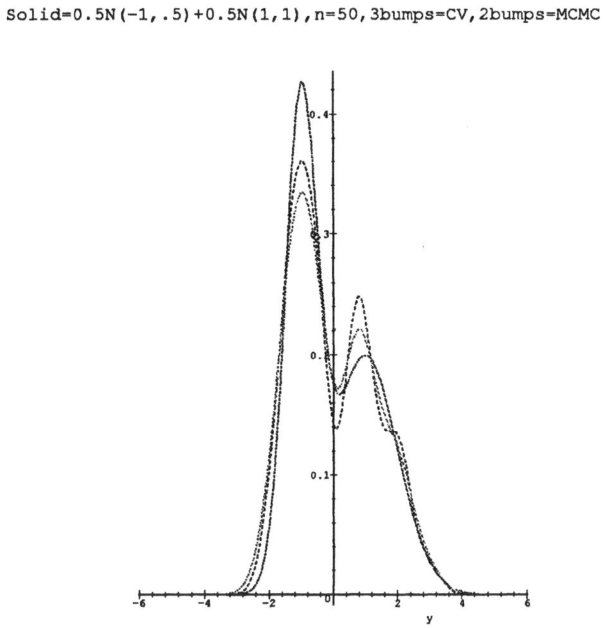

Figure 2. Posterior mean of global h vs plain cross-validation

i.e. with $\pi(h) = h^{-1}$ on mixtures of gaussians. The experiments clearly indicate that the global value of h provided by the MCMC algorithm produce a kernel estimator that is either identical to plain likelihood cross-validation or clearly superior to it depending on the experiment. A typical run is presented in figure [2] where the true density and the two estimators from 50 iid observations are shown. The MAPLE package used in the experiments is available at *http://omega.albany.edu:8008/npde.mpl.*

For comparison with other density estimators in the literature we show in figure [3] the estimate for the complete set of 109 observations of the Old Faithful geyser data. These data are the 107 observations in [2] plus the two outliers 610 and 620. This is a standard gaussian kernel with the global value of $h = 14.217$ chosen by the MCMC algorithm.

6. Conclusions

There is nothing special about dimension one. Only minor cosmetic changes (of the kind: replace h by h^d in some formulas) are needed to include the multivariate case, i.e. the case when the x_j's are d-dimensional vectors instead of real variables.

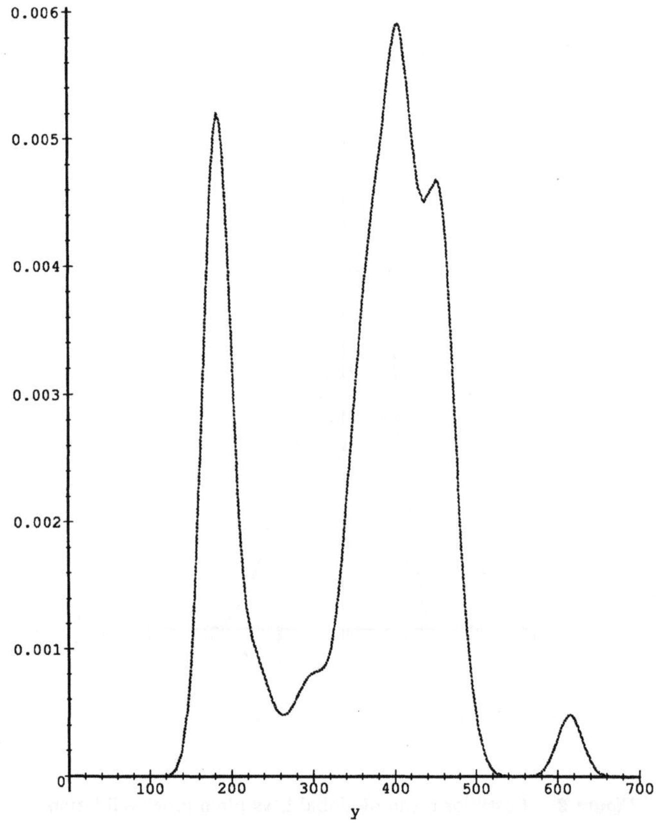

Figure 3. **Estimate for the Old Faithful geyser data, h = 14.217**

Very little is known about these estimators beyond what is presented in this paper. In particular nothing is known about rates of convergence. There are many avenues to explore with theory and with simulations but clearly the most interesting and promising open questions are those related to the performance of the QM method above.

References

1. C. C. Rodriguez, "On the estimation of the bandwidth parameter using a non informative prior," in *Proceedings of the 45th session of the ISI*, No. 1, pp. 207–208, International Statistical Institute, August 1985.
2. B. W. Silverman, *Density Estimation: for statistics and data analysis*, Monographs on Statistis and Applied Probability, Chapman and Hall, 1986.

WHICH ALGORITHMS ARE FEASIBLE? MAXENT APPROACH

D.E. COOKE, V. KREINOVICH, AND L. LONGPRÉ
Department of Computer Science
University of Texas at El Paso
El Paso, TX 79968, USA[†]

Abstract. It is well known that not all algorithms are feasible; whether an algorithm is feasible or not depends on how many computational steps this algorithm requires. The problem with the existing definitions of feasibility is that they are rather *ad hoc*. Our goal is to use the maximum entropy (MaxEnt) approach and get more motivated definitions.

If an algorithm is feasible, then, intuitively, we would expect the following to be true: *If we have a flow of problems with finite average length $\bar{l}$, then we expect the average time $\bar{t}$ to be finite as well.*

Thus, we can say that an algorithm is ***necessarily feasible*** if $\bar{t}$ is finite for *every* probability distribution for which $\bar{l}$ is finite, and ***possibly feasible*** if $\bar{t}$ is finite for *some* probability distribution for which $\bar{l}$ is finite.

If we consider *all* possible probability distributions, then these definitions trivialize: every algorithm is possibly feasible, and only linear-time algorithms are necessarily feasible.

To make the definitions less trivial, we will use the main idea of MaxEnt and consider only distributions for which the entropy is the largest possible. Since we are interested in the distributions for which the average length is finite, it is reasonable to define MaxEnt distributions as follows: we fix a number l_0 and consider distributions for which the entropy is the largest among all distributions with the average length $\bar{l} = l_0$.

If, in the above definitions, we only allow such "MaxEnt" distributions, then the above feasibility notions become non-trivial: an algorithm is possibly feasible if it takes exponential time (to be more precise, if and only if its average running time $\bar{t}(n)$ over all inputs of length n grows slower than some exponential function C^n), and necessarily feasible if it is ***sub-exponential*** (i.e., if $\bar{t}(n)$ grows slower than *any* exponential function).

Key words: maximum entropy, feasible algorithm, average computational complexity, Moore's law

[†]Emails: {dcooke,vladik,longpre}@cs.utep.edu

1. When is an algorithm feasible? The problem, and what we plan to do

1.1. SOME ALGORITHMS ARE NOT FEASIBLE

It is well known that not all algorithms are feasible (see, e.g., [3,9,10]): whether an algorithm is feasible or not depends on how many computational steps this algorithm requires.

Let us give two examples.

1.2. EXAMPLE OF A FEASIBLE ALGORITHM

First, we will give an example of a feasible algorithm. Let us consider the following simple sorting algorithm for ordering a finite list of n numbers (this algorithm is called ***selection sort***):

- First, we find the largest number. To do this, we check the elements one by one and after each check, select the largest-of-checked number. Initially, when we start the sorting, the first element is the largest-of-checked. If we have already checked the first k numbers, and we know which of them is the largest, all we have to do to find out which number is the largest among the first $k+1$ is compare the existing "champion" with the newly checked number:
  - If the champion is larger, it stays the champion.
  - If the new number is larger, it becomes a new champion.

  After we have checked all n numbers, the largest-of-checked becomes simply the largest of all of them. So, we can place this number at the right end of the desired list, and place the number, that was originally at this right end, into the champion's original location.
- After the above-described first iteration, the largest element is in its desired place. Now, we apply a similar algorithm to the remaining $n-1$ elements, find the largest of them, and place it right before the largest one.
- Then, we repeat the same procedure with $n-2$ elements, etc.

How many computational steps does this algorithm require? On each iteration, we add exactly one element to the list of already sorted ones, so, we need at most n iterations. On each of these iterations, we check at most n elements, so we need at most n computational steps per iteration. Thus, this algorithm requires a total computation time of $\leq n \times n = n^2$. This algorithm is quite realistic and thus, feasible.

In practice, for large lists, better sorting algorithms are used, with computation time $n \cdot \log(n)$; they are, clearly, also quite feasible.

1.3. EXAMPLE OF A NON-FEASIBLE ALGORITHM

Let us now give an example of a non-feasible algorithms.

There exists an algorithm for solving all problems from elementary geometry [18]. It is known that this algorithm requires, in some cases, at least doubly exponential time [2], i.e., time of order 2^{2^n}. Even for small n, this amount of time is much larger than the lifetime of the Universe divided by the smallest physically possible time interval (of $\approx 10^{-43}$ seconds). Thus, this algorithm is clearly non-feasible.

1.4. IMPORTANT COMMENT: "NON-FEASIBLE" DOES NOT NECESSARILY MEAN "USELESS"

It is important to remark that the very fact that an algorithm requires a long time for *some* examples does not necessarily make it practically unusable: it may be that these examples are rare, and in almost all practical cases, this algorithm can be successfully applied.

1.5. THE EXISTING DEFINITION OF FEASIBILITY, ITS ADVANTAGES AND DRAWBACKS

In modern theory of computation, an algorithm is usually called *feasible* if is is *polynomial-time*, i.e., if its running time is bounded by some polynomial of the length of the input. This definition has its pluses and minuses:

- This definition has many *advantages*, e.g., it is *independent* of the type of the computational device that we are using.
- On the other hand, this definition is not completely satisfactory, mainly because we are trying to formalize one of the major theoretical concepts of computing theory, but the existing definition of feasibility is rather *ad hoc.*

In the following text, we will use the maximum entropy (MaxEnt) approach [5–8] to get more motivated definitions.

2. Our main idea

If an algorithm is feasible, then, intuitively, we would expect the following to be true:

> *If we have a flow of problems with finite average length, then we expect the average time to be finite as well.*

3. Formalization: general idea

How can we formalize this statement?

- Let $l(x)$ denote the length of the word x,
- let $t_U(x)$ denote the running time of the algorithm U on the input x, and
- let $p(x)$ be the probability (frequency) of the word x in the flow.

In these notations:

- the average word length is $\bar{l} = \sum p(x) \cdot l(x)$, and
- the average time is $\bar{t} = \sum p(x) \cdot t_U(x)$.

Thus, a natural formalization of the above principle takes the following form:

- We somehow select a family $\mathcal{P}$ of *natural* probability distributions $\{p(x)\}$.
- We require that for every distribution p from the family $\mathcal{P}$ for which the average word length $\bar{l}$ is finite, the average computation time $\bar{l}$ is also finite.

4. Formalization: first approximation

The problem with this definition is that we do not know which probability distributions are natural. There are two extreme approaches to this problem:

- the *most cautious* approach is to consider all possible probability distributions, and to require that $\bar{t}$ is finite for *all* probability distributions for which $\bar{l}$ is finite;
- the *least cautious* approach is to require that $\bar{t}$ is finite for *some* probability distribution for which $\bar{l}$ is finite.

The corresponding two definitions lead to the following results:

Definition 1. *We say that an algorithm is necessarily feasible if $\bar{t}$ is finite for every probability distribution for which $\bar{l}$ is finite.*

Definition 2. *We say that an algorithm is possibly feasible if $\bar{t}$ is finite for some probability distribution for which $\bar{l}$ is finite.*

Proposition 1.

- *An algorithm U is necessarily feasible if and only if it is linear-time, i.e., if there exists a constant C for which $t_U(x) \le C \cdot l(x)$ for all inputs x.*
- *Every algorithm U is possibly feasible.*

Comment. For the readers' convenience, all the proofs are given at the end of the paper.

4.1. THE PROBLEMS WITH THIS DEFINITION

Proposition 1 is not very satisfactory:

- Linear time seems too restrictive; we would like the definition to be more restrictive.
- On the other hand, *no* restrictions seems too non-restrictive. We would like *some* restrictions.

5. MaxEnt approach

5.1. MAIN IDEA

To make the definitions less trivial, we will use the main idea of MaxEnt: instead of all possible probability distributions, we will only consider *MaxEnt distributions*, i.e., distributions for which the entropy $S = -\sum p(x) \cdot \log(p(x))$ is the largest possible.

5.2. MAIN DEFINITIONS

Since we are interested in the distributions for which the average length is finite, it is reasonable to define MaxEnt distributions as follows:

Definition 3.

- *Let l_0 be a real number. By a MaxEnt distribution corresponding to l_0 we mean the distribution for which the entropy is the largest among all distributions with the average length $\bar{l} = l_0$.*
- *By a MaxEnt distribution, we mean a distribution that is MaxEnt for some real number $l_0 > 0$.*

Definition 4. *We say that an algorithm is necessarily MaxEnt-feasible if $\bar{t}$ is finite for every MaxEnt distribution for which $\bar{l}$ is finite.*

Definition 5. *We say that an algorithm is possibly MaxEnt-feasible if $\bar{t}$ is finite for some MaxEnt distribution for which $\bar{l}$ is finite.*

5.3. AUXILIARY DEFINITIONS

To describe what algorithms are feasible in the sense of these distributions, we will need the following auxiliary definition:

Definition 6. *By an average running time of an algorithm U, we mean the function $\bar{t}_U(n)$ defined as*

$$\bar{t}_U(n) = \frac{1}{N_n} \cdot \sum_{x:l(x)=n} t_U(x),$$

where the sum is taken over all words x of length $l(x) = n$, and N_n is the total number of such words.

Now, we are ready to formulate the main result:

5.4. MAIN RESULT

Proposition 2.

- *An algorithm U is necessarily MaxEnt-feasible if and only if it is average subexponential-time, i.e., if for all $C > 0$,*

$$\lim_{n\to\infty} \frac{\bar{t}_U(n)}{\exp(C \cdot n)} = 0.$$

- *An algorithm U is possibly MaxEnt-feasible if and only if it is average exponential-time, i.e., if there exist constants C_0 and C_1 for which $\bar{t}_U(n) \le C_0 \cdot \exp(C_1 \cdot n)$ for all positive integers n.*

6. Discussion and future plans

6.1. NICE PROPERTIES OF THE ABOVE DEFINITIONS AND RESULTS

In contrast to traditional treatment of feasibility, we gain a formal distinction between the two levels of ***possible*** and ***necessary*** feasibility, the discussion intuited, e.g., by M. Gell-Mann [4], but never before properly formalized (see, e.g., [1]).

6.2. POSSIBLE PROBLEMS (?) WITH THESE DEFINITIONS AND RESULTS

Our result seemingly contradicts the traditional definition of feasible as polynomial time.

However, there is no contradiction if we take into consideration the so-called *Moore's law*, according to which the computer speed grows exponentially with time [11,12,16]. In view of this law, linear *time* means actually exponentially many computational steps, and therefore, a problem that requires exponentially many computational steps can be actually solved in reasonable time [17].

(Also, on a more speculative note, some exponential-time algorithms become quite feasible if we take the curvature of space-time into consideration [13].)

6.3. POSSIBILITIES OF MAKING THESE DEFINITIONS MORE ADEQUATE

Unless we invoke Moore's law, the above formalization is not completely adequate: e.g., an algorithm that require 2^n steps on inputs of length n is treated as possibly feasible. This may be not restrictive enough, because, such an algorithm, for an input of a reasonable length $n \approx 300$, would need more computational steps than there can be during the lifetime of our universe.

Even sub-exponential does not necessarily mean practically feasible, because, e.g., a function $2^{\sqrt{n}}$ is sub-exponential, but grows pretty fast with n.

If we want to make the above definitions more restrictive (or at least more realistic), it is desirable to find and formalize other intuitive statements about feasibility (examples of such statements are given, e.g., in [14,15]). The fact that our intuitive statement led to reasonable restriction shows that MaxEnt is indeed a reasonable technique to use.

Acknowledgments

This work was supported in part by NSF under grants No. EEC-9322370 and No. DUE-9750858, by NASA under cooperative agreement NCCW-0089, and by Future Aerospace Science and Technology Program (FAST) Center for Structural Integrity of Aerospace Systems, effort sponsored by the Air Force Office of Scientific Research, Air Force Materiel Command, USAF, under grant number F49620-95-1-0518.

The authors are thankful to all the participants of MaxEnt'97 workshop, especially to Anton Garrett and John Skilling, for valuable discussions.

References

1. D. E. Cooke, *Scientific Research: From the Particular to the General*, El Paso Energy Award for Research Excellence Presentation, University of Texas at El Paso, April 9, 1997.
2. J. H. Davenport and J. Heintz, "Real quantifier elimination is doubly exponential", *Journal of Symbolic Computations*, 1988, Vol. 5, No. 1/2, pp. 29–35.
3. M. R. Garey and D. S. Johnson, *Computers and intractability: a guide to the theory of NP-completeness*, W. F. Freeman, San Francisco, 1979.
4. M. Gell-Mann, *The Quark and the Jaguar: Adventures in the Simple and the Complex*, Freeman, N.Y., 1994.
5. K. M. Hanson and R. N. Silver (Eds.), *Maximum Entropy and Bayesian Methods*, Kluwer Academic Publishers, Dordrecht, 1996.
6. E. T. Jaynes, "Information theory and statistical mechanics", *Phys. Rev.*, 1957, Vol. 108, pp. 171–190.
7. E. T. Jaynes, "Where do we stand on maximum entropy?", In: R. D. Levine and M. Tribus (Eds.) *The maximum entropy formalism*, MIT Press, Cambridge, MA, 1979.
8. V. Kreinovich, "Maximum entropy and interval computations", *Reliable Computing*, 1996, Vol. 2, No. 1, pp. 63–79.
9. H. R. Lewis and C. H. Papadimitriou, *Elements of the Theory of Computation*, Prentice-Hall, Inc., New Jersey, 1981.
10. J. C. Martin, *Introduction to languages and the theory of computation*, McGraw-Hill, N.Y., 1991.
11. G. E. Moore, "Cramming more components onto integrated circuits", *Electronics Magazine*, 1965, Vol. 38, No. 8, pp. 114–117.
12. G. E. Moore, "Lithography and the Future of Moore's Law", In *Proceedings of the SPIE Conference on Optical/Laser Microlithography, February 1995*, SPIE Publ., Vol. 2440, 1995, pp. 2–17.
13. D. Morgenstein and V. Kreinovich, "Which algorithms are feasible and which are not depends on the geometry of space-time", *Geombinatorics*, 1995, Vol. 4, No. 3, pp. 80–97.
14. H. T. Nguyen and V. Kreinovich, "When is an algorithm feasible? Soft computing approach", *Proceedings of the Joint 4th IEEE Conference on Fuzzy Systems and 2nd IFES*, Yokohama, Japan, March 20–24, 1995, Vol. IV, pp. 2109–2112.
15. H. T. Nguyen and V. Kreinovich, "Towards theoretical foundations of soft computing applications", *International Journal on Uncertainty, Fuzziness, and Knowledge-Based Systems*, 1995, Vol. 3, No. 3, pp. 341–373.
16. R. R. Schaller, "Moore's law: past, present, and future", *IEEE Spectrum*, June 1997, pp. 53–59; see also discussion on p. 8 of the August 1997 issue of *IEEE Spectrum*.
17. D. Schirmer and V. Kreinovich, "Towards a More Realistic Definition of Feasibility", *Bulletin of the European Association for Theoretical Computer Science (EATCS)*, 1996, Vol. 90, pp. 151–153.
18. A. Tarski, *A decision method for elementary algebra and geometry*, 2nd ed., Berkeley and Los Angeles, 1951.

Appendix: Proofs

6.4. PROOF OF PROPOSITION 1

For a linear time algorithm, $t_U(x) \le C \cdot l(x)$ and therefore, for every probability distribution, $\bar{t} \le C \cdot \bar{l}$. So, if $\bar{l}$ is finite, the average time $\bar{t}$ is also finite.

Let us show that if an algorithm is not linear-time, then there exists a probability distribution for which $\bar{l}$ is finite, but $\bar{t}$ is infinite. Indeed, linear-time means that there exists a constant C for which, for all x, $t_U(x)/l(x) \le C$. Therefore, the fact that an algorithm is not linear-time means that for every possible constant C, in particular, for $C = 2^n$, there exists a word x_n for which $t_U(x_n)/l(x_n) > 2^n$. Let us take the probability distribution that is concentrated on each x_n with a probability $p_n = N \cdot 2^{-n}/l(x_n)$, where N is a normalization constant (which guarantees that $\sum p_n = 1$). Since $\sum 2^{-n} = 1 < \infty$, and $l(x_n) \geq 1$, we have

$$\sum \frac{2^{-n}}{l(x_n)} \leq \sum 2^{-n} < \infty$$

and therefore, such a normalization is indeed possible.

For this probability distribution,

$$\bar{l} = \sum p_n \cdot l(x_n) = \sum N \cdot 2^{-n} = N \cdot \sum 2^{-n} < \infty.$$

However, since $t_U(x_n)/l(x_n) \geq 2^n$, we get

$$\bar{t} = \sum p_n \cdot t_U(x_n) = \sum N \cdot 2^{-n} \cdot \frac{t_U(x_n)}{l(x_n)} \geq N \cdot \sum 1 = \infty.$$

The first part of Proposition 1 is proven.

To show that every algorithm is possibly feasible in the sense of Definition 2, it is sufficient to take a probability distribution that is concentrated on one word x_0 with probability 1. Then, $\bar{l} = l(x_0) < \infty$ and $\bar{t} = t_U(x_0) < \infty$. The proposition is proven.

6.5. PROOF OF PROPOSITION 2

Every MaxEnt probability distribution is a solution to the constrained optimization problem $S = -\sum p(x) \cdot \log(p(x)) \to \max$ under the constraints $\sum p(x) = 1$ and $\sum l(x) \cdot p(x) = l_0$. Lagrange multiplier method reduces this formula to the *unconstrained* optimization problem

$$-\sum[p(x) \cdot \log(p(x)) + \lambda \cdot p(x) + \mu \cdot p(x) \cdot l(x)] \to \max.$$

If we differentiate the resulting objective function with respect to $p(x)$ and equate the result to 0, we get $-\ln(p(x)) + 1 + \lambda + \mu \cdot l(x) = 0$, i.e., $p(x) = N \cdot \exp(B \cdot l(x))$, where B is a real number and N is the normalization constant (that guarantees $\sum p(x) = 1$).

What are the possible values of B? The value $p(x)$ depends only on the length $n = l(x)$ of the word x: $p(x) = N \cdot \exp(B \cdot n)$. For every length $l(x) = n$, there are exactly $N_n = A^n$ words of length n, where A is the number of symbols in the input alphabet. Thus, the sum p_n of the probabilities $p(x)$ for all words x of length n is equal to $p_n = N \cdot A^n \cdot \exp(B \cdot n) = N \cdot \exp(\ln(A) \cdot n) \cdot \exp(B \cdot n) = N \cdot \exp(n \cdot (B + \ln(A))$. The total probability is equal to the sum of these terms: $1 = \sum p_n$. Thus, for the sum of these terms p_n to converge, we must have $B + \ln(A) < 0$. It is easy to see that for an arbitrary positive real number $C > 0$, if we take $B + \ln(A) = -C$ (i.e., $B = -C - \ln(A)$), then we get a MaxEnt distribution.

For an arbitrary MaxEnt distribution, the average time $\bar{t} = \sum p(x) \cdot t_U(x)$, which is defined over all possible words, can also be represented as the sum of the following sub-sums: $s = s_1 + s_2 + \ldots + s_n + \ldots$, where:

- the sum s_1 is taken over all the words of length 1,
- the sum s_2 is taken over all the words of length 2,
- ...,
- the sum s_n is taken over all the words of length n,
- ...

For a fixed n, the values $p(x) = N \cdot \exp(B \cdot n)$ in the sum $s_n = \sum p(x) \cdot t_U(x)$ do not depend on x, so, $s_n = p(x) \cdot \sum t_U(x) = N \cdot \exp(B \cdot n) \cdot sumt_U(n)$. The sum of running times is, in its turn, equal to $N_n \cdot \bar{t}_U(n) = A^n \cdot \bar{t}_U(n)$. Thus, $s_n = N \cdot A^n \cdot \exp(B \cdot n) \cdot \bar{t}_U(n) = N \cdot \exp(-C \cdot n) \cdot \bar{t}_U(n)$. Thus, the convergence of the series $s_1 + \ldots + s_n + \ldots$ means that the sum $\sum \exp(-C \cdot n) \cdot \bar{t}_U(n)$ converges.

Let us show that this conclusion proves Proposition 2:

- If an algorithm U is possibly MaxEnt-feasible, then this series converges for some $C > 0$. From the convergence of the series, we can conclude that the term tends to 0, i.e., that $\bar{t}_U(n)/\exp(C \cdot n) \to 0$ as $n \to \infty$. Since this sequence tends to 0, there exists a constant C_0 that bounds it. Hence, $\bar{t}_U(n) \le C_0 \cdot \exp(C \cdot n)$, i.e., U is indeed an average exponential-time algorithm.
- Vice versa, let U be an average exponential-time algorithm, i.e., let $\bar{t}_U(n) \le C_0 \cdot \exp(C_1 \cdot n)$ for all n. Then, if we take $C = C_1 + \varepsilon$ for some $\varepsilon > 0$, we conclude that $\bar{t}_U(n) \cdot \exp(-C \cdot n) \le C_0 \cdot \exp(-\varepsilon \cdot n)$ converges, and thus, this algorithm is feasible with respect to the corresponding MaxEnt distribution.
- If an algorithm is necessarily MaxEnt-feasible, then for *every* $C > 0$, we get $\bar{t}_U(n)/\exp(C \cdot n) \to 0$ as $n \to \infty$, and thus, the algorithm U is indeed average subexponential-time.
- Vice versa, let us show that every subexponential-time algorithm is necessarily MaxEnt-feasible, i.e., that the sum $\sum \exp(-C \cdot n) \cdot \bar{t}_U(n)$ converges for all $C > 0$. Indeed, since U is an average subexponential-time algorithm, we have $\bar{t}_U(n)/\exp((C/2) \cdot n) \to 0$ and therefore,

$$\frac{\bar{t}_U(n)}{\exp((C/2) \cdot n)} \le C_0$$

  for some $C_0 > 0$ and for all n. Thus, $\bar{t}_U(n) \le C_0 \cdot \exp((C/2) \cdot n)$ and $s_n = \bar{t}_U(n) \cdot \exp(-C \cdot n) \le C_0 \cdot \exp(-(C/2) \cdot n)$. The terms s_n are bounded by a convergent sum and thus, the sum $\sum s_n$ also converges.

The proposition is proven.

MAXIMUM ENTROPY, LIKELIHOOD AND UNCERTAINTY: A COMPARISON

A. GOLAN
Dept. of Economics, American University
4400 Massachusetts Ave. NW
Washington, DC 20016, USA
E-mail: golan@are.berkeley.edu

Abstract. A framework for comparing the maximum likelihood (ML) and maximum entropy (ME) approaches is developed. Two types of linear models are considered. In the first type, the objective is to estimate probability distributions given some moment conditions. In this case the ME and ML are equivalent. A generalization of this type of estimation models to incorporate noisy data is discussed as well. The second type of models encompasses the traditional linear regression type models where the number of observations is larger than the number of unknowns and the objects to be inferred are not natural probabilities. After reviewing a generalized ME estimator and the empirical likelihood (or weighted least squares) estimator, the two are compared and contrasted with the ML. It is shown that, in general, the ME estimators use less input information and may be viewed, within the second type models, as expected log-likelihood estimators. In terms of informational ranking, if the objective is to estimate with minimum a-priori assumptions, then the generalized ME estimator is superior to the other estimators. Two detailed examples, reflecting the two types of models, are discussed. The first example deals with estimating a first order Markov process. In the second example the empirical (natural) weights of each observation, together with the other unknowns, are the subject of interest.

Key words: Empirical likelihood, Information, Maximum entropy, Maximum likelihood.

1. Introduction

A major problem in comparing the maximum entropy (ME) approach to other inference methods is that their basic goals are inherently different. In the more traditional inference world, which includes the maximum likelihood (ML) and the generalized least squares (GLS) approaches, the objective is to maximize prediction within the sample given some moment representation of the data and some distributional assumptions. In contrast, the basic objective of the ME approach is to recover a set of unknown parameters using limited information (moment conditions) or minimum assumptions, and without assuming that the analyzed sample is the most likely sample representing the population. In most cases the ME approach is applied as a consistent way to assign prior probabilities and/or to solve underdetermined problems.

In general, when comparing two or more estimators, one should take into consideration both the inference criterion used as well as the data representation. A common inference criterion is "prediction" (goodness of fit) where the objective is to minimize some function (usually quadratic) of the errors, or similarly, to maximize some likelihood function. However, one could choose "precision" ("smoothness") of the estimates as the estimation objective. A joint estimation objective leading to a dual loss function where emphasis is on both prediction and precision is also possible (e.g., Akaike, 1986; Zellner, 1994; Golan, Judge and Miller, 1996). A common data representation is simply to transform the data (e.g., y or X) into a set of moments (e.g., X'y or X'X). This is the case with all likelihood-type estimators as well as the least square type estimators. Further, it is also possible to weigh the moments according to different criteria which leads to weighted-type estimators (e.g., empirical likelihood models, weighted least squares, etc.). In contrast, it may also be possible to account for each data point (observation) directly in the estimation process (e.g., the generalized ME estimator developed by Golan, Judge and Miller, 1996).

The objective of this paper is to develop a (informational) comparison of the ME and ML (Bayesian or non-Bayesian) approaches. To avoid comparing estimators of different objectives and data representations all estimators are couched within a constraint optimization framework with similar objective function for all. With this similar objective function in the form of a generalized entropy criterion, the various estimators differ only in terms of their data representation, the constraints. These constraints are the only information used in the inference process. Using this framework, and choosing an informational based criterion, all relevant estimators are nested within a general ME estimator. The informational criterion used is the "input information" or a-priori assumptions used in each one of the different nested estimation processes. The different estimators are then ranked based on their input information (e.g., Zellner, 1988). This approach differs from Csiszar's (1991) work on an axiomatic approach to inference for linear inverse models. He compares the different selection rules under the assumption that all linear constraints may occur.

The logic of the approach developed here is as follows. Given the basic postulates that lead to Shannon's entropy as a measure of uncertainty/information, viewing the inference process as one of an inversion process (with minimum assumptions), and building on the informational theoretic approach of Levine (1980), it is possible to use the entropy measure (obeying the same postulates) within a generalized maximum entropy approach to nest the different inversion procedures, or estimators, in a certain ordering. To do so, one maximizes the joint entropy criterion, defined over the possible signal and noise outcome space, subject to different forms of the data. It is then possible to convert the entropy primal approach into an unconstrained dual one (Agmon, Alhassid and Levine, 1979; Golan, Judge and Miller, 1996). Given the common criterion function and the different scenarios under which the data are represented, a dual unconstrained/concentrated function is developed for each estimator. Having converted the different estimators into this similar form, it is possible to compare them.

The main results of this comparison are as follows. In the "classical" ME problem where the objective is to estimate a set of unknown *natural, discrete proper probabilities* given some M moments, the ME and ML are equivalent. This type of estimation problems can be generalized to allow for noisy moments. In that case, the ME is equivalent to a generalized likelihood estimator. These type of problems are discussed and formulated in Section 3. In the second group of estimation problems, all in line with the linear statistical model, the objective is to estimate a set of K unknown parameters given T data points, and possibly M side-conditions for the likelihood function. Traditionally, these type of models are estimated by least-squares type estimators, ML or Empirical Likelihood (EL) if the M additional side-conditions are assumed to be known. The "classical" ME estimators, such as the Bayesian method of moments or the classical ME estimator, can all be converted into dual unconstrained estimators. These estimators can be viewed as *expected log-likelihood* functions of the exponential distribution with M parameters (the Lagrange multipliers) to be estimated given the M moment conditions. Further, these M restrictions have to be fulfilled exactly implying the real, and inherent, objective is to *maximize prediction* within the sample. In the generalized ME estimator discussed in Section 5, each data point is taken into account, no side-conditions in terms of M moments are assumed and the maximization is done over both prediction and precision. This estimator is the one that uses the smallest set of a-priori assumptions, or input information. These types of problems and their estimators are discussed in Sections 4-5.

Following a brief background, an informational inference criterion is defined in Section 2. In Section 3 a detailed example is discussed. This example serves as the basis for comparison of the ML and ME approaches where the objects to be inferred are natural probabilities. Building on this example, an extended ME estimator is reviewed and developed. A second example is discussed in Section 4. This example serves as a basis for constructing a comparison with the Empirical Likelihood (EL) approach where the objects to be estimated are not necessarily probability distributions. This comparison is based on Zellner's Information Processing Rule ,IPR, (Zellner, 1988). The general linear regression model is discussed in Section 5. A generalized ME model, based on Golan, Judge, and Miller (1996), is reviewed and contrasted with the ML, weighted least squares, EL and Bayesian method of moments (BMOM). A ranking of these estimates, based on their a-priori assumptions, within a nested model is given in Section 6. A conclusion is given in Section 7. Finally, those readers who are not interested in the detailed examples of Sections 3-4, can jump directly from Section 2 to Section 5.

2. An Informational Comparison

2.1 Background

Given the objective of estimating a proper probability distribution where the only information is in terms of some *expected values* of this distribution, Shore and Johnson (1980) show that maximizing the entropy is the only consistent procedure that satisfies some basic four axioms (uniqueness, invariance, system and subset independence). Thus, the ME (or cross entropy, CE) principle is viewed as a correct method of inference when given new information in terms of moments. Skilling (1989) generalizes this axiomatic (subset independence, coordinate invariance, system independence and scaling) basis for the ME. He shows that this approach is a consistent way of assigning probabilities for priors, while Bayes rule should be used to update these probabilities in line with the new evidence (data). Motivated by the above work, Csiszar (1991) searches for "logically consistent" rules for selecting an element from any possible feasible set. In contrast to Shore and Johnson (1980) and Skilling (1989), in Cisszar's work the axioms do not involve the measure to be characterized but rather the inference method it leads to. His main two axioms are "regularity" and "locality" where the second one is a weaker version of the above "subset independence" axiom. Adding a third axiom of "composition consistency", Csiszar shows that the only estimators satisfying these requirements are the LS and CE.

Throughout this paper the ME approach (classical and generalized) is used to infer probability distributions and point estimates from data (and if available, priors) where those data are not necessarily in form of moments of an unknown distribution. If, within the linear regression model, no distributional assumptions are made, then all estimation problems are practically ill-posed in the sense that, in addition to the traditional K unknowns, there are T observations with T unknown (natural) weights corresponding to each observation. Thus, to convert this ill-posed, underdetermined problem into a well-posed one, a certain criterion must be chosen and optimized subject to some moment restrictions on the unknown distribution. Viewing the estimation problem in this way, the entropy criterion, satisfying the above natural axioms, can be used to estimate the unknowns. The ML or EL objectives are just other such criteria. Unfortunately, these other criteria use more a-priori information. This additional information could be completely consistent with the underlying process that generated the data, or it could have no foundation in theory or data. But, the researcher can never validate these additional requirements (information). With this in mind, we now define the comparison criterion.

2.2 The Criterion

In searching for a "logically consistent" criterion for comparison of different estimation approaches, one possibility is to define an informational based criterion. Following Bernardo (1979) and Zellner (1988) the inference problem can be treated as a maximization of some decision rule subject to a set of constraints. The decision rule can be treated as an information processing rule, IPR, where a 100% efficient IPR is an estimator (or inversion process) that has an output-to-input information ratio of one. To measure the informational content of the input and output of an estimator, the entropy measure is used. Given the decision rule, the constraints are the only information available to the researcher and include both the "hard" data (observable) and the "soft" data (unobservable such as prior knowledge or other non-measurable beliefs based on theory). Thus, the comparison criterion measures the informational content of the input information, or the a-priori assumptions used for each one of the different inversion processes.

It is, then, possible to rank efficient IPR rules in terms of their input information. This input information represents the likelihood structure as well as the constraints (the data). In the present paper, comparisons are done *only* for 100% efficient and consistent IPR rules/estimators.

3. Example 1 - Matrix Balancing and Markov Process

Consider the following model

$$\mathbf{y} = P\mathbf{x} \tag{3.1}$$

where $\mathbf{y}$ is a K-dimensional vector, $\mathbf{x}$ is a K-dimensional vector, and P is a $(K \times K)$ matrix of unknown and unobservable probabilities to be recovered where

$$\sum_i p_{ij} = 1 \quad i = 1,2,\ldots,K \tag{3.2a}$$

$$p_{ij} \geq 0 \quad . \tag{3.2b}$$

The objective is to recover P from the aggregated data $\mathbf{y}$ and $\mathbf{x}$. However, since the number of unknowns outnumbers the number of data points this problem is ill-posed (underdetermined) and one needs to choose a certain criterion to reduce the problem to a well-posed problem.

Before solving this problem, a short discussion of different applications of this type of model, are discussed. First, let $\mathbf{y}$ be a K-dimensional vector of proportions for each one of the kth states in period $(t+1)$, and let $\mathbf{x}$ be a K-dimensional vector of proportions for each state k in period t, then P is a $(K \times K)$ matrix of first order Markov transition probabilities. If more than two periods of data exist, the same framework holds with T K-dimensional vectors $\mathbf{y}_t$ and $\mathbf{x}_t$.

Second, if $\mathbf{y}$ and $\mathbf{x}$ are both K-dimensional vectors of observed aggregated data, such as row and column sums for an input-output (IO) table or social accounting matrix (SAM) for a given period, then P are just the IO or the SAM coefficients (see Golan and Vogel, 1997, for a detailed formulation and empirical examples). This type of problem is also known as "Matrix Balancing". Note that in a SAM framework, used frequently in economics, $\mathbf{y}=\mathbf{x}$. Finally, since it is simpler to work with $\mathbf{y}$ and $\mathbf{x}$ as shares in both the IO, or SAM, models a normalization of the aggregates is done by dividing each element of $\mathbf{x}$ and $\mathbf{y}$ by $Max\{x_j, y_i\}$.

3.1 Maximum Entropy Approach to the Markov Problem

Using the entropy objective (Shannon, 1948) and following Jaynes (1957a, 1957b, 1963, 1984) the estimation model is simply

$$\underset{\mathbf{P}}{Max}\, H(P) \equiv -\sum_i \sum_j p_{ij} \ln p_{ij} \tag{3.4}$$

subject to (3.1) and (3.2). This optimization yields the most uniform set of coefficients, $\hat{P}$, that satisfies the data. If prior information exists, say P^0, then the problem is to minimize the Cross-Entropy (CE) criterion (e.g., Kullback, 1959; Good, 1963)

$$I\left(P; P^0\right) \equiv \sum_i \sum_j p_{ij} \ln\left(p_{ij} / p_{ij}^0\right) . \tag{3.5}$$

Thus, one seeks to recover the unknowns p_{ij} that satisfy the data $\mathbf{y}$ and $\mathbf{x}$ and be as close as possible to p_{ij}^0. To do so, we minimize (3.5) subject to (3.1)-(3.2). The estimated coefficients are

$$\tilde{p}_{ij} = \frac{p_{ij}^0 \exp\left(\tilde{\lambda}_i x_j\right)}{\sum_i p_{ij}^0 \exp\left(\tilde{\lambda}_i x_j\right)} \equiv \frac{p_{ij}^0 \exp\left(\tilde{\lambda}_i x_j\right)}{\Omega_j\left(\tilde{\lambda}_i\right)} \quad . \tag{3.6}$$

Given these (posteriors/post-data) estimates it is possible to reformulate the problem as a dual unconstrained one. As such, it is computationally superior and allows comparison with the ML estimator. Starting with the Lagrangean, substituting for the p_{ij}'s in $I(p_{ij}; p_{ij}^0)$ the posterior (3.6) and omitting the last term of the Lagrangean (since the posteriors already satisfy the normalization), yields

$$L(\lambda) = \sum_i \sum_j p_{ij} \ln\left(p_{ij} / p_{ij}^0\right) + \sum_i \lambda_i \left(y_i - \sum_j p_{ij} x_j\right) \tag{3.7}$$

$$= \sum_i \sum_j p_{ij} \ln\left[\frac{p_{ij}^0 \exp\left(\lambda_i x_j\right)}{\Omega_j}\right] - \sum_i \sum_j p_{ij} \ln p_{ij}^0 + \sum_i \lambda_i y_i - \sum_i \sum_j \lambda_i p_{ij} x_j$$

$$= \sum_i \lambda_i y_i - \sum_j \ln \Omega_j(\lambda).$$

Maximizing $L(\lambda)$ with respect to λ and equating to zero yields $\tilde{\lambda}$, which in turn, yields the estimates $\tilde{p}_{ij}$ via Eq. 3.6.

3.2 Maximum Likelihood Approach to the Markov Problem

Given the **x** and **y** where $y_i \in [0,1]$, the likelihood function could be expressed as

$$L = \prod_{j=1}^{K} p_{1j}^{y_1} p_{2j}^{y_2} \cdots p_{Kj}^{y_K} \tag{3.8}$$

and, thus, the log-Likelihood function is

$$\ln(L) \equiv \ell = \sum_i \sum_j y_i \ln p_{ij} \tag{3.9}$$

where P is a set of K proper probability distributions. Now, since K is large it is not practical to choose the normal distribution (probit) and one is left with the logistic (exponential) distribution. That is,

$$p_{ij} = \frac{\exp\left(\beta_i x_j\right)}{1 + \sum_{i=2}^{K} \exp\left(\beta_i x_j\right)} \equiv \frac{\exp\left(\beta_i x_j\right)}{\Omega_j(\beta)} \quad \text{for } i = 2, \ldots, K \tag{3.10a}$$

and

$$p_{1j} = \frac{1}{\Omega_j(\beta)} \quad \text{for } i = 1 \tag{3.10b}$$

Substituting (3.10) into (3.9) yields

$$\ell = \sum_i \sum_j y_i \ln \left[\frac{\exp(\beta_i x_j)}{1 + \sum_{i=2}^{K} \exp(\beta_i x_j)} \right] \tag{3.11}$$

$$= \sum_i \sum_j y_i \beta_i x_j - \sum_i \sum_j y_i \ln \Omega_j(\beta) = \sum_i y_i \beta_i - \sum_j \ln \Omega_j(\beta)$$

which is just a variation of the ML-logit and is equivalent to the dual unconstrained (3.7) with $\beta = \lambda$ and with uniform priors (i.e., $p_{ij}^0 = \frac{1}{K}$ for all i and j).

The reasons for the similarity of the ME and ML estimators in this case is first, the data representation is done in terms of aggregates, or moments, and under the traditional requirement that those restrictions (moments) hold exactly and second, both IPRs use identical maximization criterion. The first requirement is relaxed and discussed in the next section. But, before proceeding, it is important to note that due to this ME-ML similarity, the usual inferential statistics hold for the ME estimator. For completeness of presentation, some summary statistics are discussed below. For more, see Golan and Vogel (1997).

3.3 Diagnostics and Inference of the Markov Problem

Within the entropy approach used here, one can investigate the amount of information in the estimated coefficients. Let the normalized entropy (information) measure be

$$S(\tilde{P}) \equiv \frac{-\sum_i \sum_j \tilde{p}_{ij} \ln \tilde{p}_{ij}}{K \ln(K)} \tag{3.12}$$

where this measure is between zero and one with one reflecting uniformity (complete ignorance) of the estimates, and zero reflects perfect knowledge. If the CE is used, the divisors should be the relevant priors $\mathbf{p}^0$. This measure, reflects the information in the whole system. Similar normalized measures to reflect the information in each one of the j's are easily defined.

Next, following our derivation of the ME estimator, it is possible to construct a log-likelihood, or entropy, ratio test. Let l_Ω be the constrained (by the data) log-likelihood (or ME), and l_ω be the unconstrained one where, say $\beta = \lambda = \mathbf{0}$. Then, the log-likelihood (entropy) ratio statistic is just $-2 \ln \frac{l_\omega}{l_\Omega}$. However, due to the equivalence between the ML and the ME, the log-likelihood value is just the value of $Max(H)$ while $L_\omega = K \ln K$. Thus, the log-likelihood ratio statistic is just

$$W(ME) = 2K \ln(K)\left[1 - S(\tilde{P})\right] \quad . \tag{3.13a}$$

Under the null hypothesis, W(ME) converges in distribution to $\chi^2_{(K-1)}$. Finally, a *Pseudo-* R^2 measure (e.g., McFadden, 1974) can be constructed as

$$Pseudo-R^2 \equiv 1-\frac{l_\Omega}{l_\omega}=1-S(\tilde{P}). \tag{3.13b}$$

These measures are easily adjusted for the CE case. We now turn to the issue of "relaxing" the exact moment requirements.

3.4 Maximum Entropy Approach to the Markov Problem - Revisited

3.4.1 Generalization

Due to the aggregated level of the data, or due to possible noisy observations, it is quite likely that the constraints (3.1) will not hold exactly. Further, the constraints represent a relationship between an observed sample of data and the population moments but the two may be different. Unlike other theoretical approaches, the constraints (say, in the Social Sciences) here do not represent exact conservation laws; they represent noisy and aggregated data. Thus, one can reformulate the constraints (the data) as

$$y_i=\sum_j p_{ij}x_j+e_i \tag{3.14}$$

where each error $e_i \in[-1,1]$. To construct this type of generalized model within the ME formalism, the e_i's need to be transformed into proper probabilities. Following Golan, Judge, and Miller (1996) the constraints are transformed to

$$y_i=\sum_j p_{ij}x_j+\sum_h v_h w_{ih} \tag{3.15}$$

where $\mathbf{w}_i$ is an H-dimensional vector of weights satisfying

$$\sum_h w_{ih}=1 \tag{3.16a}$$

and

$$\sum_h v_h w_{ih}\equiv e_i\ . \tag{3.16b}$$

The vector $\mathbf{v}$ is an H-dimensional support space with $H\geq 2$, symmetric around zero and with end points of $-1/K$ and $1/K$. We note here that even though Csiszar (1991) acknowledges the need to incorporate the possibility of errors in the constraints, the reformulation and approach taken here is different. In Csiszar's discussion of the errors, he suggests treating the model as a system of inequalities if the bounds on the magnitude of the errors are known.

Building on the CE model of Section 3.1, the new generalized model is

$$\underset{P,\mathbf{w}}{Min}\quad I\left(P,\mathbf{w};P^0,\mathbf{w}^0\right)=\left\{\sum_i\sum_j p_{ij}\ln\left(p_{ij}/p_{ij}^0\right)+\sum_i\sum_h w_{ih}\ln\left(w_{ih}/w_{ih}^0\right)\right\} \tag{3.17}$$

subject to (3.15), (3.2), and (3.16a).

The optimization yields

$$\tilde{p}_{ij} = \frac{p_{ij}^0 \exp\left(\tilde{\lambda}_i x_j\right)}{p_{1j}^0 + \sum_{i=2}^{k} p_{ij}^0 \exp\left(\tilde{\lambda}_i x_j\right)} \equiv \frac{p_{ij}^0 \exp\left(\tilde{\lambda}_i x_j\right)}{\Omega_j\left(\tilde{\lambda}\right)} \tag{3.18}$$

and

$$\tilde{w}_{ih} = \frac{w_{ih}^0 \exp\left(\tilde{\lambda}_i v_h\right)}{\sum_h w_{ih}^0 \exp\left(\tilde{\lambda}_i v_h\right)} \equiv \frac{w_{ih}^0 \exp\left(\tilde{\lambda}_i v_h\right)}{\Psi_i\left(\tilde{\lambda}\right)} \tag{3.19}$$

where w_{ih}^0 are the prior probabilities defined over the support space **v** and are taken to be uniform. Given this solution, the dual unconstrained problem can be derived as before. Further, the same diagnostic and inferential statistics developed earlier apply here as well.

Under this generalized criterion function the objective is to minimize the joint entropy distance between the data and the priors. It is a dual-loss objective function that assigns equal weights to prediction and precision. Equivalently, it can be viewed as a shrinkage estimator that simultaneously shrinks the data to the priors (or uniform distributions in the ME case) and the errors to zero. Next, to show the possibilities of this new ME estimator, the example is generalized further to accommodate for *(i)* possible non-stationarity in the data and for *(ii)* more data.

3.4.2 Extension to Non-stationary Cases

In many cases, the researcher may have aggregated data for more than a single period. Given these data it is possible to generalize the previous formulation as follows. Let x_{tj} and y_{ti} be the aggregated data per period $t = 1,2,\ldots,T$. Further, let $\mathbf{z}_t$ be an N-dimensional vector of additional covariates per period. These additional covariates represent information that *is not captured* by x or y. Thus, the available information may be represented as

$$\sum_t y_{ti} z_{tn} = \sum_t \sum_j p_{ij} x_{tj} z_{tn} + \sum_t \sum_h w_{tih} z_{tn} v_h \tag{3.20}$$

where, now, the error bounds **v** are specified as $\pm \frac{1}{K\sqrt{T}}$. Assuming we have no priors, the generalized non-stationary maximum entropy model is just

$$\underset{P,\mathbf{w}}{Max}\ H(P,\mathbf{w}) \equiv \left\{ -\sum_i \sum_j p_{ij} \ln p_{ij} - \sum_t \sum_i \sum_h w_{tih} \ln w_{tih} \right\} \tag{3.21}$$

subject to (3.20) and the normalization requirements for P and **w**.

This optimization yields

$$\tilde{p}_{ij} = \frac{\exp\left(-\sum_t \sum_n \tilde{\lambda}_{in} x_{tj} z_{tn}\right)}{1 + \sum_{i=2}^{K} \exp\left(-\sum_t \sum_n \tilde{\lambda}_{in} x_{tj} z_{tn}\right)} \equiv \frac{\exp\left(-\sum_t \sum_n \tilde{\lambda}_{in} x_{tj} z_{tn}\right)}{\Omega_j\left(\tilde{\lambda}_n\right)} \tag{3.22}$$

and

$$\tilde{w}_{ith} = \frac{\exp\left(-\sum_n \tilde{\lambda}_{in} z_{tn}\right)}{\sum_h \exp\left(-\sum_n \tilde{\lambda}_{in} z_{tn}\right)} \equiv \frac{\exp\left(-\sum_n \tilde{\lambda}_{in} z_{tn}\right)}{\Psi_{it}\left(\tilde{\lambda}_n\right)} \tag{3.23}$$

where the $\tilde{p}_{ij}$ are "constant" over the T periods and the non-stationarity is captured through the corresponding $\mathbf{z}_t$ and $\mathbf{x}_t$. The direct effects of $\mathbf{x}$ and $\mathbf{z}$ are captured through the marginals $\frac{\partial p_{ij}}{\partial \bar{x}_j}$ and $\frac{\partial p_{ij}}{\partial \bar{z}_n}$ where $\bar{x}_j$ and $\bar{z}_n$ are the mean values (over *t=1, 2, ..., T*) are the mean values (over $t = 1,2,\ldots T$ of $\mathbf{x}_j$ and $\mathbf{z}_n$. These marginals are

$$\frac{\partial p_{ij}}{\partial \bar{x}_j} = \tilde{p}_{ij}\left[\sum_n \tilde{\lambda}_{in}\bar{z}_n - \sum_i\sum_n \tilde{p}_{ij}\tilde{\lambda}_{in}\bar{z}_n\right] \tag{3.24a}$$

and

$$\frac{\partial p_{ij}}{\partial \bar{z}_n} = \tilde{p}_{ij}\bar{x}_j\left[\tilde{\lambda}_{in} - \sum_i \tilde{p}_{ij}\tilde{\lambda}_{in}\right] \tag{3.24b}$$

evaluated at the means.

Finally, the unconstrained non-stationary model is

$$L(\lambda) = \sum_t\sum_i\sum_n \lambda_{in} y_{ti} z_{tn} + \sum_j \ln\Omega_j(\lambda) + \sum_t\sum_i \ln\Psi_{it}(\lambda) \tag{3.25}$$

where the same diagnostics and inferential statistics apply here as well.

3.5 Maximum Likelihood Approach to the Markov Problem - Revisited

Following the same logic as before, and using the logit (exponential) distribution for both P and w, yields the generalized log-likelihood function

$$\ell = \sum_t\sum_n\sum_i y_{ti} z_{tn}\beta_{in} - \sum_j \ln\Omega_j(\beta) - \sum_t\sum_i \ln\Psi_{ti}(\beta) \tag{3.26}$$

with $\beta = -\lambda$ and with uniform priors for both P and $\mathbf{w}$.

To see the relationship with the more traditional ML, let the end-points of the support space v be $\pm\left(\frac{1}{K\sqrt{T}}\right)$, thus as $T \to \infty$, $\mathbf{v} \to \mathbf{0}$, the generalized ML-logit converges to the "traditional" ML-logit model. However, since the whole logic here is to estimate the parameters of a small data set, this argument is just a consistency one.

3.6 Uncertainty

Since under this reformulation the ME and the traditional ML are different, it is of great interest to compare the two. Given our earlier discussion, the comparison criterion used is the amount of input information, or similarly a ranking based on the (a-priori) assumptions used.

In this example, however, the comparison is simpler as the traditional ML is just *nested* within the extended ME model. Given the joint dual-loss objective function (3.21), or its cross-entropy counter-part, the extended ME model is just Maximizing $H(P,w)$ subject to (3.20) and the normalization requirements, while the traditional-ML IPR is achieved by maximizing $H(P,\mathbf{w})$ subject to (3.20) and the normalizations for the P's *but* with **v=0**. This is equivalent to maximizing $H(P,\mathbf{w})$ subject to the *pure* version of Eq. (3.20), which is just

$$\sum_t y_{ti} z_{tm} = \sum_t \sum_j p_{ij} x_{tj} z_{tm} \, . \tag{3.20a}$$

This "pure" representation of the data is similar to requiring that the population moments and the sample moments be equal, thereby introducing the implicit assumption that this is the most likely sample to represent the population. Thus, the objective of the estimation process reduces back to "prediction" instead of both *prediction* and *precision.* In other words, under the generalized entropy formulation, the common "conservation laws" are relaxed since with these noisy data, it seems inappropriate to require that the moments of a noisy subset of the population be exactly equal to the (unknown) population's moments. This "relaxation" is especially important for small samples of non-experimental data.

Finally, looking at the constraints in a different way, one sees that (3.20) is just a *relaxed* version of (3.20a), allowing for variations around the sample moments. This implies that the Lagrange multipliers, λ, which are the basic unknowns of interest, are of a smaller magnitude which, in turn, yields more uniform **p**'s. The implication is that always $H(\hat{\mathbf{p}})_{ML} \le H(\hat{p})_{ME}$. Since both estimators are 100% efficient, it follows that the generalized entropy approach uses *fewer* a-priori assumptions.

4. Example 2 - A Simple Linear Model

Consider the sample $y_1,\ldots,y_T$ which are i.i.d. observations from an unknown distribution F_0. There is a ***p***-dimensional parameter θ associated with F_0. The only information about θ and F_0 is in the form of some M moments. Further, assume that a unique relationship between these M moments (of the observed data) and the population is known, and is represented as

$$\sum_t p_t g_m(y_t;\theta) = 0 \qquad m = 1,\ldots, M \tag{4.1}$$

The objective is to estimate θ.

If the researcher can live happily with the assumption that she/he knows the function F_0, then the likelihood function is specified and the ML estimator is used. If, on the other hand, the distribution F_0 is unknown then a correct approach would be to search for the *natural weights* of each one of the T observations (data points). However, such an approach implies that the problem is, again, underdetermined (ill-posed) since $M<T$. Thus, again one needs to choose a certain criterion for eliminating all but a single probability vector **p** that is consistent with the data.

4.1 Maximum Empirical Likelihood for the Simple Linear Model

One such criterion is the Empirical (discrete) Likelihood (EL) criterion used in the recent statistical literature. Under this criterion one maximizes the empirical (weights) likelihood of **p** subject to the M moment restrictions. Following Hall (1990), Owen (1990, 1991), Qin and Lawless (1994), and DiCicco, Hall and Romano (1991) the EL approach, or IPR, is just to

$$\underset{\mathbf{p}}{Max}\left\{\frac{1}{T}\sum_{t=1}^{T}\ln p_t\right\} \tag{4.2}$$

subject to the M moment restrictions (4.1) and the usual normalization and positivity restrictions

$$\sum_t p_t = 1 \quad and \quad p_t \geq 0 \quad . \tag{4.3}$$

Solving for the first order conditions yields the optimal set of empirical weights

$$\hat{p}_t(EL) = T^{-1}\left[\sum_m \hat{\lambda}_m g_m(y_t;\theta) + 1\right]^{-1} \tag{4.4}$$

where $\hat{\lambda}$ is the M-dimensional optimal Lagrange multipliers associated with (4.1). If $\hat{p}_t = \frac{1}{T}$ for all t, the $EL{=}ML$. For inferential properties, other statistics, and proof of consistency see for example Qin and Lawless (1994).

Why should one use the EL criterion? Under this criterion, one maximizes the discrete-empirical likelihood of the sample. This is equivalent to choosing the unknown weights, satisfying the properties of a proper distribution, under the assumption that the observed sample is the *most probable sample that satisfies the M constraints.* Given that the true underlying mechanism generating the data is unknown, and given our ignorance with regard to the population's moments, it seems that this criterion may be somewhat restrictive.

4.2 Maximum Entropy for the Simple Linear Model

Instead of using (4.2), one can use an *expected empirical log-likelihood* measure:

$$\underset{\mathbf{p}}{Max}\ H(\mathbf{p}) \equiv -\sum_t p_t \ln p_t \tag{4.5}$$

subject to (4.1) and (4.3). Thus, instead of maximizing the EL criterion the discrete entropy, or the "expected log-likelihood", criterion is maximized to yield

$$\hat{p}_t(ME) = \frac{\exp\left(-\sum_m \hat{\lambda}_m g_m(y_t;\theta)\right)}{\sum_t \exp\left(-\sum_m \hat{\lambda}_m g_m(y_t;\theta)\right)} \equiv \frac{\exp\left(-\sum_m \hat{\lambda}_m g_m(y_t;\theta)\right)}{\Omega(\hat{\boldsymbol{\lambda}})} \tag{4.6}$$

If instead of the discrete entropy criterion one uses the continuous criterion subject to the M side-conditions, the Zellner's BMOM estimator results. This estimator is discussed in section 5.

4.3 Uncertainty

Having estimated the weights via these two different methods it is possible to compare them under the criterion of Section 2. Following Golan and Judge (1996) it is possible to show that the EL IPR uses more input information than the ME IPR. The main argument of the proof is now discussed.

Zellner (1988) developed a way to evaluate estimators via IPRs where he defines a measure for capturing the informational content of the inputs and outputs of an estimator. The two inputs are the data density, or likelihood function, and the prior distribution. The two outputs are the post-data distribution $p(\theta|\mathbf{y})$ and the marginal data density (the partition function). Thus, for example, the two input information measures for the ME (or CE with non-uniform priors q_t) are

$$I_1(\hat{\mathbf{p}}) \equiv -\sum_{t=1}^{T} \hat{p}_t \ln q_t$$

$$I_2(\hat{\mathbf{p}}) \equiv -\sum_t \hat{p}_t \ln\left[\exp\left(-\sum_m \hat{\lambda}_m g_m(y_t;\theta)\right)\right] = \sum_t \sum_m \hat{p}_t \hat{\lambda}_m g_m(y_t;\theta).$$

The two output information measures are

$$O_1(\hat{\mathbf{p}}) \equiv -\sum_t \hat{p}_t \ln \hat{p}_t$$

$$O_2(\hat{\mathbf{p}}) \equiv -\sum_t \hat{p}_t \ln\left[\Omega(\hat{\boldsymbol{\lambda}})\right] = -\ln \Omega(\hat{\boldsymbol{\lambda}}) \quad .$$

Similarly, one can define the input and output information for the *EL*. For simplicity of comparison, the uniform priors are always assumed for all estimators. This way, the comparison could also be done with estimators that are not constructed to include priors. Next, since the prior information, I_1, is equal for both estimators, and since both are 100% efficient it is sufficient to show that $I_2(ME) > I_2(EL)$. But, since the EL results always with estimates $\hat{\mathbf{p}}(EL)$ that have lower entropy than $\hat{\mathbf{p}}(ME)$ it follows that $O(EL) < O_1(ME)$ for all $\boldsymbol{\lambda} \neq \mathbf{0}$. Finally, since both are 100% efficient IPR's, it must be that $I_2(EL) < I_2(ME)$, thereby showing that the *EL* approach uses more input information, or simply uses more a-priori assumptions. Translating this result into "uncertainty" terms implies that *(i)* the ME estimator results in a more uniform **p** than the *EL*, and *(ii)* that the variances of the real unknown parameters $(\boldsymbol{\lambda})$ are smaller for the ME estimator, and that *(iii)* the relevant R^2, or $Pseudo\text{-}R^2$, for the ME is smaller and thus, reflecting a higher level of uncertainty.

5. The General Linear Regression Model.

Consider the noisy inverse model

$$\mathbf{y} = X\boldsymbol{\beta} + \mathbf{e} \tag{5.1}$$

where $\boldsymbol{\beta}$ is a *K*-dimensional vector of unobserved parameters, $\mathbf{y} = (y_1, y_2, \ldots, y_T)'$ is a *T*-dimensional vector of observed data, and *X* is a $(T \times K)$ design matrix.

5.1 Maximum Entropy

Given this linear model, the objective is to recover the unknown vector β with minimum assumptions. To retain this objective, no distributional assumptions are used and no assumptions regarding an exact relationship between sample and population moments are imposed. Thus, both the signal, β, and the noise (unknown error distribution), **e**, are assumed to be unknown. The objective is, then, to simultaneously recover β and **e**.

To achieve this goal, Golan, Judge, and Miller (1996, Chapter 6) developed the generalized ME estimator. As a first step, the parameters of Eq. 5.1 reparameterized as

$$\mathbf{y} = X\boldsymbol{\beta} + \mathbf{e} \equiv \mathbf{X}\mathbf{Z}\boldsymbol{\pi} + \mathbf{v}'\mathbf{w} . \tag{5.2}$$

In this reformulation each β_k is defined as

$$\beta_k \equiv \sum_m z_{km}\pi_{km} \tag{5.3a}$$

with

$$\sum_m \pi_{km} = 1 \tag{5.3b}$$

and where $\mathbf{z}_k = (z_{k1}, z_{k2}, \ldots, z_{kM})'$ is an $M \geq 2$ dimensional vector. This vector serves as a discrete support space, for each one of the K unknowns, and is specified such that it will span the possible (unknown) value of β_k. Thus, under this formulation, the unknown β are converted from the real line to the [0,1] interval with the properties of K proper probability distributions, π_k, defined over the K support spaces $\mathbf{z}_k$. How should one specify $\boldsymbol{Z}$? If the researcher posses no knowledge as to the possible value of β_k, $\mathbf{z}_k$ should be specified to be *symmetric around zero* and with large negative and positive boundaries. For example, $z_{k1} = -z_{kM} = -10^6$. However, in a large number of cases the researcher has some knowledge regarding the possible values of β and may use this knowledge in specifying $\boldsymbol{Z}$.

In a similar way, the e_t's may be transformed into T proper probability distributions. Define a discrete support space v of dimension $J \geq 2$, equally spaced and *symmetric around zero* and associate with it a set of proper weights, w_{tj}, such that

$$e_t \equiv \sum_j v_j w_{tj} \tag{5.4a}$$

and

$$\sum_j w_{tj} = 1 \quad . \tag{5.4b}$$

Unlike the previous Markov model, the end points (v_1 and v_J) are chosen to be $-3\sigma_y$ and $3\sigma_y$ where σ_y is the *empirical* standard deviation of y.

Having converted the two sets of unknowns into probability distributions, the estimation problem becomes

$$\underset{\pi,\mathbf{w}}{Max}\ H(\pi,\mathbf{w}) = \left\{-\sum_k\sum_m \pi_{km}\ln\pi_{km} - \sum_t\sum_j w_{tj}\ln w_{tj}\right\} \tag{5.5}$$

subject to

$$y_t = \sum_k\sum_m x_{tk}z_{km}\pi_{km} + \sum_j v_j w_{tj} \tag{5.6}$$

$$\sum_m \pi_{km} = 1 \tag{5.7a}$$

$$\sum_j w_{tj} = 1 \quad . \tag{5.7b}$$

The optimal solution is

$$\hat{\pi}_{km} = \frac{\exp\left(-\sum_t \hat{\lambda}_t x_{tk} z_{km}\right)}{\sum_m \exp\left(-\sum_t \hat{\lambda}_t x_{tk} z_{km}\right)} \equiv \frac{\exp\left(-\sum_t \hat{\lambda}_t x_{tk} z_{km}\right)}{\Omega_k(\hat{\lambda})} \tag{5.8a}$$

and

$$\hat{w}_{tj} = \frac{\exp\left(-\hat{\lambda}_t v_j\right)}{\sum_j \exp\left(-\hat{\lambda}_t v_j\right)} \equiv \frac{\exp\left(-\hat{\lambda}_t v_j\right)}{\Psi_t(\hat{\lambda})} . \tag{5.8b}$$

Finally, the point estimates $\hat{\beta}$ and $\hat{\mathbf{e}}$ are just $\hat{\beta}_k = \sum_m z_{km}\hat{\pi}_{km}$ and $\hat{e}_t = \sum_j v_j \hat{w}_{tj}$.

Before proceeding we note that if prior information regarding the possible location (within the support $\mathbf{z}_k$) of the point estimates β_k exists it could be easily formulated within the CE objective.

Finally, following Golan, Judge and Miller (1996), the dual, concentrated model can be constructed. Substituting into the first element of the RHS of the Lagrangean the post-data $\hat{\pi}'s$ and $\hat{\mathbf{w}}'s$, yields

$$L(\lambda) = \sum_t y_t\lambda_t + \sum_k \ln\Omega_k(\lambda) + \sum_t \ln\Psi_t(\lambda) . \tag{5.9}$$

Minimizing $L(\lambda)$ with respect to λ and equating to zero yields $\hat{\lambda}$ which, in turn, yields $\hat{\beta}$ and $\hat{\mathbf{e}}$.

5.2 Maximum Likelihood

Given the linear model (5.1) together with the traditional assumptions leads to the OLS IPR and with the normality assumption inference can be made. This is just the traditional ML. If, however, the objective remains to infer without distributional assumptions one may proceed via the EL approach to, first, recover the unknown weights of each observation (as was discussed in Section 4.1) and then, recover the set of unknown β given the weighted observations. That is, the profile EL for β is

$$R(\beta) = Max\left\{\frac{1}{T}\sum_t \ln p_t \middle| p_t \geq 0, \sum_t p_t = 1, \sum_t p_t \mathbf{x}_t'(y_t - \mathbf{x}_t\beta) = 0\right\} \tag{5.10}$$

where $-2\ln R(\beta) \rightarrow \chi^2_{(K)}$.

This estimator is somewhat similar to the "weighted" least squares approach (e.g. Imbens, 1993; Imbens and Hellerstein, 1994) where

$$\hat{\beta}_{WLS} = \left[\sum_t \hat{p}_t \mathbf{x}_t \mathbf{x}_t'\right]^{-1}\left[\sum_t \hat{p}_t \mathbf{x}_t y_t\right] \tag{5.11}$$

and the weights $\hat{p}_t$ are the solution to

$$\underset{\mathbf{p}}{Max}\ \frac{1}{T}\sum_t \ln p_t \tag{5.12}$$

subject to

$$\sum_t p_t = 1, p_t \geq 0, \text{ and } \sum_t p_t g(y_t, \mathbf{x}_t) = 0 . \tag{5.13}$$

If $\hat{p}_t = \frac{1}{T}$ for all t, then $\hat{\beta}_{WLS} = \hat{\beta}_{LS} = \hat{\beta}_{ML}$.

5.3 Discussion

Having developed and reviewed both the likelihood and the ME type IPRs, it is useful to summarize their main features, advantages and disadvantages based on their basic statistical assumptions. These assumptions are embodied in the objectives and data representations of these different inversion procedures.

5.3.1 The Objectives

Given a sample of T observations and no knowledge of the underlying distribution and mechanism that generated the data, the basic estimation problem is ill-posed due to the T unknown basic weights. Thus, there is a need for a criterion function to choose among the possible **p**'s satisfying the data. The EL model, or weighted LS model (5.11)-(5.13), both maximize the discrete log-likelihood function. Thus, the objective is to estimate the *most probable set of (discrete) weights* **p** given the data represented by a set of M moments. In a similar way, one can choose to optimize the *expected log-likelihood*, the entropy objective (The ME model of Section 4.2 or, in the BMOM case, the continuous entropy measure). This ensures that the recovered **p**'s are the most uniform or uninformed. However, this is *not* the approach taken in the generalized ME estimator of Section 5.

Unlike, all other ML, EL or BMOM criteria where the objective is defined directly on the unknown probability distribution **p** supported on the data, under the generalized ME IPR for the linear regression model (5.1), the objective is not defined directly on **p**. In fact, **p** does not enter in the objective (5.5) nor in the constraints (5.6). The objective function contains both π and **w** where the π's are defined over the support space for the possible β's, while the **w**'s are defined over the symmetric around zero support **v**. The empirical weights (of each observation) are determined *indirectly* via the T Lagrange multipliers. There is a unique multiplier associated with each observation, and as such, these multipliers capture the observations' natural weights. Thus, the **p**'s can be recovered *indirectly* via

$$\hat{p}_t \equiv \frac{\exp\left(-\hat{\lambda}_t\right)}{\sum_t \exp\left(-\hat{\lambda}_t\right)} \tag{5.14}$$

See Golan and Judge (1996) for further discussion.

5.3.2 The Data Representation

In all likelihood and BMOM type models the data are represented in its moment formulation. Further, under the EL and BMOM approaches, in addition to the moment representation of the T observations, there are also M additional moment restrictions while in the traditional ML there are distributional assumptions. These M side-conditions define the exact form of the M-1 moments of the (unknown) distribution under the assumption that a unique, and a well defined, relationship with the population moments must hold.

In contrast, under the generalized ME approach each observation is directly taken into account in the estimation process. Thus, no strict requirements on the direct relationship between sample and population moments are used. Each observation has to be directly accounted for in the optimization, while the noise is "pushed" toward zero, but is not required to be exactly zero on average.

5.3.3 Maximum Likelihood and Classical Maximum Entropy

A Major motivation for using ME is for evaluating underdetermined problems and for constructing priors within a Bayesian context. The main intuitive argument is that one should choose the most conservative, or uninformed, probability distribution out of the infinitely many satisfying some M moments. The axiomatic motivation was discussed in Section 2.1. Within a more traditional inference context, the ME (in continuous form) has been used when assumptions regarding the likelihood function are not desired. Such a method is the Bayesian method of moments (BMOM). See Zellner (1996, 1997) and Tobias and Zellner (1997).

The BMOM was developed with the objective of computing post-data densities for parameters, and for future observations, when insufficient information regarding a likelihood function is available. Thus, maximizing the continuous entropy (of a density function) subject to the M moments and normalization, yields the average log height of the density function which is the least informative density given these M side-conditions. Substitute the continuous entropy measure for the EL objective in model (5.10) or (5.11)-(5.13) yields estimates that are closely related to the BMOM.

Following the derivations discussed here, the ME-ML (or EL) relationship can be split into two distinct groups. The first, is the "classical" ME where the objective is to estimate a set of unknown *natural, discrete proper probabilities* given some M moments. In such problems the ME and ML are equivalent whenever the likelihood function is defined over the multinomial distribution. This is because the multinomial coefficient (multiplicity factor) and the entropy are, practically, the same. The generalized classical ME estimator for the "relaxed" moments is thus just a generalized likelihood IPR. These type of problems were discussed in Section 3.

The second group of estimators is the more interesting one and reflects the common linear statistical models. In these models the objective is to estimate a set of K unknown parameters given T data points, and possibly M side-conditions for the likelihood function. Traditionally, these type of models are estimated by least-squares type estimators, ML or EL if the M additional side-conditions are assumed to be known. The "classical" ME estimators, such as the BMOM or the discrete model of Section 4, can all be converted into dual unconstrained IPRs. These IPRs can be viewed as *expected log-likelihood* functions of the exponential distribution with M parameters (the Lagrange multipliers) to be estimated given the M moment conditions. Further, this M restrictions have to be fulfilled exactly implying the real, and inherent, objective is to *maximize prediction* within the sample. Finally, unlike the EL or ML IPRs for the linear model (5.1), the discrete or continuous entropy maximization results in an expected log-likelihood IPR which is more conservative than the EL or ML. These type of problems were discussed in Sections 4-5.

6. Maximum Entropy, Likelihood, and Uncertainty

Having discussed the generalized ME IPR and contrasted it with the ML, EL, and BMOM, it is now possible to nest these models in a certain informational order. This order is based on their input-information. Two possible routes can be taken. The first is to develop a generic entropy criterion and differentiate the various IPRs by their *hard* and *soft* data representation. The second, and somewhat similar approach, is to form a general dual unconstrained ME IPR and investigate the conditions under which it approaches each one of the other IPRs. The first approach, which seems to be more intuitive, is taken here.

6.1 The General Objective Function

A general entropy objective that encompasses all these linear IPRs (in discrete form) is

$$H(\mathbf{p},\pi,\mathbf{w}) \equiv H(\mathbf{p}) + H(\pi,\mathbf{w}) = -\sum_t p_t \ln p_t - \sum_k \sum_m \pi_{km} \ln \pi_{km} - \sum_t \sum_j w_{tj} \ln w_{tj} \quad (6.1)$$

which is just the generalized ME objective (5.5) and the entropy of the unknown (empirical) weights of the data. For simplicity, and because the sample size T is finite, $H(\mathbf{p})$ is in discrete form.

6.2 The Data Representation

As discussed earlier, within the likelihood, or BMOM, type IPRs all (hard data) information is represented in terms of moments. Thus, let the sample be represented as

$$\sum_t p_t \mathbf{x}_t y_t = \left[\sum_t p_t \mathbf{x}_t \mathbf{x}_t'\right] Z\pi + \left[\sum_t p_t \mathbf{x}_t\right] V\mathbf{w}. \quad (6.2)$$

where $\beta = Z\mathbf{p}$ and $\mathbf{e} = V\mathbf{w}$.

Maximizing (6.1) subject to (6.2), the M moment requirements $\sum_t p_t g_m(y_t, \mathbf{x}_t) = \mathbf{0}$, and the three normalization requirements yields $\hat{\mathbf{p}}_G, \hat{\boldsymbol{\pi}}_G$, and $\hat{\mathbf{w}}_G$ which in turn yields $\hat{\boldsymbol{\beta}}_G$.

6.3 Nested Models

Assumption 1. Let Z span the possible outcomes of β.

This assumption is trivial since if the β_k's have natural bounds, the $\mathbf{z}_k$ will have the same bounds. If, on the other hand, the researcher is in a state of complete ignorance regarding β, then each $\mathbf{z}_k$ is specified such that $\mathbf{z}_{k1} = -\mathbf{z}_{kM}$ with large $\mathbf{z}_{kM}$.

Case1. Given Assumption 1 and for $\hat{p}_t = \frac{1}{T}$ for all t, as $\mathbf{v} \to \mathbf{0}$, $\hat{\boldsymbol{\beta}}_G \to \hat{\boldsymbol{\beta}}_{ML}$. This result is trivial, since the RHS element of (6.2) vanishes and thus, for equal weights p_t, (6.2) is just the FOC for the ML. Intuitively speaking, letting $\mathbf{v} \to \mathbf{0}$, as $T \to \infty$ (for example $v_1 = -v_J$ and $v_J = \frac{3\sigma_y}{T}$) yields this result which is equivalent to showing consistency of estimates.

Case 2. Given Assumption 1, as $\mathbf{v} \to \mathbf{0}$, $\hat{\boldsymbol{\beta}}_G \to \hat{\boldsymbol{\beta}}_{EEL}$ where $\hat{\boldsymbol{\beta}}_{EEL}$ is the weighted, expected ML (or WLS) model (5.11) *but with the entropy objective H(p)* instead of the EL objective (5.12).

Case 3. Given Assumption 1, substituting $H(\mathbf{p})$ and (6.2) with their continuous counterparts, as $\mathbf{v} \to \mathbf{0}$, the BMOM density function for β results, which in turn yields $\hat{\boldsymbol{\beta}}_{BMOM}$.

Case 4. Given Assumption 1 and substituting the data representation (6.2) from weighted moments to weighted data (i.e. each element of 6.2 is not premultiplied by $\mathbf{x}_t$), $\hat{\boldsymbol{\beta}}_G \to \hat{\boldsymbol{\beta}}_{GME}$ where $\hat{\boldsymbol{\beta}}_{GME}$ are the generalized ME estimates of Section 5.1. Note that in this case the p_t's just *cancel out* and thus including them in the objective function (6.1) has no effect on the outcome.

6.4 Uncertainty

Building on the nested (but incomplete) model it is possible to rank the IPRs in terms of their input information. But before doing so it is important to recall that the basic EL, or WLS of Section 5.2, is not a special case of the general (nested) model. This is because the objective in these EL type IPRs is the discrete likelihood instead of $H(\mathbf{p})$. Nevertheless, following Section 4, the EL estimator can be compared to the ME general model, under an informational criterion.

Based on the informational criterion, (input information, or equivalently, the a-priori assumptions underlying each IPR) the ranking of the different IPRs, starting with the least informed model, are:

i) The generalized ME model - Case 4.
ii) The general model (6.1)-(6.2).
iii) The EEL and BMOM models - Cases 2-3.
iv) The EL or WLS models (5.10) or (5.11)-(5.13).

This ordering has the following intuition. In the first Case the only basic assumption used is the support spaces Z and v. In terms of the errors this implies that each $e_t \in [v_1, v_J]$. In addition, *all* the T data-points (observations) are evaluated explicitly and *no* assumptions regarding the higher M moments are made. In terms of the dual, one can view it as an expected empirical likelihood function defined on a composite of exponential distributions with T unknowns.

In the second case two basic characteristics are changed. First, the sample is converted into a moment representation. Thus, instead of T observations of the form y_t and $\mathbf{x}_t$, Eq. 6.2 is in terms of $X'\mathbf{y}$ and $X'X$ and weighted by p_t. From informational content point of view $\mathbf{y}$ and X can never have less information than their moment transform $X'\mathbf{y}$ and $X'X$ which are of a lower dimension. Finally, recall that in (6.2) the restrictions are not forced to hold exactly; there is an additional error term for each moment. In terms of the dual unconstrained model, this expected log-likelihood (discrete or continuous) is defined over a mix of exponential distributions but with $M<T$ parameters.

The third case does not allow for the additional freedom in (6.2) and forces $\mathbf{v} = \mathbf{0}$. Thus, its entropy value must be lower. In the fourth case, the (discrete or continuous) entropy objective *H(p)* is substituted by the EL objective and therefore, following Section 4, the input information increases relative to Case *(iii)*. Finally, the traditional ML is just a special case of (*iii*) or (*iv*) for uniform $\boldsymbol{p}$'s.

What are the implications of this ranking? All analyzed estimators are consistent and 100% efficient IPRs. If, given a unique and finite (small) data set, one estimator incorporates more input information than another, the output information is also higher and improves the estimates. But, if this additional information (assumptions) cannot be confirmed, than the choice of a consistent and 100% efficient estimator that uses less assumptions, incorporates each data point directly and without strict distributional requirements seems to be logical.

Further, if one chooses to represent each data point (observation) then, the GME inversion procedure is the only one satisfying the basic four axioms of Shore and Johnson (1980) and Skilling (1989). Intuitively speaking, in that case, the probabilities $\boldsymbol{p}$ and $\boldsymbol{w}$ are defined over the support spaces Z and $\boldsymbol{v}$ where the data points represent expected values of these distributions over these two supports. Hence, within these support spaces, as defined and chosen by the researcher, the GME is the *only* inversion procedure satisfying these four axioms. What is left, is to investigate the informational content of those supports and compare it with the informational content of the distributional, or side-conditions, of the other estimators. This is a topic for further research.

7. Conclusions

This paper compares the ME to the likelihood-type estimators under certain informational criterion. Two basic estimation problems are discussed. In the first one, the objects to be recovered are natural probabilities and the classical ME is equivalent to the ML. A generalized ME (and ML), where the constraints are relaxed, is developed, reviewed and compared with the traditional ML. The second type of estimation problem discussed is the linear statistical model. A generalized ME approach is reviewed and contrasted with the EL (WLS), ML and BMOM estimators. These estimators are then ranked according to their input information.

The comparison basis discussed and developed here is not complete and no formal proofs are given. In future research, a more complete axiomatic comparison of the estimators will be done.

8. References

1. Akaike, H (1986), "The Selection Smoothness Priors for Distributed Lag Estimation," in *Bayesian and Decision Techniques: Essays in Honor of Bruno de Finetti*, eds. P. K. Goel and A. Zellner (Amsterdam, North-Holland) 109-118.
2. Agmon, N., Y. Alhassid, and R. D. Levine (1979), "An Algorithm for Finding the Distribution of Maximal Entropy," *Journal of Computational Physics*, Vol. 30, pp. 250–259.
3. Bernardo, J.M. (1979), "Expected Information as Expected Utility," *The Annals of Statistics*, 7,686-690.
4. Csiszar, I. (1991) "Why Least Squares and Maximum Entropy? An Axiomatic Approach to Inference for Linear Inverse Problems," *The Annals of Statistics*, 19, 2032-2066.
5. DiCiccio, T., P. Hall, and J. Romano (1991), "Empirical Likelihood is Bartlett-Correctable," *The Annals of Statistics*, 19, 1053-1061.
6. Golan, A., and G. G. Judge and D. Miller (1996), *Maximum Entropy Econometrics: Robust Estimation With Limited Data*, John Wiley & Sons, New York.
7. Golan, A., and G. Judge (1996), "A Maximum Entropy Approach to Empirical Likelihood Estimation and Inference," (Working paper), UC Berkeley.
8. Golan, A., and S. J. Vogel (1997), "Estimation of Stationary and Non-Stationary Social Accounting Matrix Coefficients With Structural and Supply-Side Information," (Working paper).
9. Good, I. J. (1963), "Maximum Entropy for Hypothesis Formulation, Especially for Multidimensional Contingency Tables," *Annals of Mathematical Statistics*, Vol. 34, pp. 911–934.
10. Hall, P. (1990), "Pseudo-Likelihood Theory for Empirical Likelihood," *The Annals of Statistics*, 18, 121-140.
11. Imbens, G.W. (1993), "A New Approach to Generalized Method of Moments Estimation," (mimeo), Harvard University.

12. Imbens, G.W. and J.K. Hellerstein (1994), "Imposing Moment Restrictions by Weighting," (mimeo), Harvard University.
13. Jaynes, E.T. (1957a), "Information Theory and Statistical Mechanics," *Physics Review*, 106, 620-630.
14. Jaynes, E.T. (1957b), "Information Theory and Statistical Mechanics II," *Physics Review*, 108, 171-190.
15. Jaynes, E.T. (1963), "Information Theory and Statistical Mechanics II," in K.W. Ford (ed.), *Statistical Physics*, W.A. Benamin, Inc., New York, 181-218.
16. Jaynes, E.T. (1984), "Prior Information and Ambiguity in Inverse Problems," in D.W. McLaughlin (ed.), *Inverse Problems*, SIAM Proceedings, American Mathematical Society, Providence, RI, 151-166.
17. Kullback, J. (1959), *Information Theory and Statistics,* New York: John Wiley & Sons.
18. Levine, R.D. (1980), "An Information Theoretical Approach to Inversion Problems," *Journal of Physics, A*, 13, 91-108.
19. McFadden, D. (1974), "The Measurement of Urban Travel demand," *Journal of Public Economics*, Vol. 3, pp. 303-328.
20. Owen, A. (1990), "Empirical Likelihood Ratio Confidence Regions," *The Annals of Statistics*, 18, 90-120.
21. Owen, A. (1991), "Empirical Likelihood for Linear Models," *The Annals of Statistics*, 19, 1725-1747.
22. Qin, J. and J. Lawless (1994), "Empirical Likelihood and General Estimating Equations," *The Annals of Statistics*, 22, 300-325.
23. Shannon, C.E. (1948), "A Mathematical Theory of Communication," *Bell System Technical Journal*, 27, 379-423.
24. Shore, J.E. and R.W. Johnson (1980), "Axiomatic Derivation of the Principle of Maximum Entropy and the Principle of Minimum Cross-Entropy," *IEEE Transactions on Information Theory*, IT-26(1), 26-37.
25. Skilling, J. (1989), "The Axioms of Maximum Entropy," in J. Skilling (ed.), *Maximum Entropy and Bayesian Methods in Science and Engineering*, Kluwer Academic, Dordrecht, 173-187.
26. Tobias, J., and A. Zellner (1997), *Further Results on Bayesian Method of Moments Analysis of the Multiple Regression Model.* H.G.B. Alexander Research Foundation, University of Chicago.
27. Zellner, A. (1988), "Optimal Information Processing and Bayes Theorem," *American Statistician*, 42, 278-284.
28. A. Zellner. Bayesian method of moments/ instrumental variable (BMOM/iv) analysis of mean and regression models. In J.C. Lee, W.C. Johnson, and A. Zellner, editors, *Modeling and Prediction: Honoring Seymour Geisser*, pages 61-75. Springer- Verlag, 1996.
29. A. Zellner (1997). The Bayesian method of moments (BMOM): theory and applications. In T. Fomby and R.C. Hill, editors, *Advances in Econometrics*.

PROBABILISTIC METHODS FOR DATA FUSION

Ali Mohammad–Djafari

Laboratoire des Signaux et Systèmes (CNRS–SUPELEC–UPS)
École Supérieure d'Électricité
Plateau de Moulon, 91192 Gif–sur–Yvette Cedex, France.
E_mail: djafari@lss.supelec.fr

Abstract. The main object of this paper is to show how we can use classical probabilistic methods such as Maximum Entropy (ME), maximum likelihood (ML) and/or Bayesian (BAYES) approaches to do microscopic and macroscopic data fusion. Actually ME can be used to assign a probability law to an unknown quantity when we have macroscopic data (expectations) on it. ML can be used to estimate the parameters of a probability law when we have microscopic data (direct observation). BAYES can be used to update a prior probability law when we have microscopic data through the likelihood. When we have both microscopic and macroscopic data we can use first ME to assign a prior and then use BAYES to update it to the posterior law thus doing the desired data fusion. However, in practical data fusion applications, we may still need some engineering feeling to propose realistic data fusion solutions. Some simple examples in sensor data fusion and image reconstruction using different kind of data are presented to illustrate these ideas.

key words: Data fusion, Maximum entropy, Maximum likelihood, Bayesian data fusion, EM algorithm.

1. Introduction

Data fusion is one of the active area of research in many applications such as non destructive testing (NDT), geophysical imaging, medical imaging, radioastronomy, etc. Our main object in this paper is not to focus on any of these applications. Indeed, we want to show how we can use classical probabilistic methods such as Maximum Entropy (ME), maximum likelihood (ML) and/or Bayesian (BAYES) approaches to do data fusion.

First, we consider these three methods separately, and we describe briefly each method. Then we will see some interrelations between them.

We will see that ME can be used to assign a probability law to an unknown quantity X when we have macroscopic data (expectations) on it. ML can be used when we have assigned a parametric probability law, before getting the data, on X and we want to estimate this parameter from some microscopic data (samples of X). BAYES can be used to update probability laws, going from priors to posteriors.

When we have both microscopic and macroscopic data we can use first ME to assign a prior and then use BAYES to update it to the posterior law, doing thus the desired data fusion. In practical data fusion applications, however, we may still need some engineering feeling to propose realistic data fusion solutions.

2. Short description of the methods

2.1. Maximum Entropy (ME)

ME can be used to assign a probability law to an unknown quantity when we have macroscopic data (expectations) on it. To see this let note by X a quantity of interest and try to see when and how we can use ME. We do this through a given problem.

Problem P1: We have L sensors giving us L values $\{\mu_l, l = 1, \ldots, L\}$, representing the mean values of L known functions $\{\phi_l(X), l = 1, \ldots, L\}$ related to the unknown X:

$$\mathrm{E}\{\phi_l(X)\} = \int \phi_l(x) p(x) \,\mathrm{d}x = \mu_l, \quad l = 1, \ldots, L. \tag{1}$$

The question is then how to represent our partial knowledge of X by a probability law.

Obviously, this problem has not a unique solution. Actually these data define a class of possible solutions and we need a criterion to select one of them. The ME principle can give us this criterion and the problem then becomes:

$$\text{maximize} \qquad S(p) = -\int p(x) \ln p(x) \,\mathrm{d}x$$

$$\text{subject to} \qquad \int \phi_l(x)\, p(x) \,\mathrm{d}x = \mu_l, \quad l = 1, \ldots, L.$$

The solution is given by

$$p(x) = \frac{1}{Z(\boldsymbol{\theta})} \exp\left[-\sum_{l=1}^{L} \theta_l \phi_l(x)\right] = \frac{1}{Z(\boldsymbol{\theta})} \exp\left[-\boldsymbol{\theta}^t \boldsymbol{\phi}(x)\right], \tag{2}$$

where

$$Z(\boldsymbol{\theta}) = \int \exp\left[-\sum_{l=1}^{L} \theta_l \phi_l(x)\right] \mathrm{d}x \tag{3}$$

is the partition function and $\{\theta_1, \ldots, \theta_l\}$ are determined by the following system of equations:

$$\boxed{-\frac{\partial \ln Z(\boldsymbol{\theta})}{\partial \theta_l} = \mu_l, \quad l = 1, \ldots, L,} \tag{4}$$

See [1, 2] for more discussions.

2.2. Maximum Likelihood (ML)

Problem P2: Assume now that we have a parametric form of the probability law $p(x;\boldsymbol{\theta})$ and a sensor gives us N values $\boldsymbol{x} = [x_1, \ldots, x_N]$ of X. How to determine the parameters $\boldsymbol{\theta}$?

Two classical methods for solving this problem are:

- Moments Method (MM): The main idea is to write a set of equations (at least L) relating the theoretical and empirical moments, and solve them to obtain the solution:

$$G_l(\boldsymbol{\theta}) = \mathrm{E}\left\{X^l\right\} = \int x^l\, p(x;\boldsymbol{\theta})\, \mathrm{d}x = \frac{1}{N}\sum_{j=1}^{N} x_j^l,\; l = 1,\ldots,L \tag{5}$$

- Maximum Likelihood (ML): Here, the main idea is to consider the data as N samples of X. Then, writing the expression of $p(\boldsymbol{x};\boldsymbol{\theta})$ and considering it as a function of $\boldsymbol{\theta}$, the ML solution is defined as

$$\widehat{\boldsymbol{\theta}} = \arg\max_{\boldsymbol{\theta}} \{l(\boldsymbol{\theta}|\boldsymbol{x})\} \;\text{ with } l(\boldsymbol{\theta}|\boldsymbol{x}) = p(\boldsymbol{x};\boldsymbol{\theta}) = \prod_{j=1}^{N} p(x_j;\boldsymbol{\theta}) \tag{6}$$

It is interesting to note that, in the case of the generalized exponential families:

$$p(x;\boldsymbol{\theta}) = \frac{1}{Z(\boldsymbol{\theta})}\exp\left[-\sum_{l=1}^{L}\theta_l\,\phi_l(x)\right] = \frac{1}{Z(\boldsymbol{\theta})}\exp\left[-\boldsymbol{\theta}^t\,\boldsymbol{\phi}(x)\right] \tag{7}$$

we have

$$l(\boldsymbol{\theta}) = \prod_{j=1}^{N} p(x_j;\boldsymbol{\theta}) = \frac{1}{Z^N(\boldsymbol{\theta})}\exp\left[-\sum_{j=1}^{N}\sum_{l=1}^{L}\theta_l\,\phi_l(x_j)\right] \tag{8}$$

Then, it is easy to see that the ML solution is the solution of the following system of equations:

$$\boxed{\frac{\partial \ln Z(\boldsymbol{\theta})}{\partial \theta_l} = \frac{1}{N}\sum_{j=1}^{N}\phi_l(x_j), \quad l = 1,\ldots,L} \tag{9}$$

Comparing equations (4) & (9), we can remark an interesting relation between these two methods. See also [3] for more discussions.

2.3. ML and incomplete data: EM Algorithm

Problem P3: Consider the ***problem P2***, but now assume that the sensor gives us M values $\boldsymbol{y} = [y_1, \ldots, y_M]$ related to the N samples $\boldsymbol{x} = [x_1, \ldots, x_N]$ of X by a non invertible relation, $\boldsymbol{y} = \boldsymbol{A}\boldsymbol{x}$ with $M < N$. How to determine $\boldsymbol{\theta}$?

The solution here is still based on the ML. The only difference is the way to calculate the solution. In fact we can write

$$p(\boldsymbol{x};\boldsymbol{\theta}) = p(\boldsymbol{x}|\boldsymbol{y};\boldsymbol{\theta})\,p(\boldsymbol{y};\boldsymbol{\theta}), \qquad \forall \boldsymbol{A}\boldsymbol{x} = \boldsymbol{y}. \tag{10}$$

Taking the expectation of both sides for a given value of $\boldsymbol{\theta} = thetab'$, we have

$$\ln p(\boldsymbol{y};\boldsymbol{\theta}) = \mathrm{E}_{\boldsymbol{x}|\boldsymbol{y};\boldsymbol{\theta}'}\left\{\ln p(\boldsymbol{x};\boldsymbol{\theta})\right\} - \mathrm{E}_{\boldsymbol{x}|\boldsymbol{y};\boldsymbol{\theta}'}\left\{\ln p(\boldsymbol{x}|\boldsymbol{y};\boldsymbol{\theta})\right\} \tag{11}$$

or written differently:

$$L(\boldsymbol{\theta}) = Q(\boldsymbol{\theta};\boldsymbol{\theta}') - V(\boldsymbol{\theta},\boldsymbol{\theta}'). \tag{12}$$

Note that for a given $\boldsymbol{\theta}'$ and for all $\boldsymbol{\theta}$ we have

$$L(\boldsymbol{\theta}) - L(\boldsymbol{\theta}') = [Q(\boldsymbol{\theta};\boldsymbol{\theta}') - Q(\boldsymbol{\theta}';\boldsymbol{\theta}')] + [V(\boldsymbol{\theta};\boldsymbol{\theta}') - V(\boldsymbol{\theta}',\boldsymbol{\theta}')]. \tag{13}$$

Now, using the Jensen's inequality [4]

$$V(\boldsymbol{\theta};\boldsymbol{\theta}') \leq V(\boldsymbol{\theta}',\boldsymbol{\theta}') \tag{14}$$

an iterative algorithm, known as *Expectation-Maximization (EM)*, is derived:

$$\left\{\begin{array}{lll} \text{E:} & Q\left(\boldsymbol{\theta};\widehat{\boldsymbol{\theta}}^{(k)}\right) & = \mathrm{E}_{\boldsymbol{x}|\boldsymbol{y};\boldsymbol{\theta}^{(k)}}\left\{\ln p(\boldsymbol{x};\boldsymbol{\theta})\right\} \\ \text{M:} & \widehat{\boldsymbol{\theta}}^{(k+1)} & = \arg\max_{\boldsymbol{\theta}}\left\{Q\left(\boldsymbol{\theta};\widehat{\boldsymbol{\theta}}^{(k)}\right)\right\} \end{array}\right. \tag{15}$$

This algorithm insures to converge to a local maximum of the likelihood.

It is interesting to see that in the case of the generalized exponential families (7), the algorithm becomes:

$$\text{Step E:} \quad Q(\boldsymbol{\theta};\boldsymbol{\theta}') = \mathrm{E}_{\boldsymbol{x}|\boldsymbol{y};\boldsymbol{\theta}'}\left\{\ln p(\boldsymbol{x};\boldsymbol{\theta})\right\} = -N\ln Z(\boldsymbol{\theta}) - \sum_{j=1}^{N}\boldsymbol{\theta}^t\mathrm{E}_{\boldsymbol{x}|\boldsymbol{y};\boldsymbol{\theta}'}\left\{\boldsymbol{\phi}(\boldsymbol{x}_j)\right\}$$

$$\text{Step M:} \quad \boxed{-\frac{\partial\ln Z(\boldsymbol{\theta})}{\partial\theta_l} = \frac{1}{N}\sum_{j=1}^{N}\mathrm{E}_{\boldsymbol{x}_j|\boldsymbol{y};\boldsymbol{\theta}^{(k)}}\left\{\phi_l(\boldsymbol{x}_j)\right\}, \quad l = 1,\ldots,L} \tag{16}$$

Compare this last equation with those of (4) and (9) to see still some relations between ME, ML and the EM algorithms.

Problem P4: Consider now the same problem *P3* where we want to estimate not only $\boldsymbol{\theta}$ but also $\boldsymbol{x}$. We can still use the EM algorithm with the following modification:

$$\left\{\begin{array}{lll} \text{E:} & Q\left(\boldsymbol{\theta};\widehat{\boldsymbol{\theta}}^{(k)}\right) & = \mathrm{E}\left\{\ln p(\boldsymbol{x};\boldsymbol{\theta})|\boldsymbol{y};\widehat{\boldsymbol{\theta}}^{(k)}\right\} \\ & \widehat{\boldsymbol{x}}^{(k)} & = \mathrm{E}\left\{\boldsymbol{x}|\boldsymbol{y};\widehat{\boldsymbol{\theta}}^{(k)}\right\} \\ \text{M:} & \widehat{\boldsymbol{\theta}}^{(k+1)} & = \arg\max_{\boldsymbol{\theta}}\left\{Q\left(\boldsymbol{\theta};\widehat{\boldsymbol{\theta}}^{(k)}\right)\right\} \end{array}\right. \tag{17}$$

2.4. Bayesian Approach

Problem P5: Consider again *problems P3* or *P4* but now assume that the observations $\boldsymbol{y}$ are corrupted by noise: $\boldsymbol{y} = \boldsymbol{A}\boldsymbol{x} + \boldsymbol{b}$.

The main tool here is the Bayesian approach where, we use the data-unknown relation and the noise probability distribution to define the likelihood $p(\boldsymbol{y}|\boldsymbol{x};\boldsymbol{\theta}_1) = p_b(\boldsymbol{y} - \boldsymbol{A}\boldsymbol{x};\boldsymbol{\theta}_1)$ and combine it with the prior $p(\boldsymbol{x};\boldsymbol{\theta}_2)$ through the Bayes' rule to obtain the posterior law

$$p(\boldsymbol{x}|\boldsymbol{y};\boldsymbol{\theta}_1,\boldsymbol{\theta}_2) = \frac{p(\boldsymbol{y}|\boldsymbol{x};\boldsymbol{\theta}_1)\, p(\boldsymbol{x};\boldsymbol{\theta}_2)}{m(\boldsymbol{y};\boldsymbol{\theta}_1,\boldsymbol{\theta}_2)}, \tag{18}$$

where

$$m(\boldsymbol{y};\boldsymbol{\theta}_1,\boldsymbol{\theta}_2) = \int p(\boldsymbol{y}|\boldsymbol{x};\boldsymbol{\theta}_1)\, p(\boldsymbol{x};\boldsymbol{\theta}_2)\,\mathrm{d}\boldsymbol{x} \tag{19}$$

The posterior law $p(\boldsymbol{x}|\boldsymbol{y};\boldsymbol{\theta}_1,\boldsymbol{\theta}_2)$ contains all the information available on $\boldsymbol{x}$. We can then use it to make any inference on $\boldsymbol{x}$. We can for example define the following point estimators:

- Maximum *a posteriori* (MAP):

$$\widehat{\boldsymbol{x}} = \arg\max_{\boldsymbol{x}} \left\{p_{x|y}(\boldsymbol{x}|\boldsymbol{y};\boldsymbol{\theta}_1,\boldsymbol{\theta}_2)\right\} \tag{20}$$

- Posterior Mean (PM):

$$\widehat{\boldsymbol{x}} = \mathrm{E}_{x|y}\left\{\boldsymbol{x}\right\} = \int \boldsymbol{x}\, p_{x|y}(\boldsymbol{x}|\boldsymbol{y};\boldsymbol{\theta}_1,\boldsymbol{\theta}_2)\,\mathrm{d}\boldsymbol{x} \tag{21}$$

- Marginal Posterior Modes (MPM):

$$\widehat{\boldsymbol{x}} = \arg\max_{x_i} \left\{p(x_i|\boldsymbol{y};\boldsymbol{\theta}_1,\boldsymbol{\theta}_2)\right\}, \tag{22}$$

  where

$$p(x_i|\boldsymbol{y};\boldsymbol{\theta}_1,\boldsymbol{\theta}_2) = \int p_{x|y}(\boldsymbol{x}|\boldsymbol{y};\boldsymbol{\theta})\,\mathrm{d}x_1 \ldots \mathrm{d}x_{i-1} \ldots \mathrm{d}x_{i+1} \ldots \mathrm{d}x_n \tag{23}$$

However, in practice, we face two great difficulties:

- How to assign the probability laws $p(\boldsymbol{y}|\boldsymbol{x};\boldsymbol{\theta}_1)$ and $p(\boldsymbol{x};\boldsymbol{\theta}_2)$?
- How to determine the parameters $\boldsymbol{\theta} = (\boldsymbol{\theta}_1,\boldsymbol{\theta}_2)$?

For the first we can use either the ME principle when possible, or any other invariance properties combined with some practical, scientific or engineering sense reasoning. For the second, there are more specific tools, all based on the joint posterior probability law

$$p(\boldsymbol{x},\boldsymbol{\theta}|\boldsymbol{y}) \propto p(\boldsymbol{y}|\boldsymbol{x},\boldsymbol{\theta})\, p(\boldsymbol{x}|\boldsymbol{\theta}) p(\boldsymbol{\theta}) \propto p(\boldsymbol{x},\boldsymbol{y}|\boldsymbol{\theta})\, p(\boldsymbol{\theta}) \propto p(\boldsymbol{x}|\boldsymbol{y},\boldsymbol{\theta})\, p(\boldsymbol{\theta}),$$

The following are some known schemes:
• Joint Maximum *a posteriori* (JMAP):

$$\left(\widehat{\boldsymbol{\theta}}, \widehat{\boldsymbol{x}}\right) = \arg\max_{(\boldsymbol{\theta},\boldsymbol{x})} \{p(\boldsymbol{x}, \boldsymbol{\theta}|\boldsymbol{y})\}$$

$$\boldsymbol{y} \mapsto \boxed{\text{JMAP}} \begin{array}{l} \mapsto \widehat{\boldsymbol{x}} \\ \mapsto \widehat{\boldsymbol{\theta}} \end{array}$$

• Generalized Maximum Likelihood (GML):

$$\begin{cases} \widehat{\boldsymbol{x}}^{(k)} = \arg\max_{\boldsymbol{x}} \left\{p(\boldsymbol{x}|\boldsymbol{y}; \boldsymbol{\theta}^{(k-1)})\right\} \\ \widehat{\boldsymbol{\theta}}^{(k)} = \arg\max_{\boldsymbol{\theta}} \left\{p(\widehat{\boldsymbol{x}}^{(k)}|\boldsymbol{y}, \boldsymbol{\theta})p(\boldsymbol{\theta})\right\} \end{cases}$$

$$\begin{array}{r} \boldsymbol{y} \mapsto \\ \widehat{\boldsymbol{\theta}}^{0} \mapsto \end{array} \boxed{\text{GML}} \begin{array}{l} \mapsto \widehat{\boldsymbol{x}}^{k} \\ \mapsto \widehat{\boldsymbol{\theta}}^{k} \end{array}$$

• Marginalized Maximum Likelihood (MML):

$$\begin{cases} \widehat{\boldsymbol{\theta}} = \arg\max_{\boldsymbol{\theta}} \left\{\int p(\boldsymbol{y}|\boldsymbol{x})\, p(\boldsymbol{x}; \boldsymbol{\theta})\, \mathrm{d}\boldsymbol{x}\right\} \\ \widehat{\boldsymbol{x}} = \arg\max_{\boldsymbol{x}} \left\{p(\boldsymbol{x}|\boldsymbol{y}; \widehat{\boldsymbol{\theta}})\right\} \end{cases}$$

$$\boldsymbol{y} \mapsto \boxed{\text{ML}} \mapsto \widehat{\boldsymbol{\theta}} \mapsto \underset{\substack{\uparrow \\ \boldsymbol{y}}}{\boxed{\text{MAP}}} \mapsto \widehat{\boldsymbol{x}}$$

• MML-EM:
An analytic expression for $p(\boldsymbol{y}; \boldsymbol{\theta})$ is rarely possible. Consequently, considering $[\boldsymbol{y}, \boldsymbol{x}]$ as the complete data and $\boldsymbol{y}$ as the incomplete data, we can use the EM algorithm to obtain the following scheme:

$$\begin{cases} \text{E: } Q\left(\boldsymbol{\theta}; \widehat{\boldsymbol{\theta}}^{(k)}\right) = \mathrm{E}_{\boldsymbol{x}|\boldsymbol{y};\boldsymbol{\theta}^{(k)}} \{\ln p(\boldsymbol{x}, \boldsymbol{y}; \boldsymbol{\theta})\} \\ \text{M: } \widehat{\boldsymbol{\theta}}^{(k+1)} = \arg\max_{\boldsymbol{\theta}} \left\{Q\left(\boldsymbol{\theta}; \widehat{\boldsymbol{\theta}}^{(k)}\right)\right\} \end{cases}$$

$$\boldsymbol{y} \mapsto \overset{\substack{\widehat{\boldsymbol{\theta}}^{(0)} \\ \downarrow}}{\boxed{\text{ML-EM}}} \mapsto \widehat{\boldsymbol{\theta}}^{(k)} \qquad \widehat{\boldsymbol{\theta}} \mapsto \overset{\substack{\boldsymbol{y} \\ \downarrow}}{\boxed{\text{MAP}}} \mapsto \widehat{\boldsymbol{x}}$$

3. Data fusion

In this section we consider some simple data fusion problems and analyse the way we can use the previous schemes to solve them.

3.1. Sensors without noise

Problem P6: The sensor C1 gives N samples $\boldsymbol{x}_a = \{x_1, \ldots, x_N\}$ of X and stops. The sensor C2 gives M samples $\boldsymbol{y}_b = \{y_1, \ldots, y_M\}$ related to $\boldsymbol{x}$ by $\boldsymbol{y} = \boldsymbol{A}\boldsymbol{x} + \boldsymbol{b}$. We are asked to predict the unobserved samples $\boldsymbol{x}_b = \{x_{N+1}, \ldots, x_{N+M}\}$ of X.

	$\boldsymbol{x}_a$	$\boldsymbol{x}_b$
$C1:$	$x_1, \ldots, x_N$	$\ldots ? \ldots$
$C2:$	$\ldots$	$y_1, \ldots, y_M$
	$\boldsymbol{y}_a$	$\boldsymbol{y}_b$

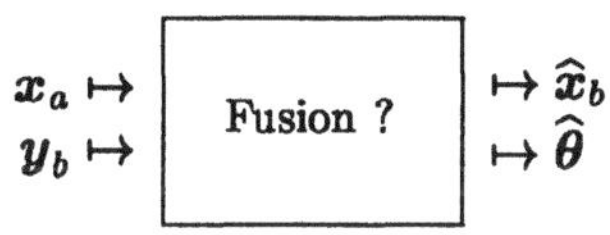

We can propose the following solutions:

- Use $\boldsymbol{x}_a$ to estimate $\boldsymbol{\theta}$, the parameters of $p(\boldsymbol{x}; \boldsymbol{\theta})$ and use it then to estimate $\boldsymbol{x}_b$ from $\boldsymbol{y}_b$:

$$\widehat{\boldsymbol{\theta}} = \arg\max_{\boldsymbol{\theta}} \{L_a(\boldsymbol{\theta}) = \ln p(\boldsymbol{x}_a; \boldsymbol{\theta})\}$$
$$\widehat{\boldsymbol{x}}_b = \arg\max_{\boldsymbol{x}_b} \left\{p\left(\boldsymbol{x}_b | \boldsymbol{y}_b; \widehat{\boldsymbol{\theta}}\right)\right\}$$

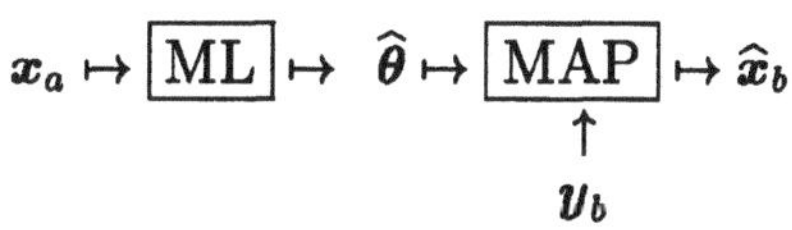

- Use both $\boldsymbol{x}_a$ and $\boldsymbol{y}_b$ to estimate $\boldsymbol{x}_b$:

$$\widehat{\boldsymbol{\theta}} = \arg\max_{\boldsymbol{\theta}} \{L_a(p(\boldsymbol{x}_a; \boldsymbol{\theta}))\}$$
$$(\widehat{\boldsymbol{x}}_b, \widehat{\boldsymbol{\theta}}) = \arg\max_{(\boldsymbol{x}_b, \widehat{\boldsymbol{\theta}})} \{p(\boldsymbol{x}_b, \boldsymbol{\theta} | \boldsymbol{x}_a, \boldsymbol{y}_b)\}$$

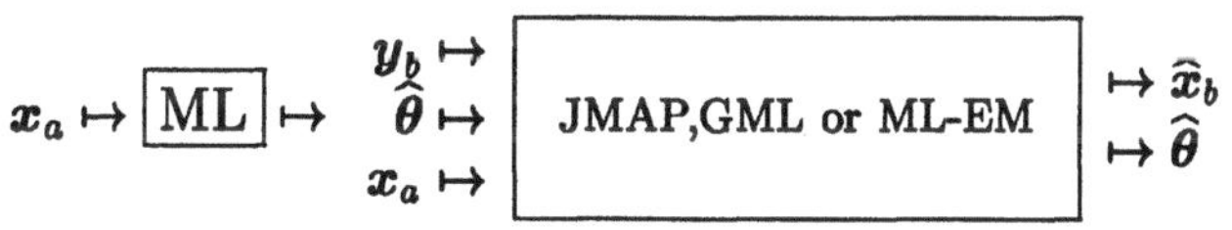

3.2. Fusion of homogeneous data

Problem P7: We have two types of data on the same unknown $\boldsymbol{x}$, both related to it through linear models:

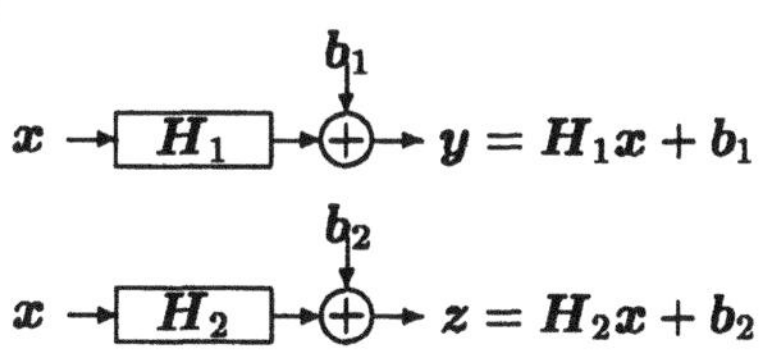

For example, consider an X ray tomography problem where $\boldsymbol{x}$ represents the mass density of the object and where $\boldsymbol{y}$ and $\boldsymbol{z}$ represent respectively a high resolution projection and a low resolution projection.

We can use directly the Bayesian approach to solve this problem:

$$p(\boldsymbol{x}|\boldsymbol{y},\boldsymbol{z}) = \frac{p(\boldsymbol{y},\boldsymbol{z}|\boldsymbol{x})\,p(\boldsymbol{x})}{p(\boldsymbol{y},\boldsymbol{z})}$$

Actually the main difficulty here is to assign $p(\boldsymbol{y},\boldsymbol{z}|\boldsymbol{x})$. If we assume that the errors associated to the two sets of data are independent then the calculation can be done more easily. For the purpose of illustration assume the following:

$$\begin{aligned} p(\boldsymbol{y}|\boldsymbol{x};\sigma_1^2) &\propto \exp\left[-\frac{1}{2\sigma_1^2}|\boldsymbol{y}-\boldsymbol{H}_1\boldsymbol{x}|^2\right] \\ p(\boldsymbol{z}|\boldsymbol{x};\sigma_2^2) &\propto \exp\left[-\frac{1}{2\sigma_2^2}|\boldsymbol{z}-\boldsymbol{H}_2\boldsymbol{x}|^2\right] \\ p(\boldsymbol{x};\boldsymbol{m},\boldsymbol{\Sigma}) &\propto \exp\left[-\frac{1}{2}[\boldsymbol{x}-\boldsymbol{m}]^t\boldsymbol{\Sigma}^{-1}[\boldsymbol{x}-\boldsymbol{m}]\right] \end{aligned}$$

Indeed, assume that the hyperparameters $(\sigma_1^2,\sigma_2^2,\boldsymbol{m},\boldsymbol{\Sigma})$ are given. Then we can use, for example, the MAP estimate, given by:

$$\widehat{\boldsymbol{x}} = \arg\max_{\boldsymbol{x}}\left\{p(\boldsymbol{x}|\boldsymbol{y},\boldsymbol{z})\right\} = \arg\min_{\boldsymbol{x}}\left\{J(\boldsymbol{x}) = J_1(\boldsymbol{x}) + J_2(\boldsymbol{x}) + J_3(\boldsymbol{x})\right\}$$

with

$$J_1(\boldsymbol{x}) = \frac{1}{2\sigma_1^2}|\boldsymbol{y}-\boldsymbol{H}_1\boldsymbol{x}|^2,$$

$$J_2(\boldsymbol{x}) = \frac{1}{2\sigma_2^2}|\boldsymbol{z}-\boldsymbol{H}_2\boldsymbol{x}|^2,$$

$$J_3(\boldsymbol{x}) = \frac{1}{2}[\boldsymbol{x}-\boldsymbol{m}]^t\boldsymbol{\Sigma}^{-1}[\boldsymbol{x}-\boldsymbol{m}]$$

However, in practical applications, the data come from different processes.

3.3. Real data fusion problems

Consider a more realistic data fusion problem, where we have two different kinds of data. As an example assume a tomographic image reconstruction problem where we have a set of data $\boldsymbol{y}$ obtained by an X ray and a set of data $\boldsymbol{z}$ obtained by an ultrasound probing system. The X ray data are related to the mass density $\boldsymbol{x}$ of the matter while the ultrasound data are related to the acoustic reflectivity $\boldsymbol{r}$ of the matter. Indeed, assume that, we have linear relations, both between $\boldsymbol{y}$ and $\boldsymbol{x}$ and between $\boldsymbol{z}$ and $\boldsymbol{r}$. Then we have:

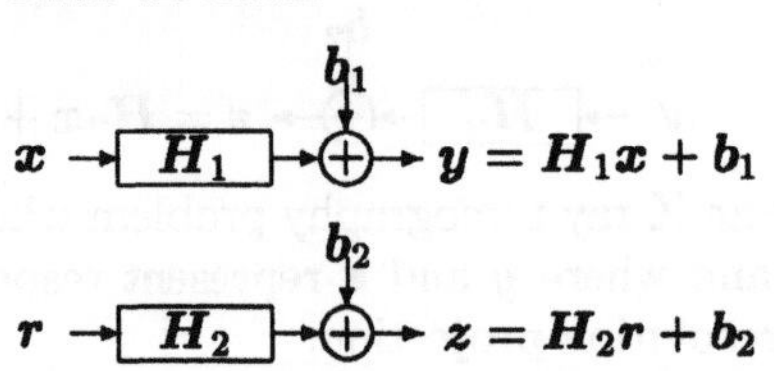

Assuming that the two sets of data are independant, we can again use the Bayes rule which now becomes

$$p(\boldsymbol{x},\boldsymbol{r}|\boldsymbol{y},\boldsymbol{z}) = \frac{p(\boldsymbol{y},\boldsymbol{z}|\boldsymbol{x},\boldsymbol{r})\,p(\boldsymbol{x},\boldsymbol{r})}{p(\boldsymbol{y},\boldsymbol{z})} = \frac{p(\boldsymbol{y}|\boldsymbol{x})\,p(\boldsymbol{z}|\boldsymbol{r})\,p(\boldsymbol{x},\boldsymbol{r})}{p(\boldsymbol{y},\boldsymbol{z})}$$

with

$$p(\boldsymbol{y},\boldsymbol{z}) = \int\!\!\int p(\boldsymbol{y}|\boldsymbol{x})\,p(\boldsymbol{z}|\boldsymbol{r})\,p(\boldsymbol{x},\boldsymbol{r})\,\mathrm{d}\boldsymbol{r}\,\mathrm{d}\boldsymbol{x}.$$

Here also the main difficulty is the assignment of the probability laws $p(\boldsymbol{y}|\boldsymbol{x})$, $p(\boldsymbol{z}|\boldsymbol{r})$, and more specifically $p(\boldsymbol{x},\boldsymbol{r})$.

Actually if we could find a mathematical relation between $\boldsymbol{r}$ and $\boldsymbol{x}$, then the problem would become the same as in the preceding case. To see this, assume that we can find a relation such as $r_j = g(x_{i+1} - x_i)$ with g a monotonic increasing function, from some physical reasons. For example, the fact that in the area where there are some important changes in the mass density of the matter both $\boldsymbol{x}$ and $\boldsymbol{r}$ change. Indeed, if g could be a linear function (an unrealistic hypothesis) then we would have

$$\left\{\begin{array}{l} \boldsymbol{y} = \boldsymbol{H}_1\boldsymbol{x} + \boldsymbol{b}_1 \\ \boldsymbol{z} = \boldsymbol{H}_2\boldsymbol{r} + \boldsymbol{b}_2 \\ \boldsymbol{r} = \boldsymbol{G}\boldsymbol{x} \end{array}\right. \longrightarrow \left\{\begin{array}{l} \boldsymbol{y} = \boldsymbol{H}_1\boldsymbol{x} + \boldsymbol{b}_1 \\ \boldsymbol{z} = \boldsymbol{G}\boldsymbol{H}_2\boldsymbol{r} + \boldsymbol{b}_2 \end{array}\right.$$

For more realistic cases we need a method which does not use a physicaly based explicit expression of g. One approach proposed and used by *Gautier et al.* [5, 6, 7, 8] is based on a compound Markovian model where the body object $\boldsymbol{o}$ is assumed to be composed of three related quantities:

$$\boldsymbol{o} = \{\boldsymbol{r},\boldsymbol{x}\} = \{\boldsymbol{q},\boldsymbol{a},\boldsymbol{x}\}$$

where $\boldsymbol{q}$ is a binary vector representing the positions of the discontinuities (edges) in the body, $\boldsymbol{a}$ a vector containing the reflectivity values such that

$$\left\{\begin{array}{l} q_j = 0 \longrightarrow r_j = 0, \\ q_j = 1 \longrightarrow r_j = a_j \end{array}\right. \quad \text{and} \quad r_j = \left\{\begin{array}{ll} g(x_{j+1} - x_j) & \text{if } |x_{j+1} - x_j| > \alpha \\ 0 & \text{otherwise} \end{array}\right.$$

and g is any monotonic increasing function.

With this model we can write

$$p(\boldsymbol{o},\boldsymbol{r}) = p(\boldsymbol{x},\boldsymbol{a},\boldsymbol{q}) = p(\boldsymbol{x}|\boldsymbol{a},\boldsymbol{q})\,p(\boldsymbol{a}|\boldsymbol{q})\,p(\boldsymbol{q})$$

and using the Bayes rule, we have

$$p(\boldsymbol{x},\boldsymbol{a},\boldsymbol{q}|\boldsymbol{y},\boldsymbol{z}) \propto p(\boldsymbol{y},\boldsymbol{z}|\boldsymbol{x},\boldsymbol{a},\boldsymbol{q})\,p(\boldsymbol{x},\boldsymbol{a},\boldsymbol{q}) = p(\boldsymbol{y},\boldsymbol{z}|\boldsymbol{x},\boldsymbol{a},\boldsymbol{q})\,p(\boldsymbol{x}|\boldsymbol{a},\boldsymbol{q})\,p(\boldsymbol{a}|\boldsymbol{q})\,p(\boldsymbol{q})$$

We illustrate this approach by making the following assumptions:

- Conditional independence of $\boldsymbol{y}$ and $\boldsymbol{z}$: $p(\boldsymbol{y},\boldsymbol{z}|\boldsymbol{x},\boldsymbol{a},\boldsymbol{q}) = p(\boldsymbol{y}|\boldsymbol{x})p(\boldsymbol{z}|\boldsymbol{a})$

– Gaussian laws for $\boldsymbol{b}_1$ and $\boldsymbol{b}_2$

$$p(\boldsymbol{y}|\boldsymbol{x};\sigma_1^2) \propto \exp\left[-\frac{1}{2\sigma_1^2}|\boldsymbol{y}-\boldsymbol{H}_1\boldsymbol{x}|^2\right]; \quad p(\boldsymbol{z}|\boldsymbol{a};\sigma_2^2) \propto \exp\left[-\frac{1}{2\sigma_2^2}|\boldsymbol{z}-\boldsymbol{H}_2\boldsymbol{a}|^2\right]$$

– Bernoulli law for $\boldsymbol{q}$: $\quad p(\boldsymbol{q}) \propto \sum_{i=1}^{n} q_i^{\lambda}(1-q_i)^{1-\lambda}$

– Gaussian law for $\boldsymbol{a}|\boldsymbol{q}$:

$$p(\boldsymbol{a}|\boldsymbol{q}) \propto \exp\left[-\frac{1}{2\sigma_a^2}\boldsymbol{a}^t\boldsymbol{Q}\boldsymbol{a}\right], \quad \boldsymbol{Q} = \operatorname{diag}[q_1,\ldots,q_n]$$

– Markovian model for $\boldsymbol{x}$: $\quad p(\boldsymbol{x}|\boldsymbol{a},\boldsymbol{q}) \propto \exp\left[-U(\boldsymbol{x}|\boldsymbol{a},\boldsymbol{q})\right]$

Then, based on

$$p(\boldsymbol{x},\boldsymbol{a},\boldsymbol{q}|\boldsymbol{y},\boldsymbol{z}) \propto p(\boldsymbol{y}|\boldsymbol{x})\,p(\boldsymbol{z}|\boldsymbol{a})\,p(\boldsymbol{x}|\boldsymbol{a},\boldsymbol{q})\,p(\boldsymbol{a}|\boldsymbol{q})\,p(\boldsymbol{q})$$

we can propose the following schemes:

– Simultaneous estimation of all the unknowns with the joint MAP estimation (JMAP):

$$(\widehat{\boldsymbol{x}},\widehat{\boldsymbol{a}},\widehat{\boldsymbol{q}}) = \arg\max_{(\boldsymbol{x},\boldsymbol{a},\boldsymbol{q})}\left\{p(\boldsymbol{x},\boldsymbol{a},\boldsymbol{q}|\boldsymbol{y},\boldsymbol{z})\right\}$$

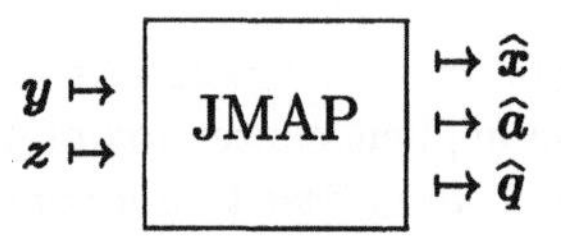

– First estimate the positions of the discontinuities $\boldsymbol{q}$ and then use them to estimate $\boldsymbol{x}$ and $\boldsymbol{a}$:

$$\begin{cases} \widehat{\boldsymbol{q}} = \arg\max_{\boldsymbol{q}}\left\{p(\boldsymbol{q}|\boldsymbol{y},\boldsymbol{z})\right\} \\ (\widehat{\boldsymbol{x}},\widehat{\boldsymbol{a}}) = \arg\max_{(\boldsymbol{x},\boldsymbol{a})}\left\{p(\boldsymbol{x},\boldsymbol{a}|\boldsymbol{y},\boldsymbol{z},\widehat{\boldsymbol{q}})\right\} \end{cases}$$

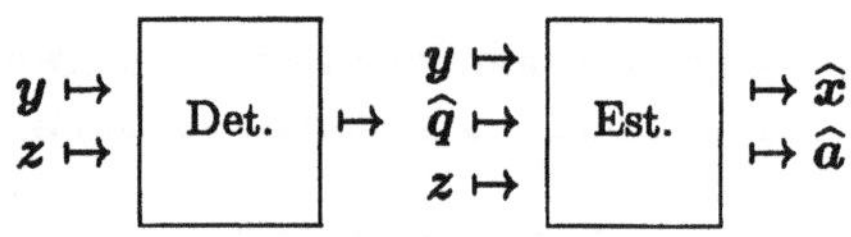

– First estimate the positions of the discontinuities $\boldsymbol{q}$ using only $\boldsymbol{z}$ and then use them to estimate $\boldsymbol{x}$ and $\boldsymbol{a}$:

$$\begin{cases} \widehat{\boldsymbol{q}} = \arg\max_{\boldsymbol{q}}\left\{p(\boldsymbol{q}|\boldsymbol{z})\right\} \\ (\widehat{\boldsymbol{x}},\widehat{\boldsymbol{a}}) = \arg\max_{(\boldsymbol{x},\boldsymbol{a})}\left\{p(\boldsymbol{x},\boldsymbol{a}|\boldsymbol{y},\boldsymbol{z},\widehat{\boldsymbol{q}})\right\} \end{cases}$$

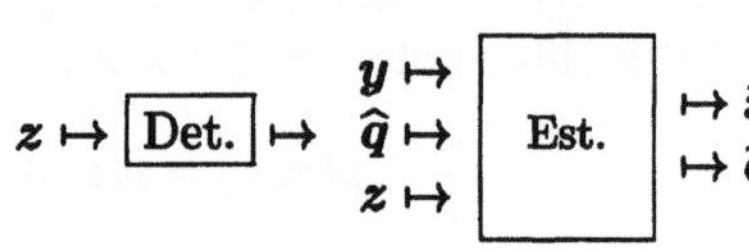

– First estimate $\boldsymbol{q}$ and $\boldsymbol{a}$ using only $\boldsymbol{z}$ and then use them to estimate $\boldsymbol{x}$:

$$\begin{cases} (\widehat{\boldsymbol{q}},\widehat{\boldsymbol{a}}) = \arg\max_{\boldsymbol{q},\boldsymbol{a}}\left\{p(\boldsymbol{q},\boldsymbol{a}|\boldsymbol{z})\right\} \\ \widehat{\boldsymbol{x}} = \arg\max_{\boldsymbol{x}}\left\{p(\boldsymbol{x}|\boldsymbol{y},\widehat{\boldsymbol{a}},\widehat{\boldsymbol{q}})\right\} \end{cases}$$

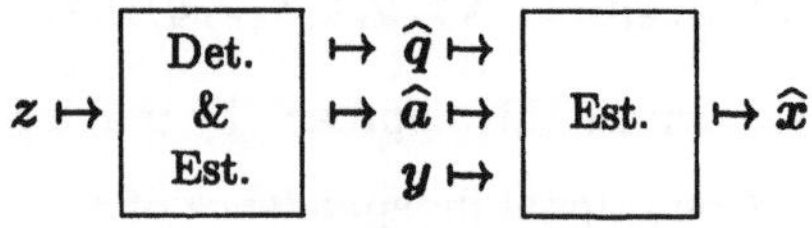

– First estimate only $\boldsymbol{q}$ using $\boldsymbol{z}$, then estimate $\boldsymbol{a}$ using $\widehat{\boldsymbol{q}}$ and $\boldsymbol{z}$, and finally, estimate $\boldsymbol{x}$ using $\widehat{\boldsymbol{q}}$, $\widehat{\boldsymbol{a}}$ and $\boldsymbol{y}$:

$$\begin{cases} \widehat{\boldsymbol{q}} = \arg\max_{\boldsymbol{q}} \{p(\boldsymbol{q}|\boldsymbol{z})\} \\ \widehat{\boldsymbol{a}} = \arg\max_{\boldsymbol{a}} \{p(\boldsymbol{a}|\boldsymbol{z},\widehat{\boldsymbol{q}})\} \\ \widehat{\boldsymbol{x}} = \arg\max_{\boldsymbol{x}} \{p(\boldsymbol{x}|\boldsymbol{y},\widehat{\boldsymbol{a}},\widehat{\boldsymbol{q}})\} \end{cases}$$

$\boldsymbol{z} \mapsto$ [Det.] $\mapsto \widehat{\boldsymbol{q}}$ $\qquad$ $\widehat{\boldsymbol{q}} \mapsto, \boldsymbol{z} \mapsto$ [Est.] $\mapsto \widehat{\boldsymbol{a}}$ $\qquad$ $\widehat{\boldsymbol{q}} \mapsto, \widehat{\boldsymbol{a}} \mapsto, \boldsymbol{y} \mapsto$ [Est.] $\mapsto \widehat{\boldsymbol{x}}$

– First estimate only $\boldsymbol{q}$ using $\boldsymbol{z}$ and then estimate $\boldsymbol{x}$ using $\widehat{\boldsymbol{q}}$ and the data $\boldsymbol{y}$:

$$\begin{cases} \widehat{\boldsymbol{q}} &= \arg\max_{\boldsymbol{q}} \{p(\boldsymbol{q}|\boldsymbol{z})\} \\ \widehat{\boldsymbol{x}} &= \arg\max_{\boldsymbol{x}} \{p(\boldsymbol{x}|\boldsymbol{y},\widehat{\boldsymbol{q}})\} \end{cases}$$

$\boldsymbol{z} \mapsto$ [Det.] $\mapsto \widehat{\boldsymbol{q}}$ $\qquad$ $\widehat{\boldsymbol{q}} \mapsto, \boldsymbol{y} \mapsto$ [Est.] $\mapsto \widehat{\boldsymbol{x}}$

Two more realistic solutions are:

Proposed method 1:
Estimate $\boldsymbol{r}$ using only $\boldsymbol{z}$ and estimate $\boldsymbol{x}$ and $\boldsymbol{q}$ using $\widehat{\boldsymbol{r}}$ and $\boldsymbol{y}$:

$\boldsymbol{z} \mapsto$ [Est.] $\mapsto \widehat{\boldsymbol{r}}$ $\qquad$ $\widehat{\boldsymbol{r}} \mapsto, \boldsymbol{y} \mapsto$ [Reconstruction] $\mapsto \widehat{\boldsymbol{x}}, \mapsto \widehat{\boldsymbol{q}}$

$p(\boldsymbol{r}|\boldsymbol{z}) \propto p(\boldsymbol{z}|\boldsymbol{r})\, p(\boldsymbol{r})$ $\qquad$ $p(\boldsymbol{x},\boldsymbol{q}|\widehat{\boldsymbol{r}},\boldsymbol{y}) \propto p(\boldsymbol{y}|\boldsymbol{x})\, p(\boldsymbol{x},\boldsymbol{q}|\widehat{\boldsymbol{r}})$

For the first part, with the assumptions made, we have

$$\widehat{\boldsymbol{r}} = \arg\max_{\boldsymbol{r}} \{p(\boldsymbol{r}|\boldsymbol{z})\} = \arg\min_{\boldsymbol{r}} \{J_1(\boldsymbol{r}|\boldsymbol{z})\}$$

with

$$J_1(\boldsymbol{r}|\boldsymbol{z}) = |\boldsymbol{z} - \boldsymbol{H}_2\boldsymbol{r}|^2 + \lambda \sum_j (r_{j+1} - r_j)^2$$

and for the second part we have

$$(\widehat{\boldsymbol{x}},\widehat{\boldsymbol{q}}) = \arg\max_{(\boldsymbol{x},\boldsymbol{q})} \{p(\boldsymbol{x},\boldsymbol{q}|\boldsymbol{y},\widehat{\boldsymbol{r}})\} = \arg\min_{(\boldsymbol{x},\boldsymbol{q})} \{J_2(\boldsymbol{x},\boldsymbol{q}|\boldsymbol{y},\widehat{\boldsymbol{r}})\}$$

with

$$J_2(\boldsymbol{x},\boldsymbol{q}|\boldsymbol{y},\widehat{\boldsymbol{r}}) = |\boldsymbol{y}-\boldsymbol{H}_1\boldsymbol{x}|^2+\lambda\sum_j(1-q_j)(x_{j+1}-x_j)^2+\alpha_1\sum_j q_j(1-\widehat{r}_j)+\alpha_2\sum_j q_j\widehat{r}_j$$

This last optimization is still too difficult to do due to the presence of $\boldsymbol{q}$ and $\boldsymbol{x}$ together. An easier solution is gien below.

Proposed method 2:
Use the ultrasound data $\boldsymbol{z}$ to detect the locations of some of the boundaries and use X ray data to make an intensity image preserving the positions of these discontinuities:

$$\boldsymbol{z} \mapsto \boxed{\text{Est.}} \mapsto \widehat{\boldsymbol{r}} \mapsto \boxed{q_j = \frac{|r_j|}{\sum_j |r_j|}} \mapsto \widehat{\boldsymbol{q}} \qquad \begin{array}{c}\widehat{\boldsymbol{q}} \mapsto \\ \boldsymbol{y} \mapsto\end{array} \boxed{\text{Est.}} \mapsto \widehat{\boldsymbol{x}}$$

Here, we made slightly different assumptions about the distributions of $\boldsymbol{r}$ and $\boldsymbol{x}$. Actually a generalized Gaussian distribution in place of Gaussian gives a good compromise of discontinuity preservation and easy implementation. A typical choice, for the first case is

$$\begin{aligned} \widehat{\boldsymbol{r}} &= \arg\max_{\boldsymbol{r}} \{p(\boldsymbol{r}|\boldsymbol{z})\} = \arg\min_{\boldsymbol{r}} \{J_1(\boldsymbol{r}|\boldsymbol{z})\} \\ \text{with} \quad J_1(\boldsymbol{r}|\boldsymbol{z}) &= ||\boldsymbol{z} - \boldsymbol{H}_2\boldsymbol{r}||^2 + \lambda_1||\boldsymbol{r}||^p, \quad 1 < p < 2 \end{aligned}$$

and for the second case is

$$\begin{aligned} \widehat{\boldsymbol{x}} &= \arg\max_{\boldsymbol{x}} \{p(\boldsymbol{x}|\boldsymbol{y}, \widehat{\boldsymbol{q}})\} = \arg\min_{\boldsymbol{x}} \{J_2(\boldsymbol{x}|\boldsymbol{y}; \widehat{\boldsymbol{q}})\} \\ \text{with} \quad J_2(\boldsymbol{x}|\boldsymbol{y}, \widehat{\boldsymbol{q}}) &= ||\boldsymbol{y} - \boldsymbol{H}_1\boldsymbol{x}||^2 + \lambda_2 \sum_j (1 - q_j)|x_{j+1} - x_j|^p, \quad 1 < p < 2 \end{aligned}$$

The aim of this paper is not to go through more details on these methods. The interested reader should refer to [8, 9].

4. Conclusions

To conclude briefly:

- ME can be used when we want to assign a probability law $p(\boldsymbol{x})$ from some expected values.
- ML can be used when we have a parametric form of the probability law $p(\boldsymbol{x}, \boldsymbol{\theta})$ and we have access to direct observations $\boldsymbol{x}$ of X, and we want to estimate the parameters $\boldsymbol{\theta}$.
- ML-EM extends the ML to the case of incomplete observations.
- When the observed data are noisy the Bayesian approach is the most appropriate.
- For practical data fusion problems the Bayesian approach seems to give all the necessary tools we need.
- Compound Markov models are convenient models to represent signals and images in a Bayesian approach of data fusion.
- The Bayesian approach is coherent and easy to understand. However, in real applications, we have still much to do to implement it:
  - Assignment or choice of the prior laws
  - Efficient optimisation of the obtained criteria
  - Estimation of the hyperparameters
  - Interpretation of the obtained results.

References

1. A. Mohammad-Djafari, *Maximum Entropy and Linear Inverse Problems; A Short Review*, pp. 253–264. Paris, France: Kluwer Academic Publishers, A. Mohammad-Djafari and G. Demoment ed., 1992.
2. A. Mohammad-Djafari, "Maximum d'entropie et problèmes inverses en imagerie," *Traitement du Signal*, pp. 87–116, 1994.
3. A. Mohammad-Djafari and J. Idier, *Maximum Likelihood Estimation of the Lagrange Parameters of the Maximum Entropy Distributions*, pp. 131–140. Seattle, USA: Kluwer Academic Publishers, C.R. Smith, G.J. Erikson and P.O. Neudorfer ed., 1991.
4. M. Miller and D. Snyder, "The role of likelihood and entropy in incomplete-data problems: Applications to estimating point-process intensities and toeplitz constrained covariances," *Proceedings of the IEEE*, vol. 75, pp. 892–906, July 1987.
5. S. Gautier, G. Le Besnerais, A. Mohammad-Djafari, and B. Lavayssière, *Data fusion in the field of non destructive testing.* Santa Fe, U.S.A.: Kluwer Academic Publishers, K. Hanson ed., 1995.
6. S. Gautier, G. Le Besnerais, A. Mohammad-Djafari, and B. Lavayssière, "Fusion de données radiographiques et ultrasonores, en vue d'une applicaion en contrôle non destructif," (Clermont-Ferrand), Second international workshop on inverse problems in electromagnetism and acoustic, May 1995.
7. S. Gautier, *Fusion de donnés gammagraphiques et ultrasonores. Application au contrôle non destructif.* PhD thesis, Université de Paris-Sud, Orsay, septembre 1996.
8. S. Gautier, J. Idier, A. Mohammad-Djafari, and B. Lavayssière, "Fusion de données gammagraphiques et ultrasonores," in *GRETSI 97*, (Grenoble, France), pp. 781–784, 1997.
9. S. Gautier, B. Lavayssière, G. Le Besnerais, and A. Mohammad-Djafari, "L2+lp deconvolution and ultrasound imaging for non destructive evaluation," in *QLCAV 97*, vol. 1, (Le Creusot), pp. 212–213, 1997.

WHENCE THE LAWS OF PROBABILITY?

ANTHONY J.M. GARRETT
63 High Street, Grantchester
Cambridge CB3 9NF, United Kingdom

Abstract. A new derivation is given of the sum and product rules of probability. Probability is treated as a number associated with one binary proposition conditioned on another, so that the Boolean calculus of the propositions induces a calculus for the probabilities. This is the strategy of R.T. Cox (1946), with a refinement: a formula is derived for the probability of the NAND of two propositions in terms of the probabilities of those propositions. Because NAND is a primitive logic operation from which any other can be synthesised, there are no further probabilities that the NAND can depend on. A functional equation is then set up for the relation between the probabilities and is solved. By synthesising the non-primitive operations NOT and AND from NAND the sum and product rules are derived from this one formula, the fundamental 'law of probability'.

Key words: probability, laws of probability, sum rule, product rule, Boolean algebra

1. Introduction: Probability

This paper sets out a novel derivation of the sum and product rules for probabilities – the 'laws of probability' – by deriving them from a single simpler equation (to which they are jointly equivalent); and by deriving that simpler equation from more primitive axioms. In deriving laws of probability from more fundamental ideas we have to engage with what 'probability' means. This is a notoriously contentious issue; fortunately, if you disagree with the definition that is proposed, there will be a get-out that allows other definitions to be preserved.

The probability $p(X|Y)$ is a shorthand for "the probability that X is true, supposing that Y is true" or, more formally, "the probability that proposition X has truth value TRUE, supposing that proposition Y has truth value TRUE". Here X and Y are binary propositions that have truth value TRUE or FALSE: they are assertions that are either true or false. The probability $p(X|Y)$ is taken to be a logical measure of the extent to which truth of X is implied by the (supposed) truth of Y, according to the ontological relations between the things to which the propositions refer; it is an 'if... then...' concept. For example, if Y is "there is a die with sides labelled '1' to '6'" and X is "the face labelled '2' is uppermost" then $p(X|Y)=1/6$, because it is an ontological assertion that there are six faces, between which Y discriminates solely according to (arbitrary) label; hence their probabilities are equal.

This idea of probability as a degree of partial implication, or ***implicability***, is emphasised in the textbook of probability by the economist Keynes (1921). It dates back to Leibniz, and implicitly to mediaeval law through the extent to which guilt is implied by evidence (Franklin 1991). This view is operationally equivalent to the Bayesian idea that probability is the degree to which it is appropriate to believe that X is true in the light of Y being true – the ***believability*** of X, supposing Y. The Keynesian view has the advantage that it carries less psychological baggage.

It is degree of implication to which the word 'probability' refers throughout this paper. If, however, you have a different idea of what probability means, you can just replace 'probability' by 'implicability' from here on, and reserve 'probability' for whatever use you wish. It is implicability, however named, that is required in any real problem in which 'probability theory' is deployed. For the aim is always to work out how strongly it is implied, from the information available including any experimental data – "the measured value was so-and-so" – that a parameter takes a particular value, or that a hypothesis is true, or whatever. Equally empty in operational terms is debate over the merits of words such as probability, credibility, verisimilitude, likeliness, plausibility, expectance, confidence, surprise *et cetera.*

2. The Single Law of Probability

We shall now find a single equation which implies both the sum and product rules. The product rule is

$$p(AB|I) = p(A|BI)p(B|I) \tag{1}$$

where AB denotes the logical product ('AND') of two propositions, and is formally defined by its truth table: it is true only if both of A and B are true, and is false otherwise. The sum rule is

$$p(A|I) + p(\overline{A}|I) = 1, \tag{2}$$

where $\overline{A}$ ('NOT A'; the negation of A) denotes the proposition that is false if A is true, and vice-versa (since $\overline{\overline{A}}=A$).

We use the sum and product rules to seek a decomposition of $p(A \uparrow B|I)$, where $A\uparrow B$ is the logic function defined to be false only if A and B are both true, and true otherwise. It may therefore be recognised as the operation NOT AND, or NAND. We seek this particular decomposition because every logical operation can be synthesised from NAND alone: given only enough NAND gates, any logic circuit can be constructed from them. Correspondingly we should be able to derive both of the product and sum rules from the decomposition of $p(A\uparrow B|I)$. NOT and AND are given by

$$\begin{aligned} \overline{A} &= A\uparrow A \\ AB &= \overline{A\uparrow B} = (A\uparrow B)\uparrow(A\uparrow B). \end{aligned} \tag{3}$$

(A comprehensive reference on Boolean algebra is Kuntzmann, 1967.) No other operation can be defined from NOT in isolation, so that the sum and product rules cannot be derived from each other, and the decomposition we seek will be more fundamental than either rule alone. It is easily found using both of them. From the sum rule

$$\begin{aligned} p(A\uparrow B|I) &= p(\overline{AB}|I) \qquad (4) \\ &= 1-p(AB|I) \qquad (5) \end{aligned}$$

whence, using the product rule,

$$p(A\uparrow B|I) = 1-p(A|BI)p(B|I). \qquad (6)$$

To verify that (6) is more fundamental than either the sum rule or product rule alone, we reverse the analysis and derive both of them from it. Begin by looking at the special case for which I includes a logical factor of $\overline{A}$: $I=\overline{A}J$, so that $A|I=F$ (for false), and $A\uparrow B|I=T$ (for true) whatever B. Denote by p_f the (fixed) probability of a proposition that is implied to be false, and p_t the probability of a proposition that is implied to be true. Then (6) simplifies to

$$p_t = 1-p_f p(B|\overline{A}J). \qquad (7)$$

For this to hold for all B the coefficient of $p(B|\overline{A}J)$ must vanish: $p_f=0$, whence immediately $p_t=1$. Next, the sum rule is generated by putting $B=A$ in (6); from (3), we have

$$p(\overline{A}|I) = 1-p(A|AI)p(A|I) \qquad (8)$$

which, since $p(A|AI)=p_t=1$, reduces to the sum rule

$$p(\overline{A}|I) = 1-p(A|I). \qquad (9)$$

To derive the product rule we first rewrite (6) as

$$p(\overline{AB}|I) = 1-p(A|BI)p(B|I). \qquad (10)$$

But, from the sum rule just derived, we have

$$p(\overline{AB}|I) = 1-p(AB|I). \qquad (11)$$

By comparing the last two formulae it follows immediately that

$$p(AB|I) = p(A|BI)p(B|I), \qquad (12)$$

which is the product rule. It shows how to move propositions across the conditioning solidus, and might better be called the conditionalising rule.

We have derived the end-point relations

$$p_f = 0\,, \qquad p_t = 1 \qquad (13)$$

and the sum and product rules from the single relation

$$p(A\uparrow B|I) = 1 - p(A|BI)p(B|I). \tag{14}$$

If, now, we can derive this relation from more basic ideas, we shall have a far tighter derivation of the conventional laws of probability. Such a derivation now follows.

3. Deriving the Calculus of Probability from the Calculus of Propositions

A probability is a number associated with a conditioned proposition. Truth values of propositions obey a calculus known as Boolean algebra, familiar in logic circuitry. The Boolean calculus of propositions induces a corresponding calculus for the numbers associated with the propositions, *i.e.* for the probabilities. We shall derive the calculus of probabilities from the calculus of propositions.

This strategy is due to R.T. Cox (1946), who, two centuries after the sum and product rules became commonplace, rederived them by this means and so provided formal justification that 'probability' can be used to mean implicability. The present derivation is a refinement of Cox's.

We begin by proposing a relationship for the decomposition of $p(A \uparrow B|I)$, which will be stated and then argued for. It is

$$p(A\uparrow B|I) = F\big(p(A|BI), p(B|I), p(B|AI), p(A|I)\big). \tag{15}$$

Since the conditioning proposition I represents what is unquestioned, it remains on the right of the conditioning solidus throughout. And, crucially, since all logical operations, including NOT and AND, can be synthesised from NAND (defined by its truth table – ignore the etymology), no logical operation need appear in the right-hand side apart from the logical product with I in the conditioning, which is known to be what is required. Hence (15) exhausts the possibilities. Except where (and how) it is stated, it is supposed that I does not include any logical factor of A, $\overline{A}$, B or $\overline{B}$.

Having justified (15) we proceed to reduce the number of probabilities in the right-hand side. Look first at the special case of (15) obtained when $B = AC$. There is loss of generality in B, which is false if A is false, but no loss of generality in A or C. On the left-hand side $A\uparrow B$ becomes

$$A\uparrow B = A\uparrow (AC) = \overline{AAC} = \overline{AC} = A\uparrow C. \tag{16}$$

The right-hand side simplifies, since

$$p(A|ACI) = p_t, \qquad p(AC|AI) = p(TC|AI) = p(C|AI)\ . \tag{17}$$

Equations (17) are properties we require of probability, not properties of Boolean algebra – although there is a happy coincidence with the truth value notation. We find that

$$p(A\uparrow C|I) = F\big(p_t, p(AC|I), p(C|AI), p(A|I)\big) \tag{18}$$

or, relabelling $C \to B$,

$$p(A \uparrow B|I) = F\big(p_t, p(AB|I), p(B|AI), p(A|I)\big). \tag{19}$$

This is to be compared with (15).

Look now at the special case of (15) for which $B = A$: from (3), it reduces to

$$p(\overline{A}|I) = F\big(p_t, p(A|I), p_t, p(A|I)\big), \tag{20}$$

which tells us that there is a direct relation connecting the probabilities of truth and falsehood of any proposition:

$$p(\overline{A}|I) = \chi\big(p(A|I)\big) \tag{21}$$

where

$$\chi(z) = F(p_t, z, p_t, z). \tag{22}$$

By eliminating $p(\overline{A}|I)$ between (21) and itself with $A \to \overline{A}$ and since $\overline{\overline{A}} = A$, we have (in operator notation) $\chi^2 = \mathcal{I}$, the unit operator; the function χ will be studied later. The next step is to replace A by $A \uparrow B$ in (21), giving, from (3),

$$p(AB|I) = \chi\big(p(A \uparrow B|I)\big), \tag{23}$$

which, upon substitution in the right-hand side of (19), yields

$$p(A \uparrow B|I) = F\big(p_t, \chi\big(p(A \uparrow B|I)\big), p(B|AI), p(A|I)\big). \tag{24}$$

At this point there are two possibilities: that the right-hand side of (24) is identically equal to $p(A \uparrow B|I)$ and has no dependence upon $p(B|AI)$ or $p(A|I)$, so that (24) is tautologous and (19) is merely a disguised form of (23); or that (24) expresses in implicit form a relation between $p(A \uparrow B|I)$, $p(B|AI)$ and $p(A|I)$. In that case we would have succeeded in simplifying (15) to dependence on just two probabilities. To investigate whether this is possible we shall use (15) to eliminate $p(A \uparrow B|I)$ in both sides of (24). Define for convenience the new function $\psi(\ldots) = \chi(F(\ldots))$ and

$$u = p(A|BI), \quad v = p(B|I), \quad x = p(B|AI), \quad y = p(A|I). \tag{25}$$

Then we have

$$p(AB|I) = \psi(u, v, x, y) \tag{26}$$

and, substituting (15) into both sides of (24) and taking χ of both sides, we obtain the functional equation

$$\psi(u, v, x, y) = \psi\big(p_t, \psi(u,v,x,y), x, y\big). \tag{27}$$

Inspection reveals that there is indeed a solution for $\psi(u, v, x, y)$ which is an arbitrary function of the last two arguments; this is what we were hoping for. However it is necessary to check for other solutions, and we must therefore solve (27) systematically. We must also be careful, since independence lies not in the four algebraic variables u, v, x, y that represent probabilities, but in the three logical variables A, B, I.

By defining $J(z,x,y)=\psi(p_t,z,x,y)$ we can rewrite this equation, in the spirit of (24), as

$$\psi = J(\psi,x,y). \tag{28}$$

In this form there are three probability variables, ψ, x and y; and, since there are three degrees of freedom (A, B and I), we may take ψ, x and y to vary independently. There are four possibilities: that J is a constant; that it is non-constant and varies with its first argument alone; that it is non-constant and independent of its first argument, varying with the others; or that it definitely varies with both its first and its other arguments. If it varies with its first argument alone, then for (28) to make sense the function $J(\psi',x',y')=\psi'$, where a prime attached to an algebraic variable denotes that it is a dummy and not a shorthand for any probability. This corresponds to the tautologous case in (24), but in (27) it yields a family of solutions for $\psi(u,v,x,y)$ satisfying $\psi(u'=p_t,v',x',y')=v'$. Next, if J is independent of its first argument we have $\psi(u,v,x,y)=f(x,y)$, the solution we anticipated. Since f is arbitrary this case can be taken to include that of constant J. The remaining case is when J varies with both its first argument and the others, so that (28) expresses a relation among ψ, x and y. This may always be reverted to give ψ as a function of x and y, so that $\psi(u,v,x,y)=f(x,y)$ again; even if the function $J(\psi',x',y')$ is multivalued, this arbitrariness can be absorbed in that of f.

Therefore the solution must be either the one we want, $\psi(u,v,x,y)=f(x,y)$, or a 'tautology' solution satisfying $\psi(u'=p_t,v',x',y')=v'$. To generate further constraints on solutions we look back to where we put $B=AC$ in (15), and instead put $A=BC$. The analysis runs exactly parallel and yields the further functional equation for ψ,

$$\psi(u,v,x,y) = \psi\big(u,v,p_t,\psi(u,v,x,y)\big). \tag{29}$$

Correspondingly this has solutions $\psi(u,v,x,y)=f(u,v)$ independent of x and y, and solutions satisfying $\psi(u',v',x'=p_t,y')=y'$. Valid solutions must fall into one or other of these categories. Of the solutions we have already found for (27), the class $f(x,y)$ are clearly not those independent of x and y (unless f is a constant); therefore to be valid they must satisfy $\psi(u',v',x'=p_t,y')=y'$. It remains to distribute those solutions of (27) satisfying $\psi(u'=p_t,v',x',y')=v'$ between these categories; some will satisfy $\psi(u,v,x,y)=f(u,v)$, some will satisfy $\psi(u',v',x'=p_t,y')=y'$ (and some will satisfy neither and be ruled out). Collecting, solutions must be: constant; or of the form $\psi(u,v,x,y)=f(u,v)$ or $f(x,y)$ where $f(p_t,z)=z$; or must satisfy both tautology conditions $\psi(u'=p_t,v',x',y')=v'$ and $\psi(u',v',x'=p_t,y')=y'$. Constant solutions are useless, while no solutions satisfy both tautology conditions, for upon putting u' and x' equal to p_t and equating the resulting expressions for $\psi(p_t,v',p_t,y')$ we obtain $v'=y'$, which is a contradiction since these are independent variables. We are left with only the desired solutions $f(u,v)$ or $f(x,y)$.

Before proceeding, we must settle a question. Since the logical product is commutative we may equate decompositions of $p(AB|I)$ and $p(BA|I)$ to give

$$\psi(u,v,x,y) = \psi(x,y,u,v). \tag{30}$$

Upon substituting our solution $\psi(u,v,x,y) = f(u,v)$ into this equation we find that $f(u,v) = f(x,y)$; were u, v, x and y four independent variables, this would imply that f is just a useless constant. However, we have seen that independence lies in the three logical variables A, B and I, one fewer degree of freedom, with symmetry between A and B. The equation $f(u,v) = f(x,y)$ is properly understood as a relation among the four probabilities u, v, x and y. If symmetry is required we can now rewrite either of our solutions $f(u,v)$, $f(x,y)$ as $\frac{1}{2}(f(u,v) + f(x,y))$, demonstrating their equivalence.

We henceforth work with the solution $\psi(u,v,x,y) = f(u,v)$ or, in probabilities,

$$p(AB|I) = f\big(p(A|BI), p(B|I)\big). \tag{31}$$

By substituting this into (23), taking χ of both sides, and using $\chi^2 = \mathcal{I}$, it follows that

$$p(A \uparrow B|I) = F(u,v,x,y) = \chi\big(f(u,v)\big) \tag{32}$$

which, upon substitution in (22), gives

$$\chi(z) = \chi\big(f(p_t, z)\big). \tag{33}$$

Since we have shown that $f(p_t, z) = z$, this does not tell us anything more about the function χ; for that we must travel further. Our two-variable solution f allows us, as we hoped, to write $p(A \uparrow B|I)$ as a function of just $p(A|BI)$ and $p(B|I)$:

$$p(A \uparrow B|I) = F\big(p(A|BI), p(B|I)\big) \tag{34}$$

where F refers now to a new function $\chi(f)$ of just two arguments. This relation for NAND is a big simplification over the four arguments in (15).

Finally, it is a useful exercise to verify that if further probabilities, such as $p(AB|I)$, are tacked on as extra arguments of F in (15), the same simplification (34) can be effected.

4. Derivation and Solution of a Functional Equation Relating Probabilities

We now wish to find the particular function F in (34) which expresses the probability of NAND of two propositions in terms of probabilities of those propositions. To do this, we repeat the above synthesis of the probability of the logical product (AND) of two propositions and of the negation of a proposition, using the result we have established, (34). R.T. Cox (1946) has investigated the functional relations for AND and NOT, so that from there on we can borrow his analysis. By putting $B = A$ in (34) we have

$$p(\overline{A}|I) = F\big(p_t, p(A|I)\big). \tag{35}$$

The fact that $p(\overline{A}|I)$ depends only on $p(A|I)$ and (hence) vice-versa was an assumption of Cox's, but here it has been derived (originally at (20)). Define for convenience

$$\chi(v) = F(p_t, v) \tag{36}$$

so that

$$p(\overline{A}|I) = \chi\big(p(A|I)\big) \tag{37}$$

where χ is the same function defined earlier, so that $\chi^2 = \mathcal{I}$. Any inverse of χ must therefore be equal to χ itself: $\chi = \chi^{-1}$. We require the mapping between the probabilities $p(A|I)$ and $p(\overline{A}|I)$ to be unique, so that χ^{-1} exists and χ is self-inverse. Any function which is its own reflection in the line $y = x$ is self-inverse, and if $S(x, y) = 0$ where S is an exchangeable function admitting an invertible relation between x and y then these are identical self-inverse functions of each other.

For the logical product we use a short cut: NAND is NOT AND, and we use (23) to remove the NOT; upon defining as before $\psi = \chi(F)$ (only with two arguments), we have upon substituting (34) into (23),

$$p(AB|I) = \psi\big(p(A|BI), p(B|I)\big). \tag{38}$$

This functional dependence, originally derived in the previous section, was a separate assumption of Cox's; a heuristic justification of it, beginning from $p(AB|I)$ as a function of the four elemental probabilities, was given by Tribus (1969, p14).

We now bring in the innovation of Cox (1946) and apply (38) twice, to decompose the logical product of three propositions. Partitioning of the logical product can be done in more than one way but, since the logical product is associative, the results must coincide. By equating the decompositions a functional equation is generated for ψ, and from its solution information can be gained about F.

Begin by writing

$$\begin{aligned} p(XYZ|I) &= p((XY)Z|I) && (39)\\ &= \psi\big(p(XY|ZI), p(Z|I)\big) && (40)\\ &= \psi\Big(\psi\big(p(X|YZI), p(Y|ZI)\big), p(Z|I)\Big). && (41) \end{aligned}$$

Alternatively,

$$\begin{aligned} p(XYZ|I) &= p\big(X(YZ)|I\big) && (42)\\ &= \psi\big(p(X|YZI), p(YZ|I)\big) && (43)\\ &= \psi\Big(p(X|YZI), \psi\big(p(Y|ZI), p(Z|I)\big)\Big). && (44) \end{aligned}$$

By equating these two decompositions and defining

$$x = p(X|YZI), \qquad y = p(Y|ZI), \qquad z = p(Z|I), \tag{45}$$

we obtain the functional equation

$$\psi\big(x, \psi(y, z)\big) = \psi\big(\psi(x, y), z\big) \tag{46}$$

where x, y and z can clearly be taken as independent variables.

Equation (46) is called the ***associativity equation*** and has been known since Abel studied it in the 19th century (see Aczél, 1966). Its solution is a matter of mathematics, and we quote the result. There is a general solution

$$\psi(u, v) = \phi^{-1}\big(\phi(u)\phi(v)\big) \tag{47}$$

(in dummy variables u, v), where ϕ is a continuous function of a single argument and ϕ^{-1} its inverse, which are both supposed to be unique; ϕ is otherwise arbitrary. Though this solution is lengthy to derive it is easy to verify: each side of (46) is equal to $\phi^{-1}\left(\phi(x)\phi(y)\phi(z)\right)$. The solution (47) was derived in an appendix by Cox (1946) and more neatly in a later exposition (Cox 1961, chapter 3). It is derived using an iterative method, and with a discussion of the conditions of validity of the solution, by Aczél (1966, section 6.2). A summary of the iteration theory for this equation is given by Smith and Erickson (1990).

Equation (46) also has particular solutions

$$\psi(u,v) = f(u), \qquad \psi(u,v) = f(v) \tag{48}$$

where, by direct substitution in (46), $f^2 = f$ (in operator language), so that f is either the identity operator ($f(w) = w$) or a projection which may be discontinuous (an extreme case is the solution $f =$ constant). The identity solution $f(w) = w$ is deducible by Cox's method, although Cox did not state it. Back in (38), the solution $\psi(u,v) = f(v)$ cannot be valid for arbitrary A. The solution $\psi(u,v) = f(u)$ fails if $A = B$ since the left-hand side then depends on B, which is arbitrary, but the right-hand side does not. For the remaining, more general solution of the associativity equation, (47), the relation (38) may be written, dropping brackets for convenience, as

$$\phi p(AB|I) = \phi p(A|BI)\,.\,\phi p(B|I)\ . \tag{49}$$

This looks uncannily like the product rule; but let us not anticipate.

Now we undo the transformations to recover F and the relation between $p(A\uparrow B|I)$, $p(A|BI)$ and $p(B|I)$. Since $\psi = \chi(F)$ and since $\chi^2 = \mathcal{I}$ it follows that $F = \chi(\psi)$, so that from (47)

$$F(u,v) = \chi\phi^{-1}\left(\phi(u)\phi(v)\right) \tag{50}$$

where further information is obtained by substituting this result for F into (36), to give

$$\chi(v) = \chi\left(\phi^{-1}(\phi p_t\,.\,\phi(v))\right). \tag{51}$$

Since χ possesses an inverse it can be peeled away from either side; it then follows that

$$\phi p_t = 1. \tag{52}$$

The next step is to substitute (50) into (34) to give, upon taking χ of both sides and simplifying,

$$(\phi\chi\phi^{-1})\phi p(A\uparrow B|I) = \phi p(A|BI)\,.\,\phi p(B|I) \tag{53}$$

or, on defining $\Theta \equiv \phi\chi\phi^{-1}$,

$$\Theta\, \phi p(A\uparrow B|I) = \phi p(A|BI)\,.\,\phi p(B|I) \tag{54}$$

where, since $\chi = \phi^{-1}\Theta\phi$ and $\chi^2(v) = v$,

$$\Theta^2(v) = v. \tag{55}$$

It is useful to define

$$p' \equiv \phi p \tag{56}$$

so that $p'_t = 1$, and (54) becomes

$$\Theta p'(A \uparrow B|I) = p'(A|BI)p'(B|I). \tag{57}$$

The last three equations encapsulate our progress so far. Equation (49) becomes

$$p'(AB|I) = p'(A|BI)\,p'(B|I), \tag{58}$$

and is readily derived by running through the decomposition of the logical product, set out at the start of this section, using (55) and (57) rather than the original equations in which the functions were undetermined. In this process we find a relation for negation by putting $B=A$ in (57):

$$\Theta p'(\overline{A}|I) = p'(A|I). \tag{59}$$

This explains why Θ satisfies (55) and must be self-inverse.

For $A=\overline{B}$ equation (58) becomes $p'_f = p'_f p'(B|I)$, which implies that $p'_f = 0$ (or ∞, which we shall demonstrate gives a logically equivalent formalism). Hence we have shown that

$$p'_f = 0\,, \qquad p'_t = 1 \tag{60}$$

and, by putting $I=ABJ$ and $I=\overline{A}J$ respectively in (57), we find that

$$\Theta(0) = 1\,, \qquad \Theta(1) = 0. \tag{61}$$

Equations (61) anchor the function Θ at the end points of its argument.

We now use (57) to synthesise a relation for the logical sum (inclusive OR); this will generate an equation for Θ. It is more convenient to synthesise the logical sum from negation and the logical product, using de Morgan's theorem, than to derive it directly from NAND. De Morgan's theorem is

$$\overline{A+B} = \overline{A}\,\overline{B} \tag{62}$$

so that

$$p'(A+B|I) = p'(\overline{\overline{A}\,\overline{B}}|I). \tag{63}$$

Our strategy will be to remove all negations using Θ, as in (59). First,

$$p'(A+B|I) = \Theta\big(p'(\overline{A}\,\overline{B}|I)\big). \tag{64}$$

Now decompose this using the 'product rule' (58):

$$\begin{aligned} p'(A+B|I) &= \Theta\big(p'(\overline{A}|I)p'(\overline{B}|\overline{A}I)\big) && (65)\\ &= \Theta\Big(\Theta\big(p'(A|I)\big)\Theta\big(p'(B|\overline{A}I)\big)\Big). && (66)\end{aligned}$$

The remaining negation in (66) is to the right of the conditioning solidus. To undo it using Θ, it must be moved to the left. This is done by exploiting commutativity of NAND: we equate the decompositions of $\Theta p'(\overline{A}\uparrow B|I)$ and $\Theta p'(B\uparrow\overline{A}|I)$ given by (57), to find

$$p'(B|\overline{A}I) = p'(B|I)p'(\overline{A}|BI)/p'(\overline{A}|I), \tag{67}$$

which is a prototype of Bayes' theorem. When this is substituted into (66) and the further negations are undone using Θ, the result is

$$p'(A+B|I) = \Theta\left(\Theta\big(p'(A|I)\big)\Theta\Big(\frac{p'(B|I)\Theta\big(p'(A|BI)\big)}{\Theta\big(p'(A|I)\big)}\Big)\right). \tag{68}$$

Define now

$$x \equiv p'(A|I)\ , \quad y \equiv p'(B|I)\ , \quad z \equiv p'(AB|I). \tag{69}$$

These are new definitions of x, y, z unrelated to (45). Now (68) becomes

$$p'(A+B|I) = \Theta\left(\Theta(x)\Theta\Big(\frac{y\Theta(\frac{z}{y})}{\Theta(x)}\Big)\right). \tag{70}$$

Since the logical sum is commutative the left-hand side is exchangeable in A and B; but, for arbitrary Θ, the right-hand side is not. If we therefore equate this expression to itself with A and B – and hence x and y – exchanged, we will gain a new equation which Θ must satisfy. On taking Θ of both sides and using (55), this equation is

$$\Theta(x)\Theta\Big(\frac{y\Theta(\frac{z}{y})}{\Theta(x)}\Big) = \Theta(y)\Theta\Big(\frac{x\Theta(\frac{z}{x})}{\Theta(y)}\Big)\ . \tag{71}$$

This functional equation must be solved for Θ subject to the boundary conditions (61) and also (55).

The solution of this equation subject to these constraints is again a matter of mathematics. In the special case $z=0$ the equation reduces without restriction on x or y to

$$\Theta(x)\Theta\Big(\frac{y}{\Theta(x)}\Big) = \Theta(y)\Theta\Big(\frac{x}{\Theta(y)}\Big) \tag{72}$$

since $\Theta(0)=1$. Redefine $x \to \Theta(x)$ and $y \to \Theta(y)$ and use (55) to obtain

$$x\,\Theta\Big(\frac{\Theta(y)}{x}\Big) = y\,\Theta\Big(\frac{\Theta(x)}{y}\Big)\ , \tag{73}$$

which incorporates (55): to see this, put either of x, y to unity and use the boundary conditions.

Equation (73) was derived from (59) in a different way by Cox (1946), who also solved it. It has been solved without assuming differentiability by Aczél (1963); another method of solution is given by Jaynes (in preparation: see references). The solution is

$$\Theta(u) = (1-u^k)^{1/k} \tag{74}$$

where k is arbitrary but positive (to satisfy the boundary conditions). It is easy to verify that this solution also satisfies the full equation (71): each side is equal to $(1-x^k-y^k+z^k)^{1/k}$. (It is possible to show without solving either equation that the solution of the reduced equation satisfies the full equation, subject to the boundary conditions; a proof will be given elsewhere.) The solution remains real – allowing an ordering as we require – provided that its domain of validity is $[0,1]$, the interval between the boundary conditions. The solution of a functional equation often depends on the domain, which may in turn depend on that solution since in the equation the function may itself appear as an argument. Here the domain is determined along with the solution by the equation and boundary conditions.

The irregular solution $p'_f = \infty$ referred to following equation (59) corresponds to $k < 0$ and domain $[1, \infty)$; this solution transforms into the present one under $k \to -k$, or equivalently $p' \to 1/p'$, and it is therefore a matter of convention which to prefer; we stick with the usual one for familiarity.

Upon substituting the solution for Θ into (57) and rearranging, we have

$$p'(A{\uparrow}B|I)^k = 1 - p'(A|BI)^k p'(B|I)^k \tag{75}$$

while (59) becomes

$$p'(A|I)^k + p'(\overline{A}|I)^k = 1 \tag{76}$$

and (70) reduces to

$$p'(A+B|I)^k = p'(A|I)^k + p'(B|I)^k - p'(AB|I)^k. \tag{77}$$

By defining

$$p'' \equiv p'^k = \phi(p)^k \tag{78}$$

these equations become

$$p''(A{\uparrow}B|I) = 1 - p''(A|BI)p''(B|I) \tag{79}$$

and

$$p''(A|I) + p''(\overline{A}|I) = 1 \tag{80}$$

(which is the sum rule) and

$$p''(A+B|I) = p''(A|I) + p''(B|I) - p''(AB|I), \tag{81}$$

which, in combination with the sum and product rules, tells us that $p''(A+\overline{A}|I)=1$ whatever I has to say about A. Equation (81) is easily derived by running through the derivation of (70) with the given form of Θ. When advance is made in expressing the solution of a functional equation it is often useful to rerun the analysis up to that point with the new version of the solution in place; this trick has has been used more than once in the present analysis. By moving the logical product to the left-hand side, equation (81) has the happy phrasing that the probability of the logical sum of two propositions and the probability of their logical product add up to the sum of probabilities of the propositions. This relation can be used to eliminate logical sums in favour of logical products or vice-versa. Finally, (58) is invariant under power-law transformations, so that

$$p''(AB|I) = p''(A|BI)p''(B|I), \tag{82}$$

which is the product rule. Hence p'' satisfies the laws of probability: the sum and product rules. Both follow from (79).

5. Uniqueness

We have already absorbed arbitrariness of the function ϕ and the index k by defining $p'' \equiv p'^k = \phi(p)^k$. Before discussing the significance of this freedom we must learn whether there is any more arbitrariness in the system. This is done by investigating the invariance properties of equation (79). Suppose that a transformed function $\Lambda(p'')$ satisfies the same equation:

$$\Lambda\big(p''(A\uparrow B|I)\big) = 1 - \Lambda\big(p''(A|BI)\big)\Lambda\big(p''(B|I)\big). \tag{83}$$

Then, by eliminating $p''(A\uparrow B|I)$ using (79) and (re)defining $x \equiv p''(A|BI)$, $y \equiv p''(B|I)$,

$$\Lambda(1-xy) = 1 - \Lambda(x)\Lambda(y). \tag{84}$$

By construction $\Lambda(z)=z$ is a solution; but are there others? Define

$$\Pi(z) \equiv 1-\Lambda(1-z), \tag{85}$$

so that (84) becomes

$$\Pi(xy) = \Lambda(x)\Lambda(y) \tag{86}$$

which for the permitted value $y=1$ reduces to

$$\Pi(x)=\Lambda(1)\Lambda(x). \tag{87}$$

By writing xy for x in this expression and using (86) to eliminate $\Pi(xy)$, we have

$$\Lambda(1)\Lambda(xy) = \Lambda(x)\Lambda(y) \tag{88}$$

which is a functional equation of Cauchy type. To solve it, differentiate it with respect to y and then put $y=1$, giving

$$\Lambda(1)x\Lambda'(x) = \Lambda'(1)\Lambda(x) \tag{89}$$

(a prime here denotes differentiation), which integrates to give solutions proportional to an arbitrary power of x. (The Cauchy functional equation is solved without assuming differentiability in Aczél, 1966, chapter 2.) Back in (84) the only solutions that survive are $\Lambda(z)=z$ and the (useless) constant roots of the quadratic $\Lambda^2 + \Lambda - 1 = 0$. Hence we have proved that there is no further arbitrariness.

The choice of ϕ and k remains arbitrary. If p is a numerical representation of the extent to which truth of one proposition is implied on supposing the truth of another, then so also is any monotonic increasing function of p. (Monotonicity is necessary to preserve ordering of probabilities.) We have learned that it is useful to write such a function as ϕ^k, where the function ϕ and the index k reflect the structure of the theory. We shall choose to work directly with p'' (and drop the primes) in order to retain a representation which satisfies the familiar two laws of probability rather than a transformed version of them. That this is possible confirms we may indeed interpret probability as a partial degree of implication. The systematic derivation of these rules from the Boolean calculus of propositions reveals that, in any problem involving logical relations between propositions, they (or their transformed equivalents) are compulsory. Anything inequivalent is wrong.

We mention some specific transformations of the laws. The theory may be re-expressed using percentages, on a scale from 0 to 100, by defining $q = 100p$. Next is the transformation $q \propto -\ln p$; the function $-\ln p$ has been interpreted as representing the extent to which we should be surprised that one proposition is true supposing that another is true (Tribus 1969). The product rule transforms to additive form

$$q(AB|I) = q(B|AI) + q(A|I). \tag{90}$$

Conversely, under $q \propto \exp(-p)$, the sum rule transforms to product form. Since we are free to transform at will, the conventional justification for the names 'product rule' and 'sum rule', that they involve products and sums of p respectively, is weak. Thankfully there is a better justification: the product rule decomposes the logical product of propositions, and the sum rule (through de Morgan's theorem) logical sums. Finally, a useful transformation is to the *odds*, defined by

$$q \equiv p/\overline{p} = p/(1-p). \tag{91}$$

Since the sum rule and product rule separately are less general than (79) they may have their own separate invariances, and we have seen that the product rule is invariant under power-law transformations of p. The sum rule is invariant under any transformation of $(p-\frac{1}{2})$ into an odd function of itself; to see this, write $q = p - \frac{1}{2}$ so that the sum rule becomes $q(A|I) = -q(\overline{A}|I)$, and demand that the transformed function $Q(q)$ satisfy the same relation, so that

$$Q\big(q(A|I)\big) = -Q\big(q(\overline{A}|I)\big) = -Q\big(-q(A|I)\big), \tag{92}$$

whence Q is odd.

6. Conclusion

The idea of deriving the calculus of probability from the calculus of propositions, since the arguments of probabilities are propositions, is due to R.T. Cox (1946). The sum and product rules have been known for centuries to apply to proportions (relative frequencies), but this derivation places them on a different, deeper foundation. Our theory is a generalisation of deductive, Boolean logic, the new ingredient being the notion of the extent to which truth of one proposition is implied by truth of another, and its quantification on a continuous scale.

The present starting point uses NAND to reduce the assumptions in the analysis, since any logic function can be constructed from NAND. Upon extending this idea to probabilities from propositions, the result is the equation

$$p(\overline{AB}|I) = 1 - p(A|BI)p(B|I), \tag{93}$$

where we have now written NAND as NOT AND. From this equation both the sum and product rules can be derived, as in section 2. This equation deserves the title 'the law of probability'. It may also be derived via NOR since, like NAND, NOR is a primitive operation in the sense of section 2.

By using (93) we are able to generate relations between probabilities of propositions from the logical relations between the propositions themselves. In combination with techniques for assigning probabilities $p(X|Y)$ from the interrelation of X and Y, this technique provides a complete theory of probability. We have seen from the example of the die in the introductory section that symmetry principles provide a method of assigning probabilities. When that symmetry is broken by 'testable information' – information about a probability distribution, such as "the mean is 3.9" for throws of a standard six-sided die – then a variational principle known as maximum (information) entropy generates the distribution; the entropy functional is Claude Shannon's (1948) and is applied to probability assignment by Jaynes (1983, and in preparation). However probability assignment is not fully understood yet.

We finish by bringing our work closer to application and deriving Bayes' theorem, which through its relation between $p(A|BI)$ and $p(B|AI)$ is the central theorem in tackling inverse problems. Since NAND is commutative we have immediately from (93)

$$p(A|BI)p(B|I) = p(B|AI)p(A|I). \tag{94}$$

When this is added to itself with $A \to \overline{A}$, and combined with the sum rule, the result is

$$p(B|I) = p(A|I)p(B|AI) + p(\overline{A}|I)p(B|\overline{A}I) \tag{95}$$

so that we may rewrite (94) in the form

$$p(A|BI) = K^{-1}p(A|I)p(B|AI), \qquad K = p(A|I)p(B|AI) + p(\overline{A}|I)p(B|\overline{A}I). \tag{96}$$

This is Bayes' theorem. For an exclusive and exhaustive set of propositions $\{A_i\}$ it may correspondingly be derived that

$$p(A_i|BI) = K^{-1}p(A_i|I)p(B|A_iI), \qquad K = p(B|I) = \sum_i p(A_i|I)p(B|A_iI). \tag{97}$$

Generalisation to the continuum proceeds routinely by defining propositions such as "the tree is between heights h and $h + dh$", and by defining a probability density function which on multiplication by dh generates the probability of this proposition. We also define densities for continuous variables to the right of the conditioning solidus, although these are not probability densities.

With these immediately applicable formulae is a good place to stop.

Acknowledgements

I thank Vladik Kreinovich and John Skilling for indicating, and the latter for helping to remedy, weaknesses in drafts of section 3.

References

Aczél, J. 1963. Remarks on probable inference. *Annales Universitatis Scientiarum Budapestinensis de Rolando Eötvös Nominatae, Sectio Mathematica* **6**, 3–11.

Aczél, J. 1966. *Lectures on Functional Equations and Their Applications.* Academic Press, New York, USA.

Cox, R.T. 1946. Probability, frequency and reasonable expectation. *American Journal of Physics* **14**, 1–13.

Cox, R.T. 1961. *The Algebra of Probable Inference.* Johns Hopkins Press, Baltimore, Maryland, USA.

Franklin, J. 1991. The ancient legal sources of seventeenth-century probability. In: *The Uses of Antiquity*, editor S. Gaukroger, Kluwer, Dordrecht, Netherlands, pp.123–144.

Jaynes, E.T. 1983. *E.T. Jaynes: Papers on Probability, Statistics and Statistical Physics.* Synthese Library **158**. Editor R.D. Rosenkrantz, Reidel, Dordrecht, Netherlands.

Jaynes, E.T. In preparation. *Probability Theory: The Logic of Science.* Cambridge University Press, Cambridge, UK. Provisional versions available on the World Wide Web at `http://bayes.wustl.edu/`.

Keynes, J.M. 1921. *A Treatise on Probability.* Macmillan, London, UK.

Kuntzmann, J. 1967. *Fundamental Boolean Algebra* (English translation). Blackie, London, UK.

Shannon, C.E. 1948. A mathematical theory of communication. *Bell System Technical Journal* **27**, 379–423 & 623–659. Reprinted in: *The Mathematical Theory of Communication*, editors C.E. Shannon & W.W. Weaver, University of Illinois Press, Urbana, Illinois, USA, 1949.

Smith, C.R. & Erickson, G.J. 1990. Probability theory and the associativity equation. In: *Maximum Entropy and Bayesian Methods, Dartmouth, USA, 1989*, editor P.F. Fougère, Kluwer, Dordrecht, Netherlands, pp.17–30.

Tribus, M. 1969. *Rational Descriptions, Decisions and Designs.* Pergamon Press, New York, USA.

BAYESIAN GROUP ANALYSIS

W. VON DER LINDEN, V. DOSE AND A. RAMASWAMI
Max-Planck-Institut für Plasmaphysik, EURATOM Association
D-85740 Garching b. München, Germany
e-mail: wvl@ipp.mpg.de

Abstract.
In many fields of research the following problem is encountered: a large collection of data is given for which a detailed theory is yet missing. To gain insight into the underlying problem it is important to reveal the interrelationships in the data and to determine the relevant input and response quantities. A central part of this task is to find the natural splitting of the data into groups and to analyze the respective characteristics. Bayesian probability theory is invoked for a consistent treatment of these problems. Due to Ockham's Razor, which is an integral part of the theory, the simplest group configuration that still fits the data has the highest probability. In addition the Bayesian approach allows to eliminate outliers, which otherwise could lead to erroneous conclusions. Simple textbook and mock data sets are analyzed in order to assess the Bayesian approach.

Key words: Auto-classification, auto-clustering, group analysis, Mahalonobis distance

1. Introduction

The type of problems we are addressing in this paper can best be explained by the simple computer vision problem depicted in fig.1. There is a collection of data points characterized by certain characteristics, here x- and y-coordinate. Our goal is to collect the data into groups, characterized by straight lines. Data in a group will be statistically distributed around a perfect straight line due to natural scatter and experimental uncertainties. A couple of questions arises immediately: How many groups (straight lines) are there? What are their characteristics, i.e. intercept and slope? Which data point belongs to which group? Are there outliers, i.e. individual points that don't fit into any of the other groups? This simple example reveals already a severe shortcoming of maximum-likelihood approaches: the perfect fit would consist of two-element groups. Obviously, this solution is far from what intuition tells us when considering fig.1. Moreover, this solution is completely useless, since we don't learn anything about the underlying problem and it is downright wrong since we know that there must be a scatter about the intrinsic straight lines. So a crucial element is missing to counteract the data-constraints: the tendency towards simplicity. In Bayesian probability theory this tendency is automatically taken care of as a result of the prior probabilities in parameter space. The gain in likelihood volume is balanced by the loss of prior volume. Why carry 100 baskets if there is only one egg to fit in?

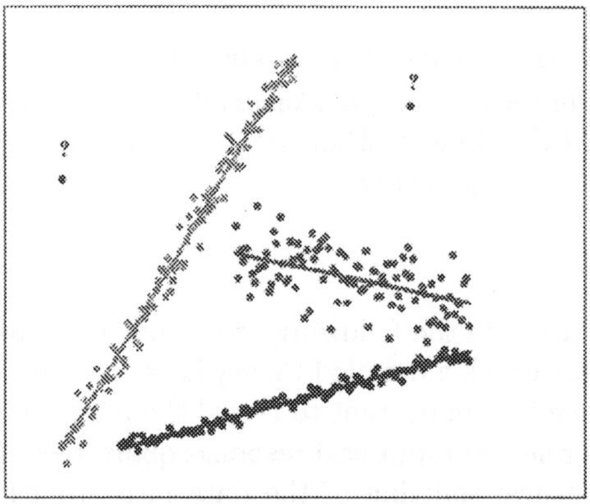

Figure 1. Simple computer vision problem for illustration.

Now let us make the problem more abstract but at the same time more realistic. Very often in science, especially in the beginning of a new field when little or nothing is known and a detailed theory is missing, data are gathered in the hope to gain insight into the underlying problem. Specifically one would like to unravel the interrelationships present in the data and to identify the relevant input and response quantities. Part of the learning process is to find the natural groupings in the data. A great deal can be learned from those characteristics that discriminate the groups.

Many – more or less ad-hoc – methods can be found in the literature for the type of problems we are discussing (see e.g. [1]). In this paper we tackle these problems by strictly following the rules of probability theory and without resorting to any ad-hoc methods or free parameters. The approach we use is a multivariate generalization of Larry Bretthorst's study *On the difference in means* [2]. In the probabilistic approach we will find two competing factors. The likelihood term, which quantifies the data-constraints, would try to find the least deviation from the model. As discussed above, this extreme case is of little use and has no predictive power. The introduction of the prior automatically enforces a tradeoff between the fit of the data and the complexity of the group, giving an automatic form of Ockham's Razor. The latter is inherent in Bayesian probability theory and favors a small number of groups.

For many scientific applications it is, however, not desirable to analyze blindly all combinatorially possible groupings to say nothing of the gigantic complexity.

A very promising approximate evaluation of the auto-classification problem using bayesian probability theory has been suggested recently [3] and implemented in a program package called AutoClass. Here we are aiming at the following type of problems: a set of data points is taken under the same experimental conditions. The external, non-scalable parameters are only changed for different sets. Due to the lack of a detailed theory it is, however, not clear from the outset, which of these parameters or which combination of changes really influences the features of the underlying problem. On the contrary, these are precisely the points one wishes to unravel. Therefore, the question to be answered is which of these data sets are different and which can be pooled into one group in order to learn about the relevant input and response quantities and there interrelationships.

For a quantitative analysis it is also necessary to automatically identify outliers in order to avoid over-rating their influence in the analysis of group characteristics. For a quantitative study, we have to specify the mathematical meaning of a group. We are not ambitious enough to head for a theory of everything and restrict our theory to linear problems. I.e. we consider groups which can be described by hyperplains in the space of the data-characteristics, like in fig.1. The restriction is actually not as severe as it looks at first glance. Many problems are globally linear, many others are locally linear in the regime for which data points are available, and many apparently nonlinear problems can be made linear by a suitable transformation. An object is characterized by m characteristics $\mathbf{d} = (d^1, \ldots, d^m)$ which are represented by real numbers, like height, length and mass. We consider only continuous characteristics. The discrete characteristics are naturally used to preselect groups. A group, ν say, is characterized – according to the above assumption – by mean vector $\mathbf{m}_\nu$ and covariance matrix C_ν. The covariance originates from the natural scatter and measurement errors. The data points $\mathbf{d}_\nu^i$ (# i in group ν) are i.i.d. multi-normal:

$$p(\mathbf{d}_\nu^i|C_\nu, \mathbf{m}_\nu, \mathcal{I}) = |2\pi C_\nu|^{-1/2} \mathrm{etr}\left\{-\frac{1}{2} C_\nu^{-1} (\mathbf{d}_\nu^i - \mathbf{m}_\nu)(\mathbf{d}_\nu^i - \mathbf{m}_\nu)^T\right\} ,$$

with the definition $\mathrm{etr}\{X\} = \exp(\mathrm{tr}(X))$. By virtue of the conditional independence the likelihood for all data-points in group ν reads

$$\begin{aligned} p(\{\mathbf{d}_\nu\}|C_\nu, \mathbf{m}_\nu, \mathcal{I}) &= |2\pi C_\nu|^{-n/2} \mathrm{etr}\left\{-\frac{1}{2} C_\nu^{-1} \overline{C_\nu}\right\} \\ \overline{C_\nu} &= \sum_{i=1}^{n} (\mathbf{d}_\nu^i - \mathbf{m}_\nu)(\mathbf{d}_\nu^i - \mathbf{m}_\nu)^T \end{aligned}$$

$\overline{C_\nu}$ is proportional to the sample covariance. The different notation has been used to simplify the expressions.

2. Bayesian Approach

Now that the problem is uniquely specified we head for the Bayesian approach. The questions raised in the introduction translate into hypotheses H_n which contain the following information

- the data belong to n different groups as specified by d^i_ν
- the number of elements in group ν is N_ν.
- group ν has mean $\mathbf{m}_\nu$ and covariance C_ν
- the means $\mathbf{m}_\nu$ of different groups are different and unknown i.e. they are nuisance parameters bound to be marginalized over.
- further specification of hypothesis H_n may consider the cases
    1. covariance matrices are the same / different in all groups
    2. covariance matrices are known / unknown.

The probability for hypothesis H_n can be determined via Bayes' theorem:

$$P(H_n|D,\mathcal{I}) \propto p(D|H_n,\mathcal{I})\, P(H_n|\mathcal{I}) \quad . \tag{1}$$

The desired probability factors into marginal likelihood $p(D|H_n,\mathcal{I})$ and prior $P(H_n|\mathcal{I})$. The latter cannot be simplified further and has to be assigned according to the problem under consideration. The proportionality constant drops out in model comparison and needs no further discussion.

2.1. MARGINAL LIKELIHOOD $P(D|H_N,\mathcal{I})$

We assume for the moment that the covariance matrices are known. In this case the marginal likelihood entering Eq.1 is $p(D|H_n,\{C_\nu\},\mathcal{I})$. One bit of information is, however, still missing to make the likelihood complete, namely the true means. They are introduced via the marginalization rule:

$$p(D|H_n,\{C_\nu\},\mathcal{I}) \;=\; p(\{\mathbf{m}_\nu^{\mathrm{ML}}\}|H_n,\mathcal{I}) \prod_{\nu=1}^{n} \int d^L\mathbf{m}_\nu p(D_\nu|H_n,C_\nu,\mathbf{m}_\nu,\mathcal{I}) \tag{2}$$

D_ν are all data points in group ν. Several remarks are in order to explain Eq.2. We employed the approach first introduced by Larry Bretthorst for the bayesian model selection [2,4–6]. Although the mathematical route he takes is different the idea behind it is identical to what we discuss next. We assumed, and this is actually part of our background information $\mathcal{I}$, that the prior $p(\mathbf{m}_\nu|H_n,\mathcal{I})$ as function of $\mathbf{m}_\nu$ is slowly varying as compared to the likelihood $p(D_\nu|H_n,C_\nu,\mathbf{m}_\nu,\mathcal{I})$. The latter is dominated by a single peak, the maximum likelihood solution $\mathbf{m}_\nu^{\mathrm{ML}}$. If this is so, we can take the prior out of the integral like in Eq.2. In other words we consider only situations where the effort put into the experiment to measure the data pays off. As far as the parameters $\mathbf{m}_\nu$ are concerned, the prior has no influence in this approach but it is utterly important for model comparison since it provides the multiplicities for Occam's razor. Obviously for ill-posed problems the assumptions need some revision to allow for the small eigenvalues of the transformation from parameter to data space and to introduce a means for regularization. For the present type of problems, however, the assumptions are perfectly valid, otherwise the maximum-likelihood approaches used in orthodox statistics would fail badly. In the assumed situation of weak prior knowledge there is naturally a wide variety of functional forms conceivable and the result must not depend on the details of the prior. We construct the prior as outlined by Jaynes [7] and extensively used by Bretthorst[4–6,2] upon using testable information and the maximum entropy principle. Reasonable testable information in our case is the average value of the characteristics over all individuals irrespective of their group affiliation, i.e. $\overline{\mathbf{m}}$. Since the prior is conditional on H_n, the number of groups n as well as the number N_ν of points in group ν is known. We can write

$$\overline{\mathbf{m}} = \int p(\{\mathbf{m}_\nu\}|H_n, I) \left\{ \sum_\nu N_\nu \mathbf{m}_\nu \right\} d\mathbf{m}^{Ln} \quad . \tag{3}$$

The maximum entropy principle [7] yields

$$p(\{\mathbf{m}_\nu\}|\lambda, I) = \prod_{i=1}^{L} \lambda_i^n \prod_{\nu=1}^{n} N_\nu^L \exp(-\lambda^T \sum_{\nu=1}^{n} N_\nu \mathbf{m}_\nu) \quad , \tag{4}$$

where λ_i are Lagrange parameters to fulfill the constraint Eq.3. They are hyperparameters of the theory and are marginalized over to obtain

$$p(\{\mathbf{m}_\nu\}|H_n, I) \propto \left(\prod_{i=1}^{L} \frac{\sum_{\nu=1}^{n} N_\nu m_\nu^i}{(\Gamma(n) \prod_{\nu=1}^{n} N_\nu)^{1/n}} \right)^{-n} = \left(\prod_{i=1}^{L} L_i^{\text{eff}} \right)^{-n} \quad . \tag{5}$$

Jeffreys' prior has been used for the positive hyperparameters. The proportionality constant is independent of the number of groups n and the dimension L, respectively and deserves no further attention. The main purpose of the prior is to count the multiplicity. In this respect it can be cast into the form of an effective flat as indicated by the last equality in Eq.5 with an uncertainty in the i-th characteristics of L_i^{eff}.

2.2. POSTERIOR PROBABILITY FOR H_N IF $C_\nu = C$ IS KNOWN

We restrict the following discussion to the case that all groups have the same covariance matrix. A generalization to differing covariances for different groups is merely a matter of more indices and thorough book-keeping. If the common covariance matrix is even known then the derivation of the posterior probability for hypothesis H_n is completed

$$\begin{aligned} P(H_n|D, C, \mathcal{I}) \quad &\propto \quad P(H_n|\mathcal{I})\, P(\{\overline{d}_\nu\}|H_n, \mathcal{I}) \prod_{\nu=1}^{n} N_\nu^{-\frac{L}{2}} \\ &\times \quad (2\pi)^{\frac{n}{2}}\, |C|^{-\frac{n}{2}}\, \text{etr}\left\{ -\frac{1}{2} C^{-1} \overline{C}^{(n)} \right\} \end{aligned} \tag{6}$$

The definition of the sample mean $\overline{d_\nu}$ and sample covariance $\overline{C}_\nu$ for group ν as well as that for the global sample covariance $\overline{C}^{(n)}$ are

$$\overline{d_\nu} = \frac{1}{N_\nu}\sum_{i=1}^{N_\nu} d_\nu^i \tag{7}$$

$$\overline{C}_\nu = \sum_{i=1}^{N_\nu}(d_\nu^i - \overline{d_\nu})(d_\nu^i - \overline{d_\nu})^T \tag{8}$$

$$\overline{C}^{(n)} = \sum_{\nu=1}^{n}\overline{C}_\nu \tag{9}$$

To avoid clumsy expressions the normalization of the covariance matrices has again been chosen different from the standard form found in text-books. The sample covariances depend on the particular specifications in H_n. The number of indices is minimized as much as possible still allowing an unambiguous interpretation. We see how Occam's factor enters automatically, favoring a small number of groups. The data-driven factors vote for complexity. If the common covariance is known this is the final result and we can start analyzing datasets. Unfortunately, only in very rare cases the covariance matrix is known experimentally and it has to be treated as a hyperparameter.

2.3. POSTERIOR PROBABILITY IF C IS NOT KNOWN

The marginal likelihood $p(D|H_n,\mathcal{I})$ entering Eq.1 is obtained upon marginalization over the true covariance matrix C:

$$p(D|H_n,\mathcal{I}) = \int \mathcal{D}C\, p(D|H_n,C,\mathcal{I})p(C|\mathcal{I})$$

The integral extends over the space of covariance matrices. Fortunately, we can borrow most of the results from what is known about the Wishart distribution [8]. In order to proceed we need the prior $p(C|I)$. Due to the lack of prior knowledge we seek a functional form for the prior which is invariant under all transformations that do not change our state of ignorance. These are all possible similarity transformations. Jaynes' transformation group approach [7] leads to a kind of multivariate Jeffreys' prior

$$p(C|\mathcal{I}) \propto |C|^{-(L+1)/2} \tag{10}$$

The proportionality constant drops out in the odds ratio since it is the same for all hypotheses under consideration. This is the place where the extension to the case of different covariances for different groups needs a little extra thought. The prior (Eq.10) reduces to the well-known Jeffreys' prior in the univariate case. The type of integral, needed for the marginalization in Eq.10, is known from the inverse Wishart distribution [8] and yields

$$P(H_n|D,\mathcal{I}) \propto P(H_n|\mathcal{I})\, P(\{\overline{d}_\nu\}|H_n,\mathcal{I}) \prod_{\nu=1}^{n} N_\nu^{-\frac{L}{2}} \times \pi^{\frac{Ln}{2}} \prod_{j=1}^{L} \Gamma(\frac{N-n-j+1}{2})\, |\overline{C}^{(n)}|^{-\frac{N-n}{2}} \quad (11)$$

This is the key formula to answer the questions we raised in the introduction. It allows to assess preselected groups, and serves as merit function for unsupervised learning (e.g. via simulated annealing). Moreover it allows to address questions like: do we have to split a group? or are there outliers?

2.4. DO WE HAVE TO SPLIT A GROUP?

Here we will consider the question whether the data favor the splitting of a group, η say, into two pieces as depicted schematically in fig. 2. This is summarized in hypothesis H_{n+1}, specifying precisely which data of group η go into the new sub-groups η_1 and η_2. Accordingly, the number of elements change $N_\eta \rightarrow \{N_{\eta,1}, N_{\eta,2}\}$. Eventually we will compare hypotheses H_{n+1} and H_n, with and without the split-

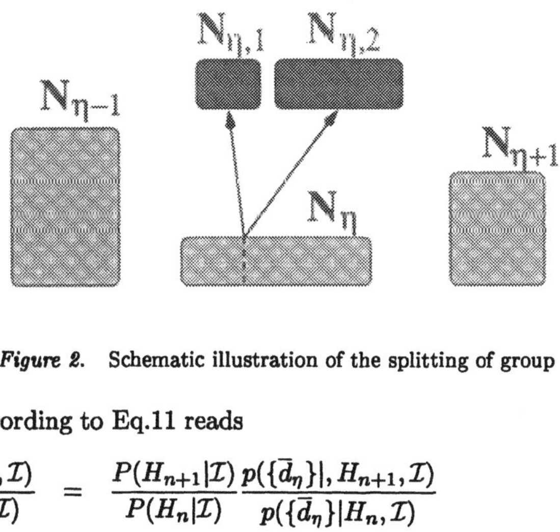

Figure 2. Schematic illustration of the splitting of group η.

ting, which according to Eq.11 reads

$$\frac{P(H_{n+1}|D,\mathcal{I})}{P(H_n|D,\mathcal{I})} = \frac{P(H_{n+1}|\mathcal{I})}{P(H_n|\mathcal{I})} \frac{p(\{\overline{d}_\eta\}|, H_{n+1},\mathcal{I})}{p(\{\overline{d}_\eta\}|H_n,\mathcal{I})} \times \left(2\pi\frac{N_{\eta,1}+N_{\eta,2}}{N_{\eta,1}N_{\eta,2}}\right)^{L/2} |\tilde{C}^{(n)}|^{1/2} \frac{\Gamma(\frac{N-n-L}{2})}{\Gamma(\frac{N-n}{2})} (\frac{N-1}{2})^{L/2} \times \left(1-\frac{N_{\eta,1}N_{\eta,2}}{(N-1)N_\eta}\Delta d_\eta{}^T (\tilde{C}^{(n)})^{-1} \Delta d_\eta\right)^{-\frac{N-n-1}{2}} \quad (12)$$

Here we switch to the standard definition of the covariance $\tilde{C}^{(n)} = \overline{C}^{(n)}/(N-1)$ and write the difference in sample means of the new sub-groups as $\Delta d_\eta = \overline{d}_{\eta,2} - \overline{d}_{\eta,1}$.

The odds ratio contains the multivariate Student-t distribution (Hotelling-T) which depends essentially on the Mahalonobis distance $\Delta d_\eta{}^T(\tilde{C}^{(n)})^{-1}\Delta d_\eta$ [9].

3. Applications

In order to get a feel for the behavior of Eq.12 we apply it to simple test examples.

3.1. UNIVARIATE EXAMPLE

We start out with two sets of data with 100 data points each, drawn from unit-variance normal distributions with means m_1 and m_2, respectively. Now we want to check whether the bayesian approach recognizes that the data actually belong to two different groups. To this end we calculate the odds ratio for the two hypotheses there are two groups (H_2) versus there is only one group (H_1). To make the test more realistic, we pretend not to know the common noise level and to have equal prior expectation for the two hypotheses,i.e. $P(H_n) = 1/2$. For simplicity we use in the present example a flat prior for $\{m_\nu\}$, i.e. $p(\{m_\nu\}|\mathcal{I}) = \prod_\nu \{\frac{1}{V}\theta(|m_\nu| - \frac{V}{2})\}$. V represents the prior uncertainty in the means m_ν. The data are generated for $m_1 = 0$ fixed and m_2 varied between -0.2 and 0.2. Actually, the result depends only on the difference $m_2 - m_1$. For the numerical analysis we use an prior uncertainty $V = 1$. Fig. 3 shows the various parts of the log-odds-ratio

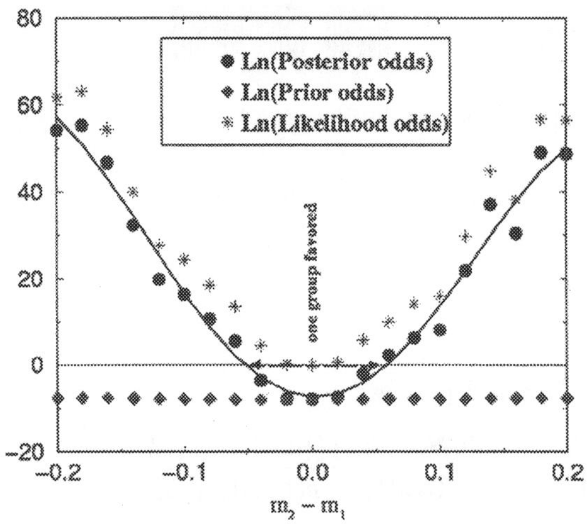

Figure 3. Contributions to the odds ratio for a mock dataset as described in the text, as function of the difference in the true means. Note that posterior and likelihood differ only by a constant prior.

$\log\{P(H_2|D,\mathcal{I})/P(H_1|D,\mathcal{I})\}$, namely prior, likelihood and posterior. In the region of positive log-odds the bayesian approach favors decisively – by several orders of magnitude – two groups if the difference in the true means exceeds the break-even point $\Delta m_{be} = 0.055$. The analytic expression for the break-even point according to Eq.12 yields for $N \gg 1$:

$$(\Delta m_{be})^2 = \frac{4\sigma^2}{N} \log(\frac{NV^2}{8\pi\sigma^2}) \tag{13}$$

As a matter of fact, the data belong to only one group if $m_2 - m_1 = 0$. However, due to the finite sample size and the statistical error in the data, the sample means have an uncertainty which is reflected in the finite value of the break-even point. If on the other hand N – the size of the two samples – tends to infinity, the sample means approach the true means and the break-even point shrinks towards zero.

3.2. OUTLIER DETECTION

An interesting application of Eq.12 concerns outlier detection. This is a special case of Eq.12, namely group η is split into: $N_{\eta,1} = 1$ and $N_{\eta,2} = N_\eta - 1$. A numerical example is studied which consists of 3 groups with 100 data points each. The data of the three groups are distributed according to unit-variance normal distributions with means $m_1 = 2$, $m_2 = 3$ and $m_3 = 4$, respectively. The position of an extra

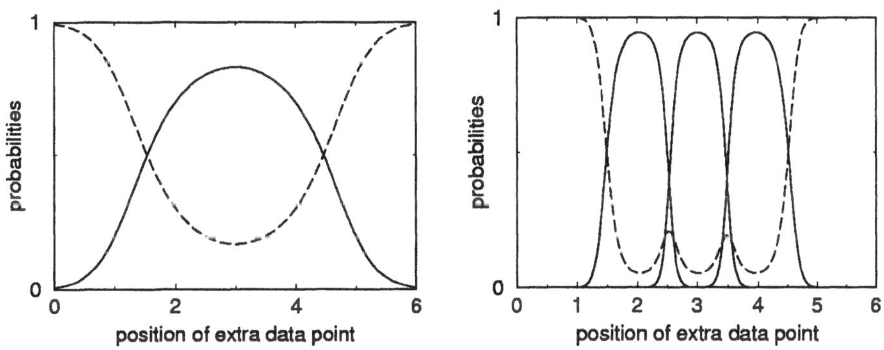

Figure 4. Outlier analysis: data belong to one group (left) or three groups (right) as solid lines. The outlier probabilities are depicted as dashed lines.

test-data point is varied between 0 and 6 and the probabilities are depicted in fig. 4 as function of the position of the test-data point. First, we consider the hypothesis "*all data belong to one group*" and ask for the probabilities that the extra point belongs to that group or that it is an outlier. The result is depicted in the left panel. It corroborates qualitatively our intuition, the outlier probability increases monotonically away from the center of the group. On top of that, Bayesian analysis indicates a failure of the assumed grouping, since the outlier probability is always rather high, even in the center of the group. Alternatively, the probability that the test-data point belongs to the group is always significantly smaller than 1, even for the "best member" of the group. The situation changes if we start from hypothesis H_3 "*there are three groups*". In this case the minimum outlier probability is much smaller. The right panel of fig. 4 shows the probabilities that the extra point belongs to group 1,2,3 or it is an outlier.

An interesting quantity in this respect is the average outlier probability in group ν

$$\overline{p_\nu} = \frac{1}{N_\nu} \sum_{i \in \text{group}_\nu} p(d_i \text{ is an outlier} | D\mathcal{I}) \tag{14}$$

The results are listed in table 1. We see that the average outlier probability is large in the case of the wrong grouping while it is about 5% for the correct grouping. We

	All In One	Three Separate Groups
Group 1	0.30	0.05
Group 2	0.17	0.06
Group 3	0.30	0.05

TABLE 1. Outlier example

see that the Bayesian analysis reveals many interesting features of the problem.

4. Key difference between the groups

Once the optimal grouping has been found the next interesting question arises as to what discriminates the groups. To this end we consider a linear combination $q = \mathbf{a}^T\mathbf{d}$ of the data points to summarize the key features of a group. $\mathbf{a}$ is the discriminating vector. A linear combination is all one can expect in a linear problem. The idea is illustrated in fig. 5. Two data clouds are represented by ellipses. In the original representation in characteristics x_1 and x_2 they can not be distinguished. Projection on the discriminating vector a_2 reveals the existence of two separate groups. We are seeking that vector $\mathbf{a}$ that discriminates the groups best, i.e. for which the odds ratio $P(H_n|\mathbf{a}, D, \mathcal{I})/P(H_1|\mathbf{a}, D, \mathcal{I})$ reaches a maximum. The conditioning on $\mathbf{a}$ means we consider projected, univariate data $\tilde{d}^i_\nu = \mathbf{a}^T\mathbf{d}^i_\nu$. The maximum odds ratio is obtained for

$$\frac{\overline{C}^{(1)}}{\mathbf{a}^T\overline{C}^{(1)}\mathbf{a}}\mathbf{a} = \left(\frac{N-n}{N-1}\frac{\overline{C}^{(n)}}{(\mathbf{a}^T\overline{C}^{(n)}\mathbf{a})} + \frac{n-1}{N-1}\mathbb{1}\right)\mathbf{a} \tag{15}$$

For $N \gg n$ we recover the conventional result[10]:

$$\begin{aligned} B\mathbf{a} &= \lambda\,\overline{C}^{(n)}\mathbf{a} \\ B &= \sum_{\nu=1}^{n} N_\nu(\overline{d}_\nu - \overline{d}))(\overline{d}_\nu - \overline{d}))^T \quad . \end{aligned} \tag{16}$$

The covariance of the sample means (covariance between groups) is denoted by B, while $C^{(n)}$ represents an average covariance within the groups. The eigenvalue λ

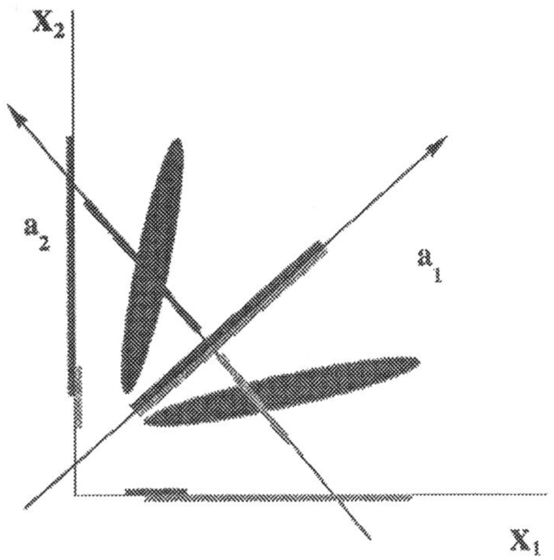

Figure 5. Illustration of the discriminating vector. Ellipses represent clouds of data. Bars along $x_1, x_2, a_1, and a_2$ stand for the projection of the data onto the respective directions.

in Eq.16 is a measure of the ratio $(\frac{\text{distance between the groups}}{\text{distance within the groups}})^2$ for the data projected onto direction **a**. For a sensible grouping the ratio will strongly exceed 1.

5. Fisher's Iris Data

Now we want to apply the theory to a realistic multi-variate problem, namely Fisher's iris data listed and analyzed in [11]. The data consist of 3 pre-formed groups according to the species: iris setosa, iris versicolor and iris virginica. The flowers are characterized only by four measurements: sepal length, sepal width, petal length and petal width. The results are given in table 2. The log-base-10

Choice	$\text{Log}_{10}(P\text{grouping})$	$\overline{P}$(outlier)
(123)	-199	0.41%
(12)(3)	-82	0.15%
(13)(2)	-36	0.12%
(1)(23)	-103	0.19%
(1)(2)(3)	0	0.09%

TABLE 2. Analysis of Fisher's iris data

probabilities (relative to the best model) are given for all different permutations of the pre-formed data groups. The meaning of the notation should be apparent. The integers in parentheses indicate which pre-formed groups are combined to one group. The data corroborate impressively the botanists' classification into 3 separate species. The odds-ratio between the best and next-to-best grouping is 10^{36}! Again we see that the average outlier-probability is much smaller for the best grouping. The smallest value (9%) is, however, about twice as large as in the mock-data set which probably points out that a further sub-classification might be reasonable. The largest eigenvalues of Eq.16 are $\sqrt{\lambda_1} = 5.7$ and $\sqrt{\lambda_2} = 0.53$. There is only one direction along which the groups are well separated. Fig. 6 shows the projected data in the sepal-length – sepal-width subspace (left panel). The data look similar in the other subspaces. The projected data clouds are never separated in the original characteristics. The projection onto the two dominant eigenvectors of Eq.16, however, clearly separates the groups as can be seen in the right panel. The scatter of the data within the groups is much smaller than that between the groups as far as the projection onto a_1 is concerned. The data point enclosed by

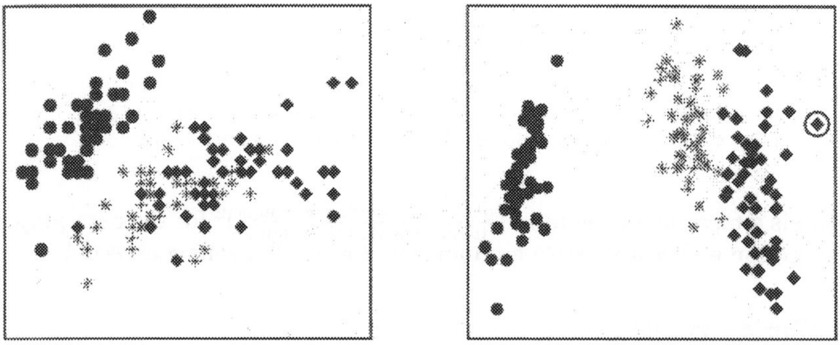

Figure 6. left: sepal width versus sepal length, right:data projected onto the two most dominant directions

a circle is identified to be an outlier.

References

1. W. Krzanowski, ed., *Principles of Multivariate Analysis*, Oxford Science Publications, Oxford, 1988.
2. G. Bretthorst, "On the difference in means," in *Physics and Probability Essays in honor of Edwin T. Jaynes*, W. T. Grandy and P. W. Milonni, eds., vol. 1, pp. 177–194, Cambridge Univ. Press, Cambridge, 1993.
3. J. Stutz and P. Cheeseman, "Autoclass - a bayesian approach to classification," in *Maximum Entropy and Bayesian Methods (1996)*, J. Skilling and S. Sibisi, eds., Kluwer Academic Publishers, Dordrecht, 1995. Further references: http://ic-www.arc.nasa.gov/ic/projects/bayesian-group/autoclass/.
4. G. Bretthorst, ed., *Bayesian Spectrum Analysis and Parameter Estimation*, Springer Press, Berlin, Heidelberg, 1988.
5. G. Bretthorst, "Excerpts from bayesian spectrum analysis and parameter estimation," in *Maximum Entropy and Bayesian Methods in Science and Engineering*, G. Erickson and C. Smith, eds., vol. 1, p. 75, Kluwer Academic Publishers, Dordrecht, 1988.
6. G. Bretthorst, "An introduction to parameter estimation using bayesian probability theory," in *Maximum Entropy and Bayesian Methods in Science and Engineering*, P. Fougere, ed., p. 53, Kluwer Academic Publishers, Dordrecht, 1989.

7. E. Jaynes, "Prior probabilities," in *E. T. Jaynes: Papers on Probability, Statistics and Statistical Physics*, R. Rosenkrantz, ed., p. 114, Reidel, Dordrecht, 1983.
8. "Wishart distribution," in *Kendall's advanced theory of statistics, Bayesian Inference*, A. O'Hagan, ed., John Wiley & Sons, New York, 1st ed., 1994. p.293ff.
9. "Mahalonobis distance," in *Principles of Multivariate Analysis*, W. Krzanowski, ed., Oxford Science Publications, Oxford, 1988. p.233ff.
10. "Canonical variates," in *Principles of Multivariate Analysis*, W. Krzanowski, ed., Oxford Science Publications, Oxford, 1988. p.291ff.
11. "Fisher's iris data," in *Principles of Multivariate Analysis*, W. Krzanowski, ed., Oxford Science Publications, Oxford, 1988. p.45ff.

SYMMETRY-GROUP JUSTIFICATION OF MAXIMUM ENTROPY METHOD AND GENERALIZED MAXIMUM ENTROPY METHODS IN IMAGE PROCESSING

O. KOSHELEVA
Department of Electrical and Computer Engineering
University of Texas at El Paso
El Paso, TX 79968, USA†

Abstract. Maximum entropy method and its generalizations are very useful in image processing. In this paper, we show that these methods naturally follow from reasonable symmetry requirements.

Key words: image processing, maximum entropy, symmetry groups

1. Introduction to the Problem

The main objective of image processing. In many problems of image processing, e.g., in optical and radio astronomy, we want to reconstruct the *image*. In precise terms, we want to know the function $I(\vec{\sigma})$ that describes how brightness depends on the coordinates $\sigma = (x, y)$.

Most of the time, we are only interested in the values of brightness over a sufficiently dense rectangular *grid* that is formed by the points $\vec{\sigma}_{ij} = (x_i, y_j)$ with coordinates $x_i = x_0 + i \cdot h_x$ and $y_j = y_0 + j \cdot h_y$. The corresponding values $I(\vec{\sigma}_{ij})$ of the brightness function form a *matrix*; the components of this matrix are, for simplicity, usually denoted as I_{ij}.

The main problem of image processing: general formulation. Real-life measurements are usually inaccurate and incomplete. Due to these inaccuracies and incompleteness, there are usually *many* different images which are consistent with the same set of observations. One of the main problems of image processing is, thus, to select the *best* (most appropriate) image among all images which are consistent with the known measurement results.

Examples. The following two simple examples illustrate the above two points:

– that image reconstruction is not unique, and

†Email: olga@ece.utep.edu

- that, often, some images, which are consistent with the measurement results, are reasonable for this particular imaging problem, while other images, which are also consistent with the same measurements, are not reasonable.

The first example is from *astronomy*; it deals with the simplest possible source, that consist of only one bright point. An arbitrary not-100%-accurate observation of this single point source is consistent not only with the corresponding one-point image, but also with an image, in which, in addition to this point source, there is a very weak second source elsewhere, so weak that, due to measurement inaccuracy, its brightness is indistinguishable from 0. Since there is no physical or observational reason for this second point, we would definitely prefer the one-point image.

The second example is from *mammography*. The main objective of mammography is to detect possible nonhomogeneities that can be indications of a growing tumor. These nonhomogeneities appear as spots on the image. The idea is that if such a spot appears, then more accurate (more costly, and often painful) testing is done to check whether it actually is a malignant tumor. If the actual tissue does not contain any nonhomogeneities, then, due to the inevitable measurement inaccuracy, the measurement results are still consistent with the assumption that we have a very small tumor. However, if we always return an image with a spot, then we would have to use the more costly and painful procedure always, and this defeats the main purpose of mammography as a "gateway" painless and cheap test. It is therefore desirable to return an image with a spot only if there are indeed reasonable grounds for suspecting that there may be a nonhomogeneity in the actual tissue.

Best image: in what sense? We want to select, from all the images that are consistent with the given observations, the most appropriate (best) image. How can we do that?

In every application area, there are experts who know which images are more appropriate. So, in principle, if we only had a few possible images to compare, we could simply ask an expert to compare them. However, for each set of measurement results, there are very many possible images that are consistent with these results, and therefore, we cannot ask an expert to compare all of them, we must do this selection *automatically*. Thus, we need to formulate the expert's preferences in precise terms that a computer will be able to understand.

Situations in which expert themselves are able to formulate their preferences in precise terms are extremely rare, and definitely in both of our examples (astronomical and medical images) experts have not been able to provide us with such a formalization. Therefore, we must either somehow extract this preference relation from the experts, or use heuristic methods and hope that they will be in good accordance with the expert preferences.

Maximum entropy and generalized maximum entropy as useful heuristic methods. Among successful heuristic methods that are used to reconstruct the image are:

- the *maximum entropy* method, according to which we select, from all the images that are consistent with the observations, the image $I(\vec{\sigma})$ for which the entropy is the largest possible:

$$-\int I(\vec{\sigma}) \cdot \log(I(\vec{\sigma}))\, d\vec{\sigma} \to \max; \tag{1}$$

  and
- *generalized maximum entropy methods*, in which we choose the image $I(\vec{\sigma})$ for which

$$\int \log(I(\vec{\sigma}))\, d\vec{\sigma} \to \max \tag{2}$$

  or

$$\int (I(\vec{\sigma}))^p\, d\vec{\sigma} \to \max \tag{3}$$

  for some real number p.

Why maximum entropy? These methods often work well, so the natural question is: why? This question is caused not only by natural scientific curiosity, but we also want to know:

- whether this success is caused by the specifics of the problems to which these methods have so far been applied, and these methods may fail for other problems,
- or these methods are universally valid, and therefore, we can fearlessly use them in future applications as well.

There exist interesting and convincing justifications of the maximum entropy method based on reasonable prior distributions (see, e.g., [7]), but this justification is not that naturally applicable to *generalized* entropy methods.

It is therefore desirable to either find a new justification of these methods, or, if these methods cannot be naturally justified, provide justifiable working methods.

What we are planning to do. In this paper, we show that, if we take into consideration natural symmetries, then the description of expert preferences leads exactly to maximum entropy and generalized maximum entropy techniques.

The results of this paper partially appeared in [3,4].

2. Towards Mathematical Formulation

How to describe preferences? There exists a general formalism. The necessity to describe *preferences*, i.e., to describe the *utility* of different alternatives for different people, is extremely important in decision making, including decision making under conflict (also known under a somewhat misleading name of *game theory*). To describe these preferences (utilities), a special *utility theory* has been developed; see, e.g., [1,5,6,8].

The mathematical formalism of utility theory comes from the observation that sometimes, when a person faces several alternatives $A_1, \ldots, A_n$, instead of choosing one of these alternatives, this person may choose a ***probabilistic*** combination of them, i.e., A_1 with probability p_1, A_2 with a probability p_2, etc. For example, if two alternatives are of equal value to a person, that person will probably choose the first one with probability 0.5 and the second one with the same probability 0.5. Such probabilistic combinations are called (somewhat misleadingly) ***lotteries***. In view of this realistic possibility, it is desirable to consider the preference relation not only for the original alternatives, but also for arbitrary lotteries combining these alternatives. Each original alternative A_i can be viewed as a *degenerate* lottery, in which this alternative A_i appears with probability 1, and every other alternative $A_j \neq A_i$ appears with probability 0.

The main theorem of utility theory states that if we have an ordering relation $L \succ L'$ between such lotteries (with the meaning "L is preferable to L'"), and if this relation satisfies natural consistency conditions such as transitivity, etc., then there exists a function u from the set $\mathcal{L}$ of all possible lotteries into the set R of real numbers for which:

- $L \succ L'$ if and only if $u(L) < u(L')$, and
- for every lottery L, in which each alternative A_i appears with probability p_i, we have $u(L) = p_1 \cdot u(A_1) + \ldots + p_n \cdot u(A_n)$.

This function u is called a ***utility function***. Each consistent preference relation can thus be described by its utility function.

General formalism (cont-d): how unique is the utility function? The correspondence between preference relations and utility functions is not 1-1: different utility functions may correspond to the same preference. For example, if the preference relation is consistent with the utility function $u(L)$, then, as one can easily see, it is also consistent with the function $u'(L) = a \cdot u(L) + b$, in which a is an arbitrary positive real number, and b is an arbitrary real number.

It is known that such linear transformations form the only non-uniqueness of utility function: namely, if two utility functions $u(L)$ and $u'(L)$ describe the same preference relation, then $u'(L) = a \cdot u(L) + b$ for some real numbers $a > 0$ and b.

Application to our problem. In general, preference relations can be described by utility functions. Therefore, to describe preferences between images, we need a utility function that is defined on the set of all possible images.

In particular, if we consider realistic images I that consist of finitely many pixels and are thus described by a matrix with components I_{ij}, $1 \leq i, j \leq N$, then we need a utility function that depends on N^2 parameters I_{ij}: $u(I) = u(I_{11}, \ldots, I_{1N}, \ldots, I_{N1}, \ldots, I_{NN})$.

What function should we choose?

Localness property. An important feature of many image processing problems, including the two above problems, is their *localness*: different parts of the image are pretty much independent of each other. Let us explain what we mean in the above examples:

- In *astronomy*, the relative quality of two possible reconstructions of a part of this image (a part that contains, e.g., a galaxy), does not depend on the remaining part of the image.
- In *medicine*, the relative quality of reconstructing a part of the mammogram does not depend on whether there is anything in the rest of this mammogram.

In mathematical terms, if

- an image I_{ij} is preferable to an image I'_{ij} that differs from I_{ij} only in pixels (i,j) from some set P, and
- images $\widetilde{I}_{ij}$ and $\widetilde{I}'_{ij}$ coincide with each other for $(i,j) \notin P$ and with, correspondingly, I_{ij} and I'_{ij} for $(i,j) \in P$,

then $\widetilde{I} \succ \widetilde{I}'$.

This "localness" ("independence") is a frequent feature in practical problems, and utility theory has developed a precise description of utility functions that satisfy this property. Namely, it has been shown that when alternatives are characterized by n parameters $x_1, \ldots, x_n$, then the localness of the preference is equivalent to the utility function $u(x_1, \ldots, x_n)$ being of one of the two types [2]:

- *additive* $u(x_1, \ldots, x_n) = u_1(x_1) + \ldots + u_n(x_n)$ for some functions $u_i(x_i)$; or
- *multiplicative* $u(x_1, \ldots, x_n) = u_1(x_1) \cdot \ldots \cdot u_n(x_n)$ for some functions $u_i(x_i)$.

In utility theory, the values $u_i(x_i)$ are called *marginal utilities.*

For images, $n = N^2$, parameters $x_1, \ldots, x_n$ are the components $I_{ij} = I(\vec{\sigma}_{ij})$, and the resulting forms of the utility function are $u(I) = \sum u_{ij}(I_{ij})$ and $u(I) = \prod u_{ij}(I_{ij})$. So, to describe the utility function, we must describe how the value of the marginal utility depends on the point $\vec{\sigma}_{ij}$ and on the brightness $I_{ij} = I(\vec{\sigma}_{ij})$ at this point. We can describe this dependence explicitly if, instead of the abbreviated notation I_{ij}, we use $I(\vec{\sigma}_{ij})$, and if we describe the dependency $u_{ij}(I_{ij})$ as $U(\vec{\sigma}_{ij}, I(\vec{\sigma}_{ij}))$. In this case, the formula for the utility function takes one of the forms $u(I) = \sum U(\vec{\sigma}_{ij}, I(\vec{\sigma}_{ij}))$ or $u(I) = \prod U(\vec{\sigma}_{ij}, I(\vec{\sigma}_{ij}))$.

Continuous case. So far, we were analyzing the problem of how to compare different pixel-by-pixel images. In real-life, the object whose image we want to describe is continuous, pixels are simply a useful approximation. It is, therefore, desirable to reconstruct not just the values on a grid, but also the entire brightness distribution, i.e., the values of $I(\vec{\sigma})$ for every point $\vec{\sigma}$. To achieve this goal, we must be able to compare the quality of different functions $I(\vec{\sigma})$, i.e., we must be able to describe the value of utility $u(I)$ for different functions I.

The denser the pixels (i.e., the smaller the distances h_x and h_y between the neighboring pixels), the closer the pixel-by-pixel image to the continuous one. Therefore, as a utility $u(I)$ of a function I, we can take the *limit* of the utilities of its pixel-by-pixel representation as $h_x \to 0$ and $h_y \to 0$. How can we describe such a limit?

This limit is easy to describe for the case when utility is a *sum* of marginal utilities: in this case, the sums are, in effect, integral sums, and therefore, as the pixels get denser, the sums tend to the integral $u(I) = \int U(\vec{\sigma}, I(\vec{\sigma}))\, d\vec{\sigma}$.

For the case when utility is a *product* of marginal utilities, the limit can be obtained indirectly: indeed, since utility is a product of marginal utilities, its *logarithm* is the *sum* of logarithms of marginal utilities: $v(I) = \log(u(I)) = \sum V(\vec{\sigma}_{ij}, I(\vec{\sigma}_{ij}))$, where $V = \log(U)$. For these logarithms, we also get integral sums and therefore, a reasonable limit expression: $v(I) = \int V(\vec{\sigma}, I(\vec{\sigma}))\,d\vec{\sigma}$, and $u(I) = \exp(v(I))$.

Comments. Before we go into further details, let us make two comments:

- At first glance, the multiplicative case seems to lead to a more complicated formula than the additive one. However, our goal is to find an image $I(\vec{\sigma})$ for which $u(I) \to \max$. Since logarithm is monotonic, the condition $u(I) \to \max$ is equivalent to $v(I) = \log(u(I)) \to \max$. Therefore, in multiplicative case, we get the same problem $\int V(\vec{\sigma}, I(\vec{\sigma}))\,d\vec{\sigma} \to \max$ as in the additive case.
- Since our goal is optimization, we would like to restrict ourselves to *smooth* (differentiable) functions U and V, because for smooth functions, optimization is as easy as computing the derivatives and equating them to 0. Moreover, since many useful optimization techniques use the second derivatives as well, we will require that these functions are *twice* differentiable.
  Fortunately, we can impose this restriction without losing generality, because, as it is well known, every continuous function can be, with an arbitrary accuracy, approximated by twice differentiable functions (even by polynomials). Since we are dealing with not 100% accurate data anyway, there is no reason to represent the expert's preferences absolutely precisely. Therefore, even if the actual expert preferences are described by a non-smooth function, we can, within an arbitrary accuracy, still approximate it by a smooth function.

Geometric symmetries. The only way that we can represent a spatial point $\vec{\sigma}$ in the computer is by using numbers, i.e., by *fixing* a coordinate system. There is *no prefered* coordinate system in the sky, so the relative quality of the two images should not depend on which coordinate system we choose.

For example, if we choose a different starting point, then a point with new coordinates $\vec{\sigma}$ will have the old coordinates $\vec{\sigma}\,' = \vec{\sigma} + \vec{s}$, where $\vec{s}$ is the vector formed by old coordinates of the new starting point. Therefore, a brightness distribution, which has a form $I(\vec{\sigma})$ in the old coordinates, has a form $I(\vec{\sigma} + \vec{s})$ in the new ones. Hence, it is reasonable to require that:

IF $I(\vec{\sigma})$ is better than the image $J(\vec{\sigma})$ (i.e., $I(\vec{\sigma}) \succ J(\vec{\sigma})$),
THEN the shifted image I should still be better than the correspondingly shifted image J: $I(\vec{\sigma} + \vec{s}) \succ J(\vec{\sigma} + \vec{s})$.

Similarly, the preference relation between images should not change if we *rotate* the coordinate axis, i.e., replace $\vec{\sigma}$ by $T\vec{\sigma}$ for some rotation T.

How can we express the invariance of the preference relation in terms of utility function? The fact that the preference relation does not change does not necessarily mean that the utility function is necessarily invariant, because, as we have mentioned, different utility functions can correspond to the same preference relation. What it does mean is that the utility function corresponding to shifted (rotated) images must describe the same preference relation as the original utility function. We know that two utility functions describe the same preference relation if and only if they can be obtained from each other by a linear transformation

$u \to a \cdot u + b$; so, we arrive at the requirement that for every shift or rotation T, there exist numbers $a(T)$ and $b(T)$ for which, for all images,

$$\int U(\vec{\sigma}, I(T\vec{\sigma}))\, d\vec{\sigma} = a(T) \cdot \int U(\vec{\sigma}, I(\vec{\sigma}))\, d\vec{\sigma} + b(T).$$

Similarly to the above argument about the smoothness of U and V, we can, without losing generality, assume that the functions $a(T)$ and $b(T)$ are differentiable.

Scale-invariance. In addition to geometric transformations, there are other natural symmetries: e.g., the relative quality of the two images should not depend on the choice of the unit in which we measure brightness. If we replace the old measuring unit by a new unit that is λ times smaller, then all numerical values of brightness become λ times larger: $I(\vec{\sigma}) \to \lambda \cdot I(\vec{\sigma})$. It is therefore reasonable to require that the resulting new utility function represent the same preference relation as the old one, i.e., that

$$\int U(\vec{\sigma}, \lambda \cdot I(\vec{\sigma}))\, d\vec{\sigma} = a(\lambda) \cdot \int U(\vec{\sigma}, I(\vec{\sigma}))\, d\vec{\sigma} + b(\lambda)$$

for some functions $a(\lambda)$ and $b(\lambda)$.

A more realistic notion of invariance. In deriving the formulas that describe invariance, we implicitly assumed that for every pair of images, an expert has a certain opinion on which of these two images is better.

In reality, experts only have expertise in comparing not arbitrary pairs of images, but only images that are consistent with the same measurement results. Usually, the main problem is to reconstruct the fine structure of the image, while the total intensity $E(I) = \int I(\vec{\sigma})\, d\vec{\sigma}$ can be easily measured. Therefore, experts usually deal not with arbitrary pairs of images, but only with pairs of images that have the same value of $E(I)$, and they do not have any experience in comparing images that correspond to different values of $E(I)$.

Thus, in the above formulas that describe invariance, the values $a(T)$, $b(T)$, $a(\lambda)$, and $b(\lambda)$ may actually be different from different values of $E(I)$, i.e., in effect, we may have to use the functions $a(T, E)$, $b(T, E)$, $a(\lambda, E)$, and $b(\lambda, E)$ instead.

As a result, we arrive at the following definitions:

3. Definitions and the Main Result

Definition 1. *By an* **image**, *we will mean a smooth non-negative function $I(\vec{\sigma})$ with compact support (i.e., that is equal to 0 outside some bounded 2D area).*

Definition 2. *By the* **total intensity** *of the image $I(\vec{\sigma})$, we mean the value*

$$E(I) = \int I(\vec{\sigma})\, d\vec{\sigma}.$$

Definition 3.

- *By an* additive utility function, *we mean an expression of the type* $u(I) = \int U(\vec{\sigma}, I(\vec{\sigma}))\,d\vec{\sigma}$, *where* $\vec{\sigma}$ *is a 2D vector, and* $U(\vec{\sigma}, I)$ *is a twice differentiable function.*
- *By an* multiplicative utility function, *we mean an expression of the type* $u(I) = \exp(v(I))$, *where* $v(I) = \int V(\vec{\sigma}, I(\vec{\sigma}))\,d\vec{\sigma}$ *and* V *is a twice differentiable function.*
- *By an* image utility function $u(I)$, *we mean either an additive utility function, or a multiplicative utility function.*

Definition 4.

- *We say that an image utility function* $u(I)$ *is* shift-invariant *if there exist smooth functions* $a(T, E)$ *and* $b(T, E)$ *such that for every image* $I(\vec{\sigma})$, *and for every shift* T, *we have* $u(I(T\vec{\sigma})) = a(T, E(I)) \cdot u(I(\vec{\sigma})) + b(T, E(I))$.
- *We say that an image utility function* $u(I)$ *is* rotation-invariant *if there exist smooth functions* $a(T, E)$ *and* $b(T, E)$ *such that for every image* $I(\vec{\sigma})$, *and for every rotation* T, *we have* $u(I(T\vec{\sigma})) = a(T, E(I)) \cdot u(I(\vec{\sigma})) + b(T, E(I))$.
- *We say that an image utility function* $u(I)$ *is* scale-invariant *if there exists smooth functions* $a(\lambda, E)$ *and* $b(\lambda, E)$ *such that for every image* $I(\vec{\sigma})$, *and for every real number* $\lambda > 0$, *we have* $u(\lambda \cdot I(\vec{\sigma})) = a(\lambda, E(I)) \cdot u(I(\vec{\sigma})) + b(\lambda, E(I))$.

Definition 5. *We say that an image utility function* $u(I)$ *is* equivalent *to the functional* $F(I)$ *if for every two images* I *and* J *with the same value of total intensity (i.e.,* $E(I) = E(J)$*),* $u(I) > u(J)$ *if and only if* $F(I) > F(J)$.

Comment. Thus, if a functional $F(I)$ is equivalent to the image utility function $u(I)$, then, for every case in which we have have to select between the images with known total intensity, the selection $u(I) \to \max$ is equivalent to the selection $F(I) \to \max$.

Theorem. *If an image utility function is shift-invariant, rotation-invariant, and scale-invariant, then it is equivalent to one of the functionals*

$F(I) = \pm \int I(\vec{\sigma}) \cdot \log(I(\vec{\sigma}))\,d\vec{\sigma}$, $F(I) = \pm \int \log(I(\vec{\sigma}))\,d\vec{\sigma}$, or $F(I) = \pm \int (I(\vec{\sigma}))^p\,d\vec{\sigma}$.

Comment. Thus, natural symmetry requirements justify the use of maximum entropy method (1) and of generalized maximum entropy methods (2)–(3).

4. Proof

1. First, let us use the fact that the utility function is rotation-invariant and shift-invariant, i.e., $u(I(T\vec{\sigma})) = a(T, E(I)) \cdot u(I(\vec{\sigma})) + b(T, E(I))$ for every rotation or shift T.

1.1. Let us consider the geometric transformation (i.e., rotation or shift) T that is a composition of two other geometric transformations $T = T_1 \circ T_2$. Then, the above equality takes the form

$$u(I(T_1 \circ T_2\vec{\sigma})) = a(T_1 \circ T_2, E(I)) \cdot u(I(\vec{\sigma})) + b(T_1 \circ T_2, E(I)). \tag{4}$$

On the other hand, we can apply a similar equality first for T_2, and then for T_1. Since geometric transformations do not change the total intensity $E(I)$, we get the following formulas:

$$u(I(T_2\vec{\sigma})) = a(T_2, E(I)) \cdot u(I(\vec{\sigma})) + b(T_2, E(I)) \tag{5}$$

and

$$u(I(T_1 \circ T_2\vec{\sigma})) = a(T_1, E(I)) \cdot u(I(T_2\vec{\sigma})) + b(T_1, E(I)) \tag{6}$$

Substituting (5) into (6), we get the formula

$$u(I(T_1 \circ T_2\vec{\sigma})) =$$

$$a(T_1, E(I)) \cdot a(T_2, E(I)) \cdot u(I(\vec{\sigma})) + a(T_1, E(I)) \cdot b(T_2, E(I)) + b(T_1, E(I)). \tag{7}$$

Formulas (4) and (7) give two expressions for the same quantity $u(I(T_1 \circ T_2\vec{\sigma}))$ as linear functions of $u(I(\vec{\sigma}))$; therefore, the coefficients of these linear functions must coincide, i.e.:

$$a(T_1 \circ T_2, E(I)) = a(T_1, E(I)) \cdot a(T_2, E(I)) \tag{8}$$

and

$$b(T_1 \circ T_2, E(I)) = a(T_1, E(I)) \cdot b(T_2, E(I)) + b(T_1, E(I)). \tag{9}$$

1.2. From the equation (8), we can conclude that for every sequence of n geometric transformations $T_1, \ldots, T_n$, we have:

$$a(T_1 \circ \ldots \circ T_n, E(I)) = a(T_1, E(I)) \cdot \ldots \cdot a(T_n, E(I)). \tag{10}$$

In particular, if we take $T_1 = \ldots = T_n =$ rotation by an angle $2\pi/n$ around the same axis, we have an identity transformation id as $T_1 \circ \ldots \circ T_n$ (for which $a(id, E(I)) = 1$), and therefore, $a^n(T_i, E(I)) = 1$. Since the value $a(T_1, E(I))$ is positive, we conclude that for such rotations, $a(T_1, E(I)) = 1$.

From (9), we can now conclude that $b(id, E(I)) = 0 = n \cdot b(T_1, E(I))$ and therefore, that $b(T_1, E(I)) = 0$.

1.3. An arbitrary rotation by an angle $2\pi \cdot p/n$, with integer p and n, can be represented as a composition of p rotations by an angle $2\pi/n$. For each of these angles, as we have already shown, $b = 1$ and $a = 0$. Therefore, by applying the formulas (8) and (9), we can conclude that for their composition, we also have $a = 1$ and $b = 0$.

1.4. Let us now show that $a = 1$ and $b = 0$ for an arbitrary rotation T. Indeed, let α be this rotation's angle. The real number $\alpha/(2\pi)$ can be represented as a limit of rational numbers p/n; therefore, the angle α is equal to the limit of angles $2\pi \cdot p/n$, and hence, the rotation T can be represented as a limit of rotations by angles $2\pi \cdot p/n$. We already know that for all these rotations, $a = 1$ and $b = 0$, and we have assumed that the functions $a(T, E)$ and $b(T, E)$ are differentiable and therefore continuous.

Hence, $a(T,E)=1$ and $b(T,E)=0$ for an arbitrary rotation T. In other words, $u(I(T\vec{\sigma}))=u(I(\vec{\sigma}))$ for an arbitrary rotation T.

1.5. An arbitrary shift can be represented as a composition of two rotations: one by an angle α, and another by the opposite angle $-\alpha$ around a parallel (but different) axis. Since for each rotation, $a=1$ and $b=0$, from the formulas (8) and (9), we can conclude that $a=1$ and $b=0$ for their composition as well, i.e., for an arbitrary shift.

In other words, $u(I(T\vec{\sigma}))=u(I(\vec{\sigma}))$ for an arbitrary shift T.

2. For an additive utility function, we have $\int U(\vec{\sigma},I(T\vec{\sigma}))\,d\vec{\sigma}=\int U(\vec{\sigma},I(\vec{\sigma}))\,d\vec{\sigma}$ for an arbitrary geometric transformation T. If we change variables $\vec{\sigma}$ to new variables $\vec{\sigma}\,'=T\vec{\sigma}$, then the first integral turns into $\int U(T^{-1}\vec{\sigma}\,',I(\vec{\sigma}\,'))\,d\vec{\sigma}\,'$, where T^{-1} is the inverse geometric transformation.

Hence, for every image $I(\vec{\sigma})$, we have $\int U(T^{-1}\vec{\sigma},I(\vec{\sigma}))\,d\vec{\sigma}=\int U(\vec{\sigma},I(\vec{\sigma}))\,d\vec{\sigma}$. In particular, if we take an image that is different from 0 (and equal to some value I) only in a small vicinity of a point $\vec{\sigma}$, we conclude that $U(T^{-1}\vec{\sigma},I)=U(\vec{\sigma},I)$ for all $\vec{\sigma}$, I, and T.

Since every two points $\vec{\sigma}$ and $\vec{\sigma}\,'$ on a plane can be transformed into each other by an appropriate rotation T, we conclude that $U(\vec{\sigma}\,',I)=U(\vec{\sigma},I)$ for arbitrary two points $\vec{\sigma}$ and $\vec{\sigma}\,'$, i.e., that the function U does not depend on its first (vector) parameter $\vec{\sigma}$ at all: $U(\vec{\sigma},I)=U(I)$, and $u(I)=\int U(I(\vec{\sigma}))\,d\vec{\sigma}$.

3. For a *multiplicative* utility function, from the invariance of the utility function $u(I)$, we conclude that its logarithm $v(I)$ is also invariant, and therefore, similarly to part 2 of this proof, we conclude that $V(\vec{\sigma},I)=V(I)$.

4. Let us now use scale-invariance, i.e., the condition

$$u(\lambda\cdot I(\vec{\sigma}))=a(\lambda,E(I))\cdot u(I(\vec{\sigma}))+b(\lambda,E(I)). \tag{11}$$

4.1. To exploit this equality, we will use the differentiability assumption. Namely, let us take an image $I(\vec{\sigma})$ that is positive in a certain area $\mathcal{A}$. Then, for every other smooth function $\delta I(\vec{\sigma})$ that is equal to 0 outside this area $\mathcal{A}$, we can consider a 1-parametric family of images $I_\varepsilon(\vec{\sigma})=I(\vec{\sigma})+\varepsilon\cdot\delta I(\vec{\sigma})$ with a real parameter ε, and then use the above equality (11) for the images from this family.

Due to our definition of utility in terms of one of the functions U or V, and due to the assumption that functions U and V are differentiable, we can conclude that the expressions $E(I_\varepsilon)$, $u(I_\varepsilon)$, and $u(\lambda\cdot I_\varepsilon)$ are differentiable with respect to ε, and their derivatives at the point $\varepsilon=0$ can be explicitly computed. For example, the derivative of $E(I)$ is equal to $\int\delta I(\vec{\sigma})\,d\vec{\sigma}$, etc.

From (11), it follows that if the derivatives of $E(I)$ and of $u(I)$ are equal to 0, then the derivative of $u(\lambda\cdot I)$ must also be equal to 0.

4.2. For *additive* utility functions, this property means that if

$$\int\delta I(\vec{\sigma})\,d\vec{\sigma}=0, \tag{12}$$

and

$$\int U'(I(\vec{\sigma})) \cdot \delta I(\vec{\sigma})\, d\vec{\sigma} = 0, \tag{13}$$

where by U', we denoted the derivative of a function $U(I)$, then

$$\int U'(\lambda \cdot I(\vec{\sigma})) \cdot \delta I(\vec{\sigma})\, d\vec{\sigma} = 0. \tag{14}$$

For *multiplicative* utility functions, we get a similar property, but with V instead of U.

4.3. The above if-then property is formulated in (not very intuitive) analytical terms, but it can be re-formulated in more intuitive geometric terms if we take into consideration that all our functions are smooth and located on $\mathcal{A}$, and therefore, belong to $L^2(\mathcal{A})$. In terms of L^2, the conditions (12) and (13) mean that the function δI is orthogonal to a constant function 1 (appropriately bounded outside $\mathcal{A}$, to make it an element of L^2), and to $U'(I(\vec{\sigma}))$, and the conclusion means that δI is orthogonal to $U'(\lambda \cdot I(\vec{\sigma}))$. In these terms, the above property says that every element of L^2 that is orthogonal to 1 and to $U'(I)$ is also orthogonal to $U'(\lambda \cdot I)$ (it actually says so not about *every* element of L^2, but about every *smooth* element of L^2, but since smooth elements are everywhere dense in L^2, we can easily extend this property to all possible functions from L^2).

In geometric terms, it is easy to prove that if a vector v is orthogonal to every vector x that is orthogonal to two given vectors v_1 and v_2, then v belongs to the linear space generated by v_1 and v_2: indeed, otherwise, we could take a projection $\pi(v)$ of v on the orthogonal complement to that linear space; this projection is orthogonal to both v_i, but not to v.

Thus, for every λ, the function $U'(\lambda \cdot I)$ is a linear combination of the functions 1 and $U'(I)$, i.e., $U'(\lambda \cdot I(\vec{\sigma})) = \alpha(\lambda) + \beta(\lambda) \cdot U'(I(\vec{\sigma}))$ for some values $\alpha(\lambda)$ and $\beta(\lambda)$. This is true for all points $\vec{\sigma}$, and therefore, this equality must be true for all possible values of I. Hence, the function $u'(I)$ must satisfy the following functional equation: for every $\lambda > 0$ and for every I, we have

$$U'(\lambda \cdot I) = \alpha(\lambda) + \beta(\lambda) \cdot U'(I). \tag{15}$$

4.4. We would like to use differentiability to solve the functional equation (15).

The function U' is differentiable, so we need to prove the differentiability of the functions α and β. Let us do it.

Indeed, if we consider the equation (15) for two different values I_1 and I_2, and subtract the resulting equations, we conclude that $U'(\lambda \cdot I_1) - U'(\lambda \cdot I_2) = \beta(\lambda) \cdot (U'(I_1) - U'(I_2))$, and, therefore,

$$\beta(\lambda) = \frac{U'(\lambda \cdot I_1) - U'(\lambda \cdot I_2)}{U'(I_1) - U'(I_2)}.$$

Since the function U is twice differentiable, the right-hand side of this equality is differentiable, and so, $\beta(\lambda)$ is a differentiable function.

Now, from the equation (15), we conclude that $\alpha(\lambda) = u'(\lambda \cdot I) - \beta(\lambda) \cdot u'(I)$. Since all the terms in the right-hand side of this equality are differentiable, the function $\alpha(\lambda)$ is differentiable as well.

4.5. Now, we are ready to deduce the differential equation from the functional equation (15).

Since all three functions $U'(I)$ (we will denote it by $W(I)$), $\alpha(\lambda)$, and $\beta(\lambda)$, are differentiable, we can differentiate both sides of the equation (15) with respect to λ and substitute $\lambda = 1$. As a result, we get the following differential equation: $W'(I) \cdot I = A + B \cdot W$, where we denoted $A = \alpha'(1)$ and $B = \beta'(1)$. Hence,

$$\frac{dW}{dI} \cdot I = A + B \cdot W. \tag{16}$$

4.6. To solve the equation (16), let us first simplify it. To simplify this equation, let us separate the variables W and I by multiplying both sides by $dI/(I \cdot (A + B \cdot W))$; then, the equation takes the form

$$\frac{dW}{A + B \cdot W} = \frac{dI}{I}. \tag{17}$$

This equation is easy to integrate; the resulting solution is slightly different for $B = 0$ and $B \neq 0$.

4.6.1. If $B = 0$, then integrating both parts of (17), we get $A^{-1} \cdot W = \ln(I) + C_1$ ($C_1, C_2, \ldots$ will denote constants). Hence, $U'(I) = W = A \cdot \ln(I) + C_2$, and integrating again, we get $U(I) = A \cdot I \cdot \log(I) + C_2 \cdot I + C_3$ for some constants C_i.

If $C_3 \neq 0$, then the expression for $U(I)$ would include an infinite integral; therefore, $C_3 = 0$, and $U(I) = A \cdot I \cdot \log(I) + C_2 \cdot I$. Hence,

$$u(I) = A \cdot \int I \cdot \log(I) \, d\vec{\sigma} + C_2 \cdot \int I \, d\vec{\sigma}.$$

Since we are only comparing images with the same value of $E(I) = \int I \, d\vec{\sigma}$, the condition $u(I) > u(J)$ is equivalent to the condition $F(I) > F(J)$ for an entropy functional $F(I) = \pm \int I \cdot \log(I) \, d\vec{\sigma}$ (the sign is equal to the sign of the constant A).

4.6.2. If $B \neq 0$, then

$$\frac{dW}{A + B \cdot W} = \frac{d(W + A/B)}{A(W + A/B)},$$

and therefore, after integrating both parts of the equation (17), we get

$$A^{-1} \cdot \ln(W + A/B) = \ln(I) + C_1;$$

hence $\ln(W + A/B) = A \cdot \ln(I) + C_2$, and so, after exponentiating, we get $W + A/B = C_3 \cdot I^A$. Thence, $W = U' = C_3 \cdot I^A + C_4$.

- If $A \neq -1$, we get $U = C_5 \cdot I^{A+1} + C_4 \cdot I + C_6$. Similarly to the case $B = 0$, we can now conclude that $C_6 = 0$, and that the corresponding utility function is equivalent to a functional $F(I) = \pm \int I^p \, d\vec{\sigma}$ (for $p = A + 1$).

– If $A = -1$, we similarly get $U(I) = C_5 \cdot \ln(I) + C_4 \cdot I + C_6$, in which case the utility function $u(I)$ is equivalent to a functional $\pm \int \log(I)\, d\vec{\sigma}$.

4.7. In both cases $B = 0$ and $B \neq 0$, the utility function is equivalent to one of the three functionals $F(I)$ from the formulation of the theorem. The theorem is thus proven.

Acknowledgment

This work was supported in part by NSF grant CDA-9522207. The author is thankful to Sergio Cabrera, Vladik Kreinovich, and to all the participants of MaxEnt'97 for valuable discussions.

References

1. P.C. Fishburn, *Utility Theory for Decision Making*, John Wiley & Sons Inc., New York, 1969.
2. P.C. Fishburn, *Nonlinear preference and utility theory*, The John Hopkins Press, Baltimore, MD, 1988.
3. O.M. Kosheleva and V. Kreinovich, "A letter on maximum entropy method," *Nature*, 1979, Vol. 281, No. 5733 (Oct. 25), pp. 708–709
4. O.M. Kosheleva and V. Kreinovich, *Utility functions that describe invariant preferences*, Technical Report, Center for New Information Technology "Informatika", Leningrad, 1989 (in Russian).
5. D.R. Luce and H. Raiffa, *Games and Decisions, Introduction and critical survey*, John Wiley & Sons, Inc., New York, 1957.
6. R.B. Myerson, *Game theory. Analysis of conflict*, Harvard University Press, Cambridge, MA, 1991.
7. J. Skilling, These Proceedings.
8. P. Suppes, D.M. Krantz, R.D. Luce, and A. Tversky, *Foundations of measurement. Vol. II. Geometrical, threshold, and probabilistic representations*, Academic Press, San Diego, CA, 1989.

PROBABILITY SYNTHESIS

How to Express Probabilities in Terms of Each Other

ANTHONY J.M. GARRETT
63 High Street, Grantchester
Cambridge CB3 9NF, United Kingdon

Abstract. A probability is a number associated with one proposition conditioned on another. We give a formalism which answers whether (and how) one probability can be expressed in terms of other given probabilities, no matter how many propositions are involved in the probabilities in how complicated a manner. The analogous problem in Boolean algebra is also solved for the expression of one binary function of variables in terms of others. The analysis is based on the connection between the calculus of propositions and the calculus of probabilities, since a probability is a number associated with one proposition conditioned on another. This connection has great potential in formalising inductive logic.

Key words: probability, Boolean algebra, laws of probability

1. Introduction: the Problem

A probability is a number associated with a conditioned proposition – *i.e.* with one proposition, conditioned on the truth of another. For example, $p(X|Y)$ is the probability that proposition X is true, supposing that proposition Y is true. Propositions are either true or false and their truth values obey Boolean algebra, familiar today in logic circuitry. Our question here is how to synthesise one probability from others that are given. For example, let us enquire whether there exists a function f such that

$$p(A|BI) = f\big(p(A|I), p(B|AI), p(B|\overline{A}I)\big). \tag{1}$$

In this case f does exist, and has the form

$$f(x, y, z) = \frac{xy}{xy + (1-x)z} \,. \tag{2}$$

(Old hands will recognise this as Bayes' theorem.) We seek a formalism which will tell us whether (and how) one probability can be expressed in terms of other given probabilities, no matter how many propositions are involved and how complicated are the binary functions in which they appear, on either side of the conditioning soliduses. Clearly the laws of probability, namely the sum rule

$$p(A|I) + p(\overline{A}|I) = 1 \tag{3}$$

and the product rule

$$p(AB|I) = p(A|BI)p(B|I) \tag{4}$$

play a central role in such syntheses; our simple example above is easily verified from them. These laws are derived as the algebra of the numbers which corresponds to the Boolean algebra of the propositions, when numbers – interpreted as probabilities – are associated with propositions. This derivation is due to R.T. Cox (1946); a version that employs fewer assumptions is given by Garrett (these proceedings).

2. Results We Shall Use

As well as the logical product ('AND') which occurs in the product rule, and negation ('NOT') which occurs in the sum rule, we shall use some results concerning the logical sum (inclusive OR, denoted $A+B$). It is related to AND and NOT by de Morgan's theorem,

$$\overline{A+B} = \overline{A}\,\overline{B} \tag{5}$$

which may be verified from truth tables. (A good reference on Boolean algebra is Kuntzmann, 1967.) De Morgan's theorem enables us to use the sum and product rules to find the probability of a logical sum:

$$\begin{aligned} p(A+B|I) &= p(\overline{\overline{A}\,\overline{B}}|I) && (6)\\ &= 1-p(\overline{A}\,\overline{B}|I) && (7)\\ &= 1-p(\overline{A}|\overline{B}I)p(\overline{B}|I) && (8)\\ &= 1-[1-p(A|\overline{B}I)]p(\overline{B}|I) && (9)\\ &= p(B|I)+p(A\overline{B}|I) && (10)\\ &= p(B|I)+p(\overline{B}|AI)p(A|I) && (11)\\ &= p(B|I)+[1-p(B|AI)]p(A|I) && (12)\\ &= p(A|I)+p(B|I)-p(AB|I). && (13) \end{aligned}$$

At the heart of our method is the expansion of an arbitrary binary expression as a sum of products of all the variables involved. The resulting canonical expression is known as the disjunctive normal form, or minterm expansion. Direct verification for $A=T$ and $A=F$ (for True and False) indicates immediately that, for any binary function f,

$$f(A,\bullet) = Af(T,\bullet) + \overline{A}f(F,\bullet) \tag{14}$$

where $\bullet$ denotes any number of propositions that do not contain A. Now single a proposition B out of $\bullet$ and expand with respect to B in the same way:

$$f(A,B,\bullet) = ABf(T,T,\bullet) + A\overline{B}f(T,F,\bullet) + \overline{A}Bf(F,T,\bullet) + \overline{A}\,\overline{B}f(F,F,\bullet) \tag{15}$$

and so on; for n variables, there will be 2^n terms on the right-hand side. Clearly, only those terms for which $f=T$ contribute to the right-hand side. The resulting 'minterm' expansion of f in all its arguments is the logical counterpart to the Taylor expansion of complex functions.

The 'minterms' in the complete expansion of a binary function in all its variables are orthogonal, in the sense that the logical product of any minterm with any other is False; and complete, in the sense that any binary function can be so expanded. To verify orthogonality, observe that the logical product of distinct minterms always includes at least one factor of the form $X\overline{X}$, which is False. It follows from (13) that the probability of a minterm expansion is equal to the sum of probabilities of individual minterms: in the probability of the logical sum of two minterms, from (13), the final 'cross' term is the probability of the logical product of two orthogonal propositions, which is zero. Hence the result is established for two minterms, and extends by induction to an arbitrary number.

3. Probability Synthesis: the Solution

We now give the solution to the problem of probability synthesis. We shall tackle the simple example given at the start; it will be clear that the same principles apply to more complicated syntheses. Our basic strategy is to make minterm expansions of the propositional arguments of all the probabilities. This is useful because the probability of a minterm expansion is equal to the sum of probabilities of each term. However we must first overcome a problem: the conditioning propositions are liable to be different in the probabilities given. Moreover the conditioning alters each time the product rule is used. (The product rule might better be called the 'conditionalising rule'.) Before using the minterm method we must employ the product rule to ensure that every proposition is identically conditioned.

With this in mind we tackle the problem of seeking a function f such that

$$p(A|BI) = f\big(p(A|I), p(B|AI), p(B|\overline{A}I)\big). \tag{16}$$

In this problem the proposition I is the highest common (logical) factor to the right of the conditioning soliduses. Therefore we use the product rule to remove all other conditioning information to the left of the soliduses, with result

$$\frac{p(AB|I)}{p(B|I)} = f\left(p(A|I), \frac{p(AB|I)}{p(A|I)}, \frac{p(\overline{A}B|I)}{p(\overline{A}|I)}\right). \tag{17}$$

Now we make minterm expansions in every propositional variable to the left of the soliduses. The variables are A and B, so that

$$\frac{p(AB|I)}{p(AB+\overline{A}B|I)} = f\left(p(AB+A\overline{B}|I), \frac{p(AB|I)}{p(AB+A\overline{B}|I)}, \frac{p(\overline{A}B|I)}{p(\overline{A}B+\overline{A}\,\overline{B}|I)}\right). \tag{18}$$

We now use the result that the probability of a minterm expansion is just the sum of probabilities of each term, so that

$$\frac{p(AB|I)}{p(AB|I)+p(\overline{A}B|I)} = f\left(p(AB|I)+p(A\overline{B}|I), \frac{p(AB|I)}{p(AB|I)+p(A\overline{B}|I)}, \frac{p(\overline{A}B|I)}{p(\overline{A}B|I)+p(\overline{A}\,\overline{B}|I)}\right). \tag{19}$$

All probabilities in this expression are probabilities of minterms. Our aim is now to rewrite this equation so as to find the function f. For ease, label the probabilities of the minterms:

$$\alpha = p(AB|I), \qquad \beta = p(\overline{A}B|I), \qquad \gamma = p(A\overline{B}|I), \qquad \delta = p(\overline{A}\,\overline{B}|I), \tag{20}$$

so that our equation becomes

$$\frac{\alpha}{\alpha+\beta} = f\left(\alpha+\gamma, \frac{\alpha}{\alpha+\gamma}, \frac{\beta}{\beta+\delta}\right) . \tag{21}$$

The minterm probabilities are not all independent: by orthogonality their sum is equal to the probability of the logical sum of the minterms, which is just $(A + \overline{A})(B + \overline{B})$. It follows from the probability relations above that

$$\alpha + \beta + \gamma + \delta = 1, \tag{22}$$

and we use this normalisation condition to remove one degree of freedom – here it is convenient to eliminate δ – and obtain the equation

$$\frac{\alpha}{\alpha+\beta} = f\left(\alpha+\gamma, \frac{\alpha}{\alpha+\gamma}, \frac{\beta}{1-(\alpha+\gamma)}\right). \tag{23}$$

To find f we simply define its arguments as new variables and seek to re-express the equation in terms of them. Define therefore

$$x = \alpha + \gamma, \qquad y = \frac{\alpha}{\alpha+\gamma}, \qquad z = \frac{\beta}{1-(\alpha+\gamma)} . \tag{24}$$

This relation expresses the probabilities of the building blocks in terms of the probabilities of minterms. The relation between these sets is easily inverted for the minterm probabilities in the left-hand side of (23):

$$\alpha = xy, \qquad \beta = (1-x)z. \tag{25}$$

In terms of the new variables equation (23) now becomes

$$f(x,y,z) = \frac{xy}{xy + (1-x)z} , \tag{26}$$

as was quoted earlier. We have used only the sum and product rules in our method, so that these will always suffice as tools for synthesis.

All essential features of more complicated syntheses are included in this example. If the relation, that expresses the probabilities of those minterms on the left-hand side (α and β) in the arguments (x, y, z) of the desired function, is not invertible, then no solution exists. Since the minterms are a complete set there is no other route to a solution and the synthesis is impossible. Non-invertibility is likely when there are more terms in the minterm expansion than there are arguments of f. If a solution exists it is always a ratio of one sum of products to another, as above. If, conversely, the inversion is non-unique, there may be many syntheses from the given building blocks. This happens when, for example, $p(A|I)$ and $p(\overline{A}|I)$ are both given as building blocks, since these are related by the sum rule.

Often the failure for a solution to exist can be remedied by using as extra building blocks probabilities conditioned on just I ('prior' probabilities).

4. Synthesis of Logical Expressions

It is natural to ask whether we can solve the analogous problem for propositions: can a particular binary function of propositional variables be expressed in terms of other, given binary functions of those variables; and, if so, how? Suppose we seek to express the binary function $\Psi(A, B, C\ldots)$ in terms of the m binary functions $\Phi_1(A, B, C\ldots)$, $\Phi_2(A, B, C\ldots), \ldots \Phi_m(A, B, C\ldots)$, so that

$$\Psi(A,B,C\ldots) = f\Big(\Phi_1(A,B,C\ldots), \Phi_2(A,B,C\ldots), \ldots \Phi_m(A,B,C\ldots)\Big). \quad (27)$$

Of course some of the propositional variables $A, B, C\ldots$ may be absent in Ψ or in certain of the Φ's. Some elementary examples are

$$AB = f(ABC, AB\overline{C}, A\overline{B}C): \qquad f(X,Y,Z) = X + Y \quad (28)$$

$$AB = f(AC, BC): \qquad \text{no } f \text{ exists} \quad (29)$$

$$ABC = f(AB, AC): \qquad f(X,Y) = XY \quad (30)$$

$$ABC = f(AB, \overline{C}): \qquad f(X,Y) = X\overline{Y}. \quad (31)$$

The first example follows immediately from the minterm expansion of AB in A, B, C, since the building blocks are all logical products involving every variable defined. The general method for seeking f is: write out the minterm expansion for an arbitrary function of m variables; substitute into this the given functions $\Phi_1 \ldots \Phi_m$; reduce the result to minterm form in all the variables in the problem; compare with the corresponding minterm expansion of Ψ, the binary function in question. This comparison establishes the expansion coefficients in the minterm expansion of the required function.

For instance, in the third example we seek a function of two variables, so write it as

$$f(X,Y) = f_{tt}XY + f_{ft}\overline{X}Y + f_{tf}X\overline{Y} + f_{ff}\overline{X}\,\overline{Y} \quad (32)$$

where f_{tf} is a convenient shorthand for $f(T, F)$. Now substitute $X = AB$, $Y = AC$, to give

$$f(AB, AC) = f_{tt}ABAC + f_{ft}\overline{AB}AC + f_{tf}AB\overline{AC} + f_{ff}\overline{AB}\,\overline{AC} \quad (33)$$

and reduce this to minterm form in the variables A, B, C; if the functions given as building blocks are too complicated for this to be done by inspection, a further minterm expansion of each term is used. The result is

$$\begin{aligned} f(AB, AC) \;=\; & f_{tt}ABC + f_{ft}A\overline{B}C + f_{tf}AB\overline{C} \\ & + f_{ff}(\overline{A}BC + \overline{A}\,\overline{B}C + \overline{A}B\overline{C} + \overline{A}\,\overline{B}\,\overline{C} + A\overline{B}\,\overline{C}) \end{aligned} \quad (34)$$

$$\begin{aligned} = \; & f_{tt}ABC + f_{ff}\overline{A}BC + f_{ft}A\overline{B}C + f_{tf}AB\overline{C} \\ & + f_{ff}A\overline{B}\,\overline{C} + f_{ff}\overline{A}B\overline{C} + f_{ff}\overline{A}\,\overline{B}C + f_{ff}\overline{A}\,\overline{B}\,\overline{C} \end{aligned} \qquad (35)$$

which, on comparison with the (trivial) minterm expansion of the desired function ABC gives a set of simultaneous linear equations for the expansion coefficients. Existence of solutions may be tested, and all solutions established, using the standard formalism for solving Boolean equations (included in Kuntzmann's book); since the number of expansion coefficients is 2^m, the algorithm for finding f is non-polynomial in m. In the present simple example the equations for the expansion coefficients are direct expressions for the expansion coefficients:

$$f_{tt} = T; \qquad f_{tf} = f_{ft} = F; \qquad f_{ff} = F \text{ (five times)}, \qquad (36)$$

which, on substitution in (32), give immediately $f(X,Y) = XY$, as intuition foretells in this simple case. There are three variables – and hence $2^3 = 8$ equations – in just $2^2 = 4$ coefficients, since we seek a function of two variables, so that contradictory conditions may arise on the expansion coefficients. This is what happens in (29).

5. Conclusion

The machinery has been laid bare for answering whether one probability can be expressed in terms of others that are given; and if so, how. The same has been done for binary functions of propositions. The analysis is another example of the connection between the calculus of propositions and the calculus of probabilities, because a probability is a number associated with one proposition conditioned on another. This connection was first made quantitative in Cox's (1946) derivation of the laws of probability from Boolean algebra. The connection can undoubtedly be exploited to make further progress in the formalising of inductive logic.

References

Cox, R.T. 1946. Probability, frequency and reasonable expectation. *American Journal of Physics* **14**, 1–13.

Garrett, A.J.M. 199*. Whence the laws of probability? *These Proceedings.*

Kuntzmann, J. 1967. *Fundamental Boolean Algebra* (English translation). Blackie, London, UK.

INVERSION BASED ON COMPUTATIONAL SIMULATIONS

K.M. HANSON, G.S. CUNNINGHAM, AND S. S. SAQUIB*
Los Alamos National Laboratory, MS P940
Los Alamos, New Mexico 87545 USA

Abstract. A standard approach to solving inversion problems that involve many parameters uses gradient-based optimization to find the parameters that best match the data. We will discuss enabling techniques that facilitate application of this approach to large-scale computational simulations, which are the only way to investigate many complex physical phenomena. Such simulations may not seem to lend themselves to calculation of the gradient with respect to numerous parameters. However, adjoint differentiation allows one to efficiently compute the gradient of an objective function with respect to all the variables of a simulation. When combined with advanced gradient-based optimization algorithms, adjoint differentiation permits one to solve very large problems of optimization or parameter estimation. These techniques will be illustrated through the simulation of the time-dependent diffusion of infrared light through tissue, which has been used to perform optical tomography. The techniques discussed have a wide range of applicability to modeling including the optimization of models to achieve a desired design goal.

Key words: simulation, inversion, reconstruction, adjoint differentiation, sensitivity analysis, optimization, model validation

1. INTRODUCTION TO THE GENERAL PROBLEM

Frequently a physical situation can only be described fully by a computational model. We wish to address the general problem of finding the values of the parameters in such a model that best match a given set of data. This problem in often referred to as that of inversion. In data matching the objective function to be minimized is often the negative logarithm of the likelihood of the data given their predicted values, which yields the maximum likelihood (ML) solution. Alternative approaches include regularized versions of maximum likelihood and Bayesian methods, in which the objective function is the minus-log-posterior, yielding the maximum a posteriori (MAP) estimate.

We confine ourselves to objective functions that depend on the parameters in a continuous and differentiable fashion. We do not necessarily avoid problems for which the objective function possesses multiple minima. However, because the techniques that we present make use of gradients in the optimization process, they will work effectively only when one can easily find the basin of attraction for the global minimum, for example, by multiscale or multiresolution optimization.

Email: kmh@lanl.gov, cunning@lanl.gov, saquib@polaroid.com
WWW: http://bayes.lanl.gov/
Saquib's current address: Polaroid Corporation, 750 Main St., Cambridge, MA 02139.

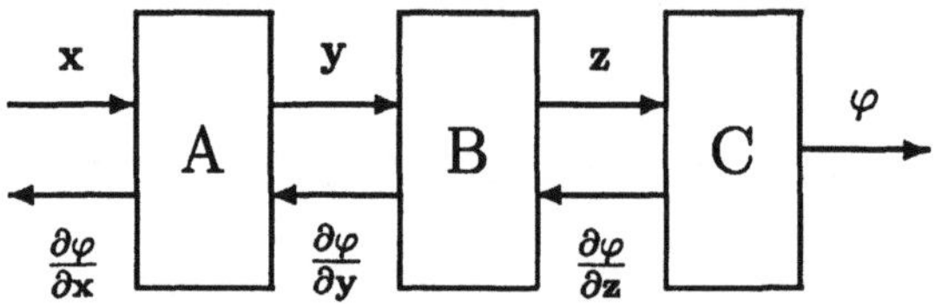

Figure 1: Data flow diagram showing a sequence of transformations, represented by the boxes A, B, and C connected by the arrows pointing to the right, starting with the data structure $\mathbf{x}$ and ending with the scalar φ. The data flow for the adjoint derivatives is indicated by the arrows pointed left.

The proposed method for solving the inverse problem is applicable to a wide variety of problems in which the measurements for the process in question are adequately described by a predictive forward computational model. We demonstrate the method in the context of time-dependent diffusion. We demonstrate the feasibility of reconstructing the diffusion constants within a region from time-dependent light intensities measured for infrared light that diffuses through the region [1]. Adjoint differentiation is already being used in ocean modeling [2, 3]. We believe it may be useful in dealing with a host of other advanced modeling situations, including geophysical interpretation of seismic data and modeling of the atmosphere, fluid flow, and shock-wave phenomena. It should be applicable to optimization of engineering designs in complex situations, such as streamlining of airplane foils and automobile bodies to reduce drag.

A number of other enabling techniques will be mentioned, including multi-scale analysis to constrain and accelerate the optimization process, deformable geometric models for describing the boundaries of objects, and the Markov Chain Monte Carlo method of sampling the uncertainty distribution of the estimated parameters.

2. ADJOINT DIFFERENTIATION

We wish to address problems that require minimizing a scalar function φ by varying the many (10^3 to 10^6 or more) variables that comprise the parameters of the object model. This optimization problem would be intractable without knowing the gradient of φ, or sensitivities, with respect to the parameters on which it depends. There exists a technique to calculate these crucial sensitivities, called adjoint differentiation [4], that is apparently relatively unappreciated. Using the adjoint differentiation technique, the calculation of all these derivatives can be done in a computational time comparable to the forward calculation.

Suppose that a calculation proceeds as a sequence of transformations as shown in Fig. 1. The independent variables in the data structures designated by the vector $\mathbf{x}$ are transformed in block A to produce the dependent variables $\mathbf{y}$. These are transformed in blocks B and C to produce the dependent data structure $\mathbf{z}$ and the final scalar φ, respectively.

We call the sequence of transformations

$$\mathbf{x} \xrightarrow{\mathrm{A}} \mathbf{y} \xrightarrow{\mathrm{B}} \mathbf{z} \xrightarrow{\mathrm{C}} \varphi ,$$

the forward calculation. We assume that the transformations are general, with the only restriction being that they are differentiable. Each transformation is self-contained; it requires only its input variables to calculate its output variables, e.g., module B uses only its input $\mathbf{y}$ to calculate its output $\mathbf{z}$. Therefore, each transformation should require nothing more than its input to implement the derivative of its output variables with respect to its input variables. The data structures are likewise general.

The derivatives of φ with respect to the ith component of $\mathbf{x}$ are obtained using the chain rule,

$$\frac{\partial \varphi}{\partial x_i} = \sum_{jk} \frac{\partial \varphi}{\partial z_k} \frac{\partial z_k}{\partial y_j} \frac{\partial y_j}{\partial x_i} . \tag{1}$$

Even if the transformations are nonlinear, this expression amounts to a product of matrices. The order of the summations can obviously be done in two different ways. If the sum over j is done before the sum over k, the calculation proceeds in the same direction as the forward model calculation. As the dimensions of $\mathbf{x}$, $\mathbf{y}$, and $\mathbf{z}$ are assumed to be large, this sequence results in very large intermediate matrices, which we would like to avoid.

On the other hand, if the sum on k is done before that on j, the sequence of calculations is

$$\mathbf{I} \xrightarrow{\mathrm{C}'^{\mathrm{T}}} \frac{\partial \varphi}{\partial \mathbf{z}}^{\mathrm{T}} \xrightarrow{\mathrm{B}'^{\mathrm{T}}} \frac{\partial \varphi}{\partial \mathbf{y}}^{\mathrm{T}} \xrightarrow{\mathrm{A}'^{\mathrm{T}}} \frac{\partial \varphi}{\partial \mathbf{x}}^{\mathrm{T}} ,$$

where, for example, B'^{T} effectively multiplies $\frac{\partial \varphi}{\partial \mathbf{z}}$ by the transpose of the matrix $\frac{\partial \mathbf{z}}{\partial \mathbf{y}}$. The symbol $\mathbf{I}$ at the beginning of the sequence represents the identity structure, indicating that the sensitivity calculation begins with the transpose of the derivative of transformation C, that is, C'^{T}. This sequence implies intermediate data structures (e.g., $\frac{\partial \varphi}{\partial \mathbf{y}}$) that mimic the normal data structures (e.g., $\mathbf{y}$) implying storage requirements identical to the forward calculation. If the forward transformations are nonlinear, the forward data may be required for the adjoint calculation. The backward flow of the adjoint derivatives is depicted in Fig. 1.

The significant conclusion is that with adjoint differentiation the derivatives of a scalar quantity, φ in the above example, with respect to all the variables in the model can be computed in a time comparable to the forward calculation.

We have coined the acronym Adjoint Differentiation In Code Technique (ADICT) [5] to describe a particular approach to adjoint differentiation. The distinctive feature of ADICT is that a separate computer code is used to compute the adjoint derivatives and that code is based on the simulation code with the explicit intent to "differentiate" the forward calculation or numerical algorithm. For optimization of a functional based on computation, it is desirable to have the gradient of the computation, not of the physics equations that the computation is supposed to approximate [6, 7]. The ADICT principle can produce more accurate derivatives [8].

For the Bayes Inference Engine (BIE) [5], we manually coded the adjoint code for each transformation based on its forward code, or the specific algorithm it employs. This task can sometimes be daunting, but one that can be learned. For example, such manual coding was employed to obtain the adjoint differentiation of a transformation from a surface, described in terms of triangles, to a 3D voxelated grid, which involves some fairly complex calculations [9]. In some cases, it may be helpful to make use of known derivatives of functions of matrices [10]. Object-oriented design and programming is tremendously advantageous for linking calculations together to form a data-flow diagram made of autonomous transformations, as in the BIE.

There are several compilers of FORTRAN code that "automatically" produce differentiation code, including the well-known ADIFOR [11] and GRESS [12]. These compilers are based on computing the derivatives of the result of each line of the simulation code during the forward calculation. In ADIFOR these derivatives are propagated along with the forward calculation, which limits the number of parameters for which it can compute derivatives in large problems. On the other hand, GRESS stores these line-by-line derivatives for a subsequent adjoint calculation, which makes it more useful for computing derivatives with respect to all the parameters. However, this approach imposes heavy memory requirements that make it unsuitable for lengthy simulations.

More promising for application to large simulation codes is a code-based approach created by Ralf Giering [13]. His compiler, dubbed TAMC for Tangent-linear and Adjoint Model Compiler, effectively implements the underlying concepts of ADICT to convert an input FORTRAN code into an enhanced forward code and adjoint code. These codes are to be run consecutively to obtain the result of the forward calculation and the derivatives with respect to the desired variables. Its significant difference from GRESS is that the adjoint code effectively reverses the forward calculations accumulating the derivatives as it goes, instead of calculating and storing the same derivatives during the forward calculation, as GRESS does. This approach allows TAMC to avoid storing a vast number of derivatives and thus opens up the possibility of handling really huge problems. Another issue that TAMC handles is the storage of the state of the forward calculation, which is typically needed for the adjoint derivative calculation. To keep these storage requirements from becoming overwhelming, TAMC allows one to insert check points in the forward calculation at which the state of calculation is stored. Then, during the adjoint calculation, the forward simulation is automatically redone to reestablish the intervening parameters that were not stored.

TAMC has successfully been used to generate adjoint differentiation code for models that simulate ocean dynamics [2, 3], which is a really large problem! We have also had success applying TAMC to 1D hydrodynamics codes [8].

3. OPTIMIZATION

The ML or MAP solution is found by minimizing a scalar functional φ with respect to all the model parameters. Given the possibly large number of parameters, it is imperative to use the derivatives of φ with respect to all parameters. Fortunately, there is a technique to efficiently calculate these gradients as described in the previous section. The best-known approaches to gradient-based optimization of functions of many parameters are steepest descent and conjugate gradient. Another set of techniques that make use of the gradient are referred to as quasi-Newton methods. The general idea is to build up an approximate expression for the Hessian (the second-derivative matrix of φ with respect to all the variables), or the inverse Hessian. Davidon pioneered the quasi-Newton movement with what he called the variable metric algorithm [14]. We have had good success using the Davidon algorithm although at times the conjugate-gradient algorithm works better.

The BFGS (Broyden-Fletcher-Goldfarb-Shanno) algorithm was an outgrowth of the Davidon algorithm, which had become known as DFP (Davidon-Fletcher-Powell). The BFGS algorithm has the advantage over the DFP in that it does not require accurate line minimizations along the quasi-Newton directions to build up the approximate Hessian [15]. Thus, BFGS potentially reduces the number of function evaluations required to complete an optimization procedure. In the kth iteration of the optimization procedure, the change in the parameter vector from its present value x_k is based on the gradient $g_k = \frac{\partial \varphi}{\partial x}|_{x_k}$ and the present estimate of the inverse Hessian[1] C_k. The position of the minimum is estimated by the Newton formula: $x^* = x_k - \alpha C_k g_k$. Starting with $\alpha = 1$, a line search is conducted to find the value of α that minimimizes $\varphi(x^*)$, which yields the new estimate $x_{k+1} = x^*$. Then, designating the change in position by $s_k = x_{k+1} - x_k$ and the corresponding change in the gradient by $y_k = g_{k+1} - g_k$, the BFGS update formula for the inverse Hessian is

$$C_{k+1} = V_k^{\mathrm{T}} C_k V_k + c_k \, s_k s_k^{\mathrm{T}} \, , \tag{2}$$

where $V_k = I - c_k \, y_k s_k^{\mathrm{T}}$ and $c_k = (s_k^{\mathrm{T}} y_k)^{-1}$.

Important to note for applications involving many variables is that the inverse Hessian does not have to be stored as a matrix. Rather, whenever the product of the inverse Hessian with a vector is desired, for example to estimate the next step, the expansion (2) may be used. Therefore, only the parameter vector and the gradient at each iteration of the optimization need to be stored.

More recently, there has appeared a limited-memory version of BFGS [16], as well as the truncated Newton [17] algorithm, both of which have the advantage that they do not require keeping the full sequence of optimization iterations. They both seem to have strengths for different kinds of problems [18]. Although we have not yet implemented either of these in the BIE, we think that the limited-memory BFGS algorithm will be very useful for large optimization problems.

[1] We avoid the prevalent confusing designation of the inverse Hessian by H.

4. DEMONSTRATION: TIME-DEPENDENT DIFFUSION

As an example of the success of the ADICT approach, we summarize our previous paper [1], in which we investigated the diffusion of infrared light through tissue. We solved the problem of inversion of time-resolved data to obtain the distribution of diffusion coefficients through which the light passed. Suppose that the intensity of diffused light at position (x, y) and time t is denoted by $U(x, y, t)$ and the source strength by $R(x, y, t)$. Then the time evolution of U through a region described by diffusion coefficients $D(x, y)$ and absorption coefficients $\mu_a(x, y)$ is given by the diffusion equation

$$\frac{\partial U}{\partial t} = \frac{\partial}{\partial x}\left(D\frac{\partial U}{\partial x}\right) + \frac{\partial}{\partial y}\left(D\frac{\partial U}{\partial y}\right) - c\mu_a U + R\,, \tag{3}$$

where the spatial and temporal dependence of the parameters has been suppressed. The quantity c is the speed of light.

We approach the computational problem in terms of discrete samples of U on a spatial grid with locations specified as a subscript s and in time with a superscript n. When the spatial position subscript is dropped, the resulting quantity is a column vector obtained by either row-ordering or column-ordering the corresponding two-dimensional field (e.g., U^n).

For simplicity we assume that the measurements are degraded by additive uncorrelated Gaussian noise. The minus-log-likelihood of the observations Y, given D and μ_a, is

$$\varphi = -\log P(Y|D, \mu_a) = \tfrac{1}{2}\chi^2 = \sum_{s,n} \frac{(Y_s^n - \tilde{U}_s^n)^2}{2\sigma_{s,n}^2}\,, \tag{4}$$

where $\sigma_{s,n}^2$ is the noise variance at spatial position s and time n and the tilde on the U indicates only those positions at which the light intensities are measured. The general strategy for estimating the parameters, D and μ_a is to minimize $\frac{1}{2}\chi^2$, or if prior information is to be incorporated by means of a Bayesian formulation, the minus-log-posterior, which is $\frac{1}{2}\chi^2$ plus the minus-log-prior.

To compute (4), given D and μ_a, we need to solve the diffusion equation (3) forward in time to obtain the diffuse intensity U_s^n for all time n and spatial positions s. We will briefly summarize the approach to this calculation employed in Ref. [1], which should be consulted for the details.

4.1. Solving the Forward Problem

The general approach taken to solve this forward problem is to use the finite-difference method in which the spatial and temporal derivatives in Eq. (3) are replaced by their finite-difference approximations. This substitution results in a difference equation that needs to be solved forward in time. When solving the difference equation for U^{n+1}, the finite-difference approximations to the spatial derivatives can be evaluated either at time index $n + 1$ or n. In the implicit method[2] for solving differential equations, the spatial derivatives are evaluated at

[2]The implicit method is presented here for clarity, although a slightly different method, the Alternating-Directions Implicit (ADI) method, was used in Ref. [1] for calculational efficiency.

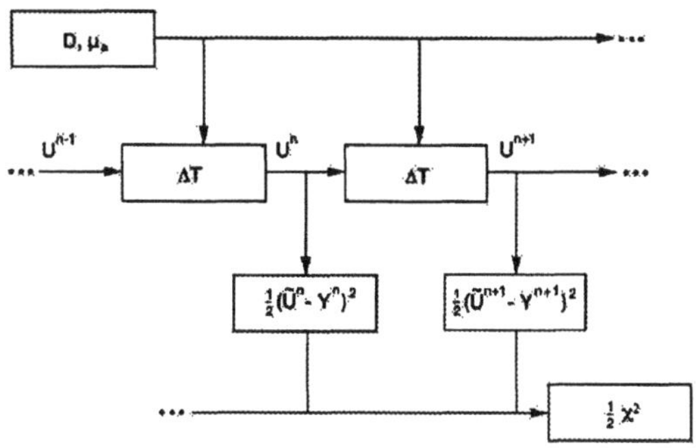

Figure 2: Data flow diagram for the forward calculation of the diffusion problem. The boxes labeled ΔT compute the update to the system for each time step based on the spatial distribution of D and μ_a provided by the topmost box. The predicted intensities are compared to the corresponding measurements Y to obtain $\varphi = \frac{1}{2}\chi^2$. In the adjoint differentiation calculation, the data flow is reversed.

the time instance $(n+1)$ when computing the diffuse intensity U^{n+1}. The implicit method is unconditionally stable for any value of Δt.

Substituting in Eq. (3), the equation to be solved to obtain U^{n+1} from U^n is, in vector notation,

$$AU^{n+1} = U^n + \bar{R}^{n+1/2} , \tag{5}$$

where A is a sparse matrix (because derivatives involve only local variables) whose elements depend on the D and μ_a values. $\bar{R}^{n+1/2}$ denotes the integrated source strength between time instances n and $n+1$.

The procedure for calculating the time-evolution of U is depicted in Fig. 2. The transformations denoted by ΔT essentially involve solving Eq. (5) to move forward by one time step. These calculations depend on the spatial distribution of the diffusion constant D and absorption coefficient μ_a. The minus-log-likelihood $(\frac{1}{2}\chi^2)$ is the accumulation of the sum of the squares of the differences between the measurements Y_s^n and their predicted values $\tilde{U}_s^n$. Thus it gets a contribution from each measurement time. The assumed time-independent distribution of D and μ_a is used in each time step calculation.

4.2. Gradient Computation

We designate the unknown parameters by the column vector $\theta = [D\ \mu_a]^T$. We need the derivative or sensitivity of $\varphi(\theta)$ with respect to θ to facilitate the solution of the inverse problem. The ADICT principle requires us to work backwards in time differentiating the same discretized equations that are used to compute the forward solution. We present a brief outline of the approach.

The sensitivity of φ with respect to U^n is obtained by using the sensitivity of

φ with respect to U^{n+1}. Application of the chain rule yields

$$\frac{d\varphi}{dU^n} = \left[\frac{dU^{n+1}}{dU^n}\right]^T \frac{d\varphi}{dU^{n+1}} + \frac{\partial\varphi}{\partial U^n}, \tag{6}$$

where $\frac{\partial\varphi}{\partial U^n}$ denotes the change in φ when only U^n is varied, keeping all other variables constant, while $\frac{d\varphi}{dU^n}$ denotes the total change in φ when U^n is varied along with all variables that depend on U^n. Differentiating Eq. (4) with respect to U_s^n, we obtain

$$\frac{\partial\varphi}{\partial U_s^n} = \frac{Y_s^n - \tilde{U}_s^n}{\sigma_s^2}, \tag{7}$$

which provide the second term in (6). As indicated in Fig. 2, similar contributions are derived from $\frac{1}{2}\chi^2$ at each time step for which there is a measurement. Differentiating the time-step update, Eq. (5), with respect to U^n, we obtain

$$\frac{dU^{n+1}}{dU^n} = A^{-1}. \tag{8}$$

Using Eqs. (7) and (8), we obtain the sensitivity of φ with respect to U^n

$$\frac{d\varphi}{dU^n} = (A^{-1})^T \frac{d\varphi}{dU^{n+1}} + \frac{\partial\varphi}{\partial U^n}. \tag{9}$$

In the diagram this result comes from each ΔT transformation and flows backwards from U^{n+1} to U^n.

Similar use of the chain rule yields the sensitivity of φ with respect to θ, which result flows out of the top of the ΔT box in Fig. 2 and gets added to the total derivative of φ with respect to the D vector.

4.3. Inversion

The problem of reconstructing the unknown parameters D and μ_a from the measurements Y_s^n is an ill-posed inverse problem. Some form of regularization is necessary to make the solution well behaved. We accomplish this by employing an image model in the reconstruction process that incorporates our *a priori* knowledge regarding the unknown fields D and μ_a. Markov random fields (MRF) have been extensively used in image processing applications. We model D as a generalized Gaussian MRF (GGMRF) [19] with an energy function of the form

$$V\left(\frac{D}{\sigma_D}\right) = \sum_{\{s,r\}\in\mathcal{N}} \frac{b_{s-r}}{p}\left|\frac{D_s - D_r}{\sigma_D}\right|^p, \tag{10}$$

where $\mathcal{N}$ is the set of all neighboring pixel pairs, b is the weight assigned to the specific neighbors, and σ_D is the scale parameter of the model. The popular choice of $p = 2$ in the signal-processing literature yields a quadratic cost function, which tends to excessively penalize large deviations resulting in blurred edges. It is possible to provide good edge preservation in the reconstructed image for $p \approx 1$

[20]. Furthermore, the form of the model facilitates the estimation of the strength of this prior directly from the data [20]. In our formalism, the above energy function is taken to the minus-log-prior, which is added to the minus-log-likelihood (4) to obtain our objective function, the minus-log-posterior. Minimization of the minus-log-posterior yields the MAP solution.

Our example consists of a simulation of time-resolved data for a 6.4-cm-square section of tissue. Figure 3a shows as a 64×64 image the original diffusion coefficients, which range in value from 0.7 to 1.4 $\mathrm{cm^2 ns^{-1}}$. The absorption coefficients are set to a constant value of 0.1 $\mathrm{cm^{-1}}$. The values of these coefficients, as well as the physical dimensions of the problem, have been chosen to reflect the properties of real tissue, since this modality is envisioned as potentially useful in medical imaging. Although the above method can be used to estimate D and μ_a simultaneously, we will restrict ourselves to the simpler case of just estimating D and assume that μ_a is known.

We assume that there are four pulsed sources placed at the midpoints of each side of the square region. There are 52 detectors evenly spaced around the perimeter, which measure the time-dependent signal in response to each pulsed source. Gaussian noise is added to the simulated signals with an rms value of 3% of the rms signal value over the 1.0 ns observation time, corresponding to a signal-to-noise ratio of 30 dB. The time step used is $\Delta t = 0.005$ ns and the detector resolution is 0.02 ns. Figure 4 shows the type of measurements obtained for one source position.

Figure 3b shows the MAP reconstruction for $p = 1.1$ obtained using 70 iterations of the conjugate gradient algorithm (taking about nine hours on an HP 9000/755). The reconstruction is remarkably good considering that effectively only four views (in the nomenclature of computed tomography) are used. This result confirms the value of incorporating ADICT into a simulation code to solve this inversion problem in which roughly 4000 parameters are determined from approximately 10000 measurements (54 detectors × 50 time samples × 4 source positions).

Inversion of these diffusion data is a truly difficult nonlinear problem. This task is much more complicated than inversion from x-ray radiographic measurements because the paths that the photons take depend on the distribution of diffusion coefficients that are being determined.

5. OTHER ENABLING TECHNIQUES

There are several other techniques that complement adjoint differentiation and optimization and help complete the suite of tools needed to conduct inference using large-scale simulations.

5.1. MULTISCALE OPTIMIZATION

When optimizing nonlinear models it is possible to encounter objective functions with multiple local minima, which can introduce difficulties for gradient-based algorithms. However, for many kinds of problems, the desired minimum can be found either by knowledgeably choosing a good starting point or by using a multiresolution approach, that is, by first finding the minimum at coarse resolution

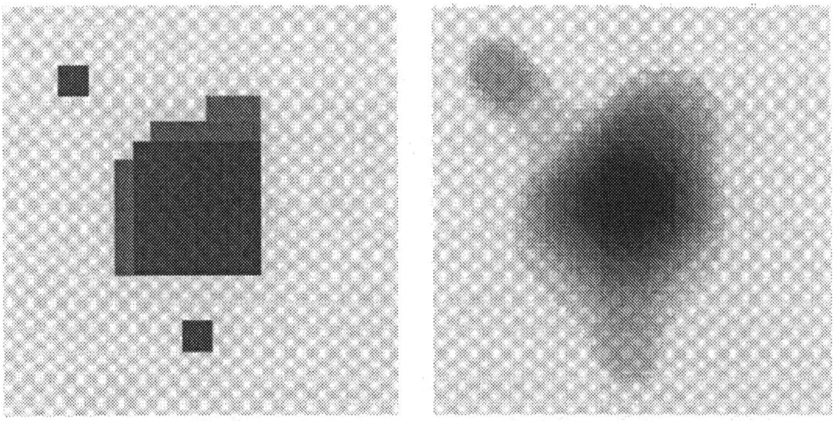

(a) (b)

Figure 3: The original distribution of diffusion coefficients (a) and their reconstructed values (b) derived from time-dependent measurements made around the periphery in response to short pulses introduced at the middle of each of the four sides.

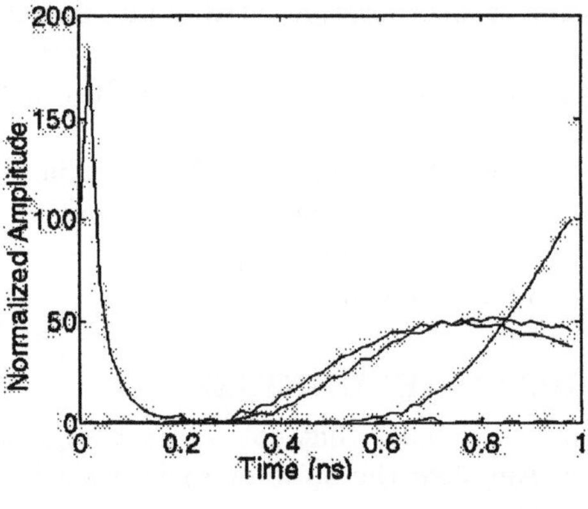

Figure 4: The simulated time-dependent measurements for a selected set of detectors presented as a function of time after a short pulse of light is introduced into the midpoint of the top edge of Fig. 3a. The first signal is from a detector located close to the source. The two signals that peak around 0.8 ns are from detectors positioned at the midpoint of the two sides and the late signal from the bottom side.

and then working toward finer resolutions [21, 22]. If this approach does not work for a particular problem, to find the global minimum it may be necessary to resort to stochastic optimization algorithms (simulated annealing or genetic algorithm), which are notoriously inefficient compared to gradient-based approaches. It might be possible to combine the best of both approaches through a hybrid algorithm in which the starting point of a gradient-based algorithm is chosen stochastically.

5.2. Deformable Geometric Models

It is often desirable to use a high-level model to describe an object or situation of interest. As an example, we have found it very useful to employ a deformable geometric model to represent the boundary of an object that we wish to reconstruct from projection data [21, 23]. In three dimensions, such a model consists of a surface, which might be represented by a grid of triangles [9, 24], as an example. An even higher-level model would use spline-based patches, which would result in fewer parameters, but not necessarily much less computation time. High-level models are often invoked to help regularize or control the inversion problem in the belief that it should be easier to solve a problem involving fewer variables. However, we have come to realize that it may be more desirable to use a very flexible description involving many parameters. The flexibility of such a model can be controlled through the use of a prior or graded constraint function, which effectively reduces the number of degrees of freedom of that model. Such constraints often take the form of an integral of the square of a derivative of some quantity, which basically acts to smooth that quantity. The advantage of this general approach is that it allows one to choose the prior that is most appropriate for the problem and even locally turn off the control when that is demanded by the data or circumstance [25].

5.3. Intermediate Elemental Representation

In constructing the Bayes Inference Engine [5], we have uncovered another basic tool for model building. In order to easily accommodate a variety of deformable geometric models in conjunction with a variety of potential measurement scenarios, we decided to employ an intermediate elemental representation for the object of interest. Every high-level model is converted to the elemental representation before the measurement process can be accomplished. For a physical object the elemental representation is a discretized density image (array of pixels) in two dimensions or a voxelated array in three dimension. This approach permits implementation of a new high-level object model without the necessity of writing a new code to calculate the result of each type of measurement.

Beyond its calculational advantage, we have come to recognize the underlying value of the elemental representation as a basic modeling tool. Its importance lies in the fact that when one uses and sees only a high-level model, deficiencies in matching the data can only be displayed in terms of the gradients of φ with respect to the model parameters. It can be difficult to recognize which aspects of the model do not accommodate the data. However, by displaying the gradients in the elemental representation, it can become evident how the high-level model needs to be changed, and possibly augmented, to better match the data [25].

5.4. Markov Chain Monte Carlo

One aspect of complex models that does not typically receive enough attention is the assessment of uncertainties in the parameters estimated from measurements. The probable reason for this neglect is that uncertainties are perceived to be difficult to estimate when there are numerous parameters. In a likelihood formulation the parameter uncertainties are described by the likelihood function. In a Bayesian formulation, they are described by the posterior probability distribution. The problem of assessing uncertainties amounts to characterizing the likelihood or the posterior. The Markov Chain Monte Carlo (MCMC) technique provides an effective means to characterize a probability distribution [26, 27, 28, 23]. MCMC generates a sequence of random samples of an arbitrary probability distribution. The only requirement of the basic algorithm is that one be able to evaluate the relative value of the probability density function for any set of parameter values, which in our context simply means doing the forward calculation to obtain ϕ plus the minus-log-prior.

This technique seems to be an essential tool to deal with the issue of uncertainties in estimated parameters. One valuable use of MCMC is to visualize the degree of model variation allowed by the uncertainties, which only requires that one has a way of displaying the model [29, 23]. Most useful is the ability to characterize the uncertainty in the model in whatever way one wishes, e.g., in terms of the variance in the parameters or of quantities derived from the model. In uncertainty assessment, it is crucially important to include the effects of correlations among the uncertainties, which MCMC does.

6. DISCUSSION

We have presented some useful tools that permit one to efficiently estimate parameters of a complicated forward model from measurements. The use of a forward model is important because many aspects of a model that describes a physical system and measurement scenario may not easily be directly inverted. This general approach allows one to construct complete models to fully account for the observations.

The primary technique in the toolkit is adjoint differentiation implemented according to the ADICT principle, which yields derivatives of a functional (objective function) based on a forward computational code with respect to all the parameters in the computational model. The derivatives of the computed functional are desirable when minimizing that functional. In the particular approach that we suggest, adjoint differentiation is accomplished through code rather than by storing derivative matrices. The resulting calculational time for the derivatives is comparable to that of the forward model calculation. ADICT can, in principle, be implemented for any code that is differentiable. Adjoint differentiation is being applied to some large simulation codes, e.g., to calculate the dynamics of the ocean and atmosphere.

ADICT permits the use of gradient-based optimization algorithms. We have had success with a quasi-Newton method of optimization and we therefore suggest using the limited-memory BFGS algorithm [16] for problems involving many parameters of mixed type. An important message that we wish to get across is that analysts should not necessarily be afraid of using models that contain large numbers of variables. With reasonable constraints on the models, one can easily accommodate many parameters. Our experience indicates that the penalties associated with using many parameters, either in terms of computational speed or ill posedness, are often not as severe as one might imagine.

Work is continuing on many fronts. The photon-diffusion work presented in Sect. 4. is progressing in several directions [30]. For example, we have demonstrated that it is feasible to reconstruction both D and μ_a from the kind of data used in this paper. Also, the diffusion simulation code has been extended to obtain three-dimensional reconstructions. These new results will be presented elsewhere [31]. Other work presently underway includes reconstruction of time-evolving 3D objects [24], improving the efficiency of the basic MCMC algorithm [32], and implementation of adjoint differentiation in a 2D hydrodynamics code [33, 7]. The current status of our developments may be found on the web: `http://bayes.lanl.gov/~kmh/publications.html`.

If one is going to rely on a simulation code to interpret data and make predictions, one needs to make certain that the simulation adequately matches real world behavior. This process is called validation [34]. While much needs to be learned about validation and how to go about it, we believe that the techniques presented here will play an important role in the methodology that evolves. Because the understanding of the uncertainties in predictions is a key component in validation, the MCMC or adjoint sensitivity techniques could become central elements in the validation process.

7. ACKNOWLEDGEMENTS

We have had helpful conversations with John Skilling, Steve Gull, Ralf Giering, Rudy Henninger, Maria Rightley, Paul Maudlin, and Bryan Travis. This work was supported by the United States Department of Energy under contract W-7405-ENG-36.

References

1. S. S. Saquib, K. M. Hanson, and G. S. Cunningham, "Model-based image reconstruction from time-resolved diffusion data," in *Medical Imaging: Image Processing*, K. M. Hanson, ed., *Proc. SPIE*, **3034**, pp. 369–380, 1997.
2. G. Burgers, R. Giering, and M. Fischer, "Construction of the adjoint of the HOPE OGCM," *Ann. Geophysicae.*, **C14**, p. 390, 1996.
3. F. C. Vossepoel, G. Burgers, and P. J. vanLeeuwen, "Radar altimeter data assimilation in an intermediate ENSO model," *Ann. Geophysicae.*, **C14**, p. 429, 1996.
4. W. C. Thacker, "Automatic differentiation from an oceanographer's perspective," in *Automatic Differentiation of Algorithms: Theory, Implementation, and Application*, A. Griewank and G. F. Corliss, eds., pp. 191–201, SIAM, Philadelphia, 1991.
5. K. M. Hanson and G. S. Cunningham, "A computational approach to Bayesian infer-

ence," in *Computing Science and Statistics* **27**, M. M. Meyer and J. L. Rosenberger, eds., pp. 202–211, Interface Foundation, Fairfax Station, VA 22039-7460, 1996.

6. R. J. Henninger, P. J. Maudlin, and E. N. Harstad, "Differential sensitivity theory applied to the MESA2D code for multi-material problems," in *Shock Compression of Condensed Matter*, S. C. Schmidt, et al., ed., pp. 283–287, Amer. Inst. Physics, Woodbury, NY, 1996.
7. R. J. Henninger, P. J. Maudlin, and M. L. Rightley, "Accuracy of differential sensitivity for one-dimensional shock problems," in *Shock Compression of Condensed Matter*, S. C. Schmidt, et al., ed., Amer. Inst. Physics, Woodbury, NY, 1998 (to be published).
8. M. L. J. Rightley, R. J. Henninger, and K. M. Hanson, "Adjoint differentiation of hydrodynamic codes," in *CNLS Newsletter*, Center for Nonlinear Studies, Los Alamos National Laboratory, Nov., 1997 (WWW: http://cnls.lanl.gov/Publications/newsletters.html).
9. X. L. Battle, G. S. Cunningham, and K. M. Hanson, "3D tomographic reconstruction using geometrical models," in *Medical Imaging: Image Processing*, K. M. Hanson, ed., *Proc. SPIE*, **3034**, pp. 346–357, 1997.
10. D. A. Harville, *Matrix Algebra from a Statistician's Perspective*, Springer, New York, 1997.
11. C. Bischof, A. Carle, P. Khademi, and A. Mauer, "The ADIFOR 2.0 system for the automatic differentiation of FORTRAN 77 programs," Tech. Rep. ANL-MCS-P481-1194, Argonne National Laboratory, 1995.
12. J. E. Horwedel, E. M. Oblow, B. A. Worley, and F. G. Pin, "GRESS 3.0 Gradient Enhanced Software System," Tech. Rep. PSR-231, RSIC Peripheral Shielding Routine Collection, Oak Ridge National Laboratory, 1994.
13. R. Giering, "Tangent linear and adjoint model compiler," Tech. Rep. TAMC 4.7, Max-Planck-Institut fuer Meteorologie, 1997 (e-mail: giering@dkrz.de).
14. W. C. Davidon, "Variable metric method for minimization," Tech. Rep. ANL-5990, Argonne National Laboratory, 1966.
15. R. Fletcher, "A new approach to variable metric algorithms," *Computer Journal*, **13**, pp. 317–322, 1970.
16. D. C. Liu and J. Nocedal, "On the limited memory BFGS methods for large scale optimization," *Math. Programming*, **45**, pp. 503–528, 1989.
17. S. G. Nash, "Preconditioning of truncated-Newton methods," *SIAM J. Sci. Stat. Comput.*, **6**, pp. 599–616, 1985.
18. S. G. Nash and J. Nocedal, "A numerical study of the limited memory BFGS method and the truncated-Newton method for large scale optimization," Tech. Rep. NAM 02, Northwestern University - Dep't. Elec. Eng. and Comp. Sci., 1989.
19. C. A. Bouman and K. Sauer, "A generalized Gaussian image model for edge-preserving MAP estimation," *IEEE Trans. on Image Proc.*, **2**, pp. 296–310, 1993.
20. S. S. Saquib, C. A. Bouman, and K. Sauer, "ML parameter estimation for Markov random fields, with applications to Bayesian tomography," Tech. Rep. TR-ECE 95-24, School of Electrical and Computer Engineering, Purdue University, 1995.
21. M. Kass, A. Witkin, and D. Terzopoulos, "Snakes: active contour models," *Inter. J. Comp. Vision*, **1**, pp. 321–331, 1988.
22. G. S. Cunningham, I. Koyfman, and K. M. Hanson, "Improved convergence of gradient-based reconstructions using multi-scale models," in *Image Processing*, M. H. Loew and K. M. Hanson, eds., *Proc. SPIE*, **2710**, pp. 145–155, 1996.
23. K. M. Hanson, G. S. Cunningham, and R. J. McKee, "Uncertainty assessment for

reconstructions based on deformable models," *Int. J. Imaging Systems and Technology*, **8**, pp. 506–512, 1997.
24. G. S. Cunningham, K. M. Hanson, and X. L. Battle, "Three-dimensional reconstructions from low-count SPECT data using deformable models," *Opt. Express*, 1998 (to be published).
25. K. M. Hanson, R. L. Bilisoly, and G. S. Cunningham, "Kinky tomographic reconstruction," in *Image Processing*, M. H. Loew and K. M. Hanson, eds., *Proc. SPIE*, **2710**, pp. 156–166, 1996.
26. J. Besag, P. Green, D. Higdon, and K. Mengersen, "Bayesian computation and stochastic systems," *Stat. Sci.*, **10**, pp. 3–66, 1995.
27. W. R. Gilks, S. Richardson, and D. J. Spiegelhalter, *Markov Chain Monte Carlo in Practice*, Chapman and Hall, London, 1996.
28. R. M. Neal, *Bayesian Learning for Neural Networks*, Springer, New York, 1996.
29. J. Skilling, D. R. T. Robinson, and S. F. Gull, "Probabilistic displays," in *Maximum Entropy and Bayesian Methods*, W. T. Grandy, Jr. and L. H. Shick, eds., pp. 365–368, Kluwer Academic, Dordrecht, 1991.
30. A. H. Hielscher, S. S. Saquib, and K. M. Hanson, "Model-based iterative image reconstruction for photon migration tomography," in *Methods for Solving Ill-Posed Inverse Imaging Problems: Medical and Nonmedical Applications*, R. L. Barbour, M. J. Carvlin, and M. A. Fiddy, eds., *Proc. SPIE*, **3171**, pp. 106–117, 1997.
31. A. H. Hielscher, A. Klose, D. M. Catarious, Jr., and K. M. Hanson, "Tomographic imaging of breast and brain tissue by time-resolved, model-based, iterative image reconstruction," in *Proc. Conf. on Advances in Optical Imaging and Photon Migration*, Optical Society of America, Orlando, 1998 (to be published).
32. K. M. Hanson and G. S. Cunningham, "Posterior sampling with improved efficiency," in *Medical Imaging: Image Processing*, K. M. Hanson, ed., *Proc. SPIE*, **3338**, 1998 (to be published).
33. R. J. Henninger, P. J. Maudlin, M. L. Rightley, and K. M. Hanson, "Application of forward and adjoint techniques to hydrocode sensitivity analysis," in *Proc. 9th Nuclear Explosives Code Developers' Conference*, F. Graviana, et al., ed., Lawrence Livermore National Laboratory, 1996 (proceedings are classified; article available from authors as report LA-CP-96-235.
34. W. L. Oberkampf, M. N. Sindir, and A. T. Conlisk, "Guide for the verification and validation of computational fluid dynamics simulations," Tech. Rep. G-077-98, Amer. Inst. Aeronautics and Astronautics, 1998.

MODEL COMPARISON WITH ENERGY CONFINEMENT DATA FROM LARGE FUSION EXPERIMENTS

R. PREUSS, V. DOSE AND W. VON DER LINDEN
Max-Planck-Institut für Plasmaphysik, EURATOM Association
D-85748 Garching b. München, Germany †

Abstract. Bayesian probability theory is applied to the model selection problem in describing the energy confinement of a fusion plasma by scaling laws. Here several possible physical scenarios lead to different models with a certain number of free parameters which are used to assign dimensional information to the scaling law. Using a recently assembled database of the W7-AS stellarator magnetic confinement device a dimensionally constrained analysis is performed. Probabilities for all models and the associated exponents of a power law scaling function are presented.

Key words: Model Comparison, Plasma Energy Confinement, Scaling Law Exponents.

1. Introduction

Controlled thermonuclear fusion is considered to be one of the most promising future energy resources. The supply with deuteron fuel contained in the oceans would last for eons. However, a ready to use power station is far from being accomplished. The technical difficulties in getting a plasma dense and hot enough to ignite and in confining it long enough to produce an energy gain are huge. Thus, temperatures in the order of tens of million degrees Kelvin are necessary to get a deuteron plasma burning. Due to the high temperatures the plasma has to be kept away from any wall material. One way to achieve this is by employing large magnetic fields to confine the plasma.

Various experimental reactors have been build to examine the physical laws of plasma behavior under such extreme conditions. One line is the stellarator type of machines in shape of a doughnut, where the confinement of the plasma is achieved by external magnetic fields alone. The data used in this article is taken from such a machine, the stellarator W7-AS. Here the plasma is characterized by the particle density n, the magnetic field B, the heating power P, the minor radius a and the rotational transform ι (which is the ratio of how often a magnetic field

†Email: preuss@ipp.mpg.de

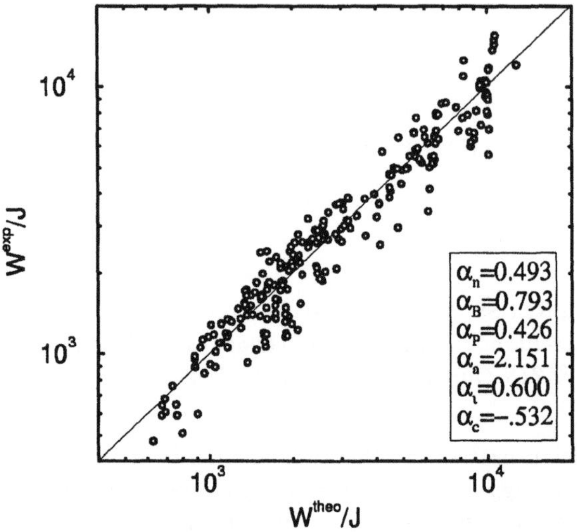

Figure 1. **Unconstrained least squares fit of the N = 250 data. The plasma energy content of a machine predicted according to this fit would be found along the diagonal line.**

line encircles the torus in poloidal and toroidal direction). Additional machine parameters like the major radius of the torus come only into play in comparisons with other stellarators of different size.

One decisive quantity in achieving a burning plasma is the energy content W which depends in a certain way on the above parameters. Since an exact description of the plasma is a highly complicated many particle problem which is impossible to solve, global scaling laws are used to determine W [1]. Introduced in the seventies scaling laws were of service for comparing different confinement regimes and inter-machine performances. Derived from small and mid sized machines, they were surprisingly successful in the performance prediction of much larger devices and are therefore of crucial importance in the planing of new machines (e.g. ITER).

Due to the fact that the exact plasma behavior is unknown one is restricted to assume a certain functional form where all the plasma determining parameters are fed into. The initial assumption of a power law dependence has survived up to now because of convenience and simplicity:

$$W_i = e^{\alpha_c} n_i^{\alpha_n} B_i^{\alpha_B} P_i^{\alpha_P} a_i^{\alpha_a} \iota_i^{\alpha_\iota} + \varepsilon_i \quad , \tag{1}$$

where W_i is the i-th measurement of the plasma energy content and ε_i its error. The constant factor has been expressed as e^{α_c} for calculational convenience.

We are out for the scaling exponents $\vec{\alpha}^* = (\alpha_n, \alpha_B, \alpha_P, \alpha_a, \alpha_\iota, \alpha_c)^T$. The simplest way to do this is to perform a least squares fit where $\tilde{\chi}^2 = \sum_{i=1}^{N}(W_i^{exp} - W_i^{theo})^2$ is minimized over $\vec{\alpha}^*$ (see Fig. 1). The obtained exponents would then be used in the plasma energy content predictions of new machines.

M_j	plasma model	degrees of freedom N_P	parameter vector $\vec{x}$
1	collisionless low-β	1	x
2	collisional low-β	2	$(x,y)^T$
3	collisionless high-β	2	$(x,z)^T$
4	collisional high-β	3	$(x,y,z)^T$
5	ideal fluid	1	x
6	resistive fluid	2	$(x,y)^T$

TABLE 1. Plasma models according to Connor and Taylor [2].

But this approach disregards conflicting dimensions of the quantities on the l.h.s. and the r.h.s. of Eq. (1). Moreover it ignores available information on the exponents from basic plasma theory. This was provided by Connor and Taylor [2] who explored the invariances of the basic equations of plasma behavior under similarity transformations and derived constraints on exponents of power law scaling expressions characteristic for different basic physics assumptions. The examined six models are listed in table 1 together with the number of the remaining free parameters. The implementation of the models leads to constraints on α_n, α_B, α_P and α_a, (further abbreviated as $\vec{\alpha}$). The number of degrees of freedom N_P in choosing $\vec{\alpha}$ have reduced according to the complexity of the physical model from originally $N_M = 4$ exponents α_i to one, two or three. The corresponding parameter vectors are also shown in table 1. As an example the explicit dependences of $\vec{\alpha}$ are shown for the third model M_3:

$$\begin{aligned} \alpha_n &= 1 - x - z \\ \alpha_B &= 2 - 3x \\ \alpha_P &= x \\ \alpha_a &= 4 - 4x - 2z \end{aligned} \qquad \vec{c} = \begin{pmatrix} 1 \\ 2 \\ 0 \\ 4 \end{pmatrix}, \qquad L = \begin{pmatrix} -1 & -1 \\ -3 & 0 \\ 1 & 0 \\ -4 & -2 \end{pmatrix}$$

with $\vec{\alpha} = \vec{c} + L\vec{x}$ and $\vec{x} = (x,z)^T$.

However, despite of the relative success in the past there is no reason to assume that a power scaling law represents the relevant plasma physics adequately [3]. Therefore we shall not require the power law exponents to satisfy the Connor/Taylor [CT] relations exactly. We therefore call our results dimensionally constrained as opposed to dimensionally exact.

In the following, Bayesian inference is employed to quantify the probability of each of the six CT models in the light of the data and to derive the respective exponent sets which approach the CT dimensional relation as close as possible at an as small as possible expense in data fit.

2. Model comparison

The probability of model M_j in the light of the data D and the prior information I is given using Bayes' theorem.

$$p(M_j|D,I) = \frac{p(M_j|I)p(D|M_j,I)}{p(D|I)} = \frac{p(M_j|I)p(D|M_j,I)}{\sum_k^{\#M} p(M_k|I)p(D|M_k,I)} \quad . \tag{2}$$

Since a priori no model is preferred, a constant uniform prior is assigned to each model and $p(M_j|I)$ cancels out

$$p(M_j|D,I) = \frac{p(D|M_j,I)}{\sum_k^{\#M} p(D|M_k,I)} \quad . \tag{3}$$

What remains to be determined is the global likelihood of the data, $p(D|M_j,I)$. This is achieved in marginalizing over all quantities which contribute to the problem in the joint probability, as there are the unknown error level represented by the hyperparameter ω, the exponent vector $\vec{\alpha}^*$ and its hyperparameter λ, and the parameter vector $\vec{x}$ and its hyperparameter μ:

$$p(D|M_j,I) = \int p(D,\omega,\vec{\alpha}^*,\lambda,\vec{x},\mu|M_j,I)\; d\omega d\vec{\alpha}^* d\lambda d\vec{x} d\mu \tag{4}$$

$$= \int d\omega d\lambda d\mu\; p(\omega|I)p(\lambda|I)p(\mu|I) \int d\vec{\alpha}^* d\vec{x}\; p(D,\vec{\alpha}^*,\vec{x}|\omega,\lambda,\mu,M_j,I) \quad . \tag{5}$$

Let us postpone the integrals over the hyperparameters. In the following we are concerned with the evaluation of the last of the upper integrals, which is

$$p(D|\omega,\lambda,\mu,M_j,I) = \int d\vec{x}\; p(\vec{x}|\mu,I) \int d\vec{\alpha}^*\; p(\vec{\alpha}^*|\vec{x},\lambda,M_j,I)p(D|\vec{\alpha}^*,\omega,M_j,I) \; . \tag{6}$$

Now the remaining quantities have to be assigned. We start with the likelihood function. Since we assume a Gaussian error statistic with constant absolute error, we get

$$p(D|\vec{\alpha}^*,\omega,I) = \left(\frac{\omega}{2\pi}\right)^{\frac{N}{2}} \exp\left\{-\frac{\omega}{2}\sum_{i=1}^{N}(D_i - W_i(\vec{\alpha}^*))^2\right\} \quad . \tag{7}$$

The hyperparameter ω is related to the noise level by $\langle\varepsilon^2\rangle = 1/\omega$. Here we identify the data vector D with the experimental measurement, and $W(\vec{\alpha}^*)$ with the theoretical power law.

(α_c, α_ι) are not determined by any model and therefore marginalized over. A direct integration is impossible and we therefore use the method of steepest descent to give

$$p(D|\vec{\alpha},\omega,I) \propto \left(\frac{\omega}{2\pi}\right)^{\frac{N-2}{2}} \exp\left\{-\frac{\omega}{2}\left(\tilde{\chi}_o^2 + \Delta\vec{\alpha}^T H \Delta\vec{\alpha}\right)\right\} \quad , \tag{8}$$

where a uniform distribution was assumed for the prior in $(\alpha_c,\ \alpha_\iota)$. Here and further on all the quantities which cancel out in the model comparison of Eq. 2 are dropped. H is the reduced Hesse matrix over the remaining $\vec{\alpha}$, and $\Delta\vec{\alpha} = \vec{\alpha} - \vec{\alpha}^{ML}$ with $\vec{\alpha}^{ML}$ the position of the minimum of $\tilde{\chi}^2$.

Next we assign the prior functions. Using finite power assumption that $\lambda^{-1} = \langle||\vec{\alpha} - \vec{c} - L\vec{x}||^2\rangle$ the principle of maximum entropy gives [4]

$$p(\vec{\alpha}|\lambda,\vec{x},M_j,I) = \left(\frac{\lambda}{2\pi}\right)^{\frac{N_M}{2}} \exp\left\{-\frac{\lambda}{2}||\vec{\alpha} - \vec{c} - L\vec{x}||^2\right\} \quad . \tag{9}$$

The same assumption is made to assign the prior for $\vec{x}$:

$$p(\vec{x}|\mu,M_j,I) = \left(\frac{\mu}{2\pi}\right)^{\frac{N_{P_j}}{2}} \exp\left\{-\frac{\mu}{2}\vec{x}^T\vec{x}\right\} \quad . \tag{10}$$

We now assume that the likelihood function is sharply peaked as compared to the prior $p(\vec{\alpha}|H,I)$ and that the prior in $\vec{\alpha}$ is sharply peaked as compared to the prior $p(\vec{x}|H,I)$ [5]. This is the normal case, where the measurement contains information considerably more detailed than the prior, and in the same way that the $\vec{\alpha}$ are far better determined than the parameters $\vec{x}$. Thereby an hierarchy in the hyperparameters

$$\omega \gg \lambda \gg \mu \tag{11}$$

is established. The integral in Eq. (6) may consequently be approximated by

$$p(D|\omega,\lambda,\mu,M_j,I) = p(\hat{\vec{x}}|\mu,M_j,I) \int d\vec{x}\ p(\hat{\vec{\alpha}}|\vec{x},\lambda,M_j,I) \int d\vec{\alpha}\ p(D|\vec{\alpha},\omega,I) \ , \tag{12}$$

where $\hat{\vec{\alpha}}$ and $\hat{\vec{x}}$ denote the vectors for which the likelihood $p(D|\vec{\alpha},M_j,I)$ and the prior $p(\hat{\vec{\alpha}}|\vec{x},\lambda,M_j,I)$ attain their maximum, respectively.

First we want to integrate over the remaining nonlinear parameters in the likelihood.

$$\int d\vec{\alpha}\ p(D|\vec{\alpha},\omega,I) \propto \left(\frac{\omega}{2\pi}\right)^{\frac{N-N_M-2}{2}} \exp\left\{-\frac{\omega}{2}\tilde{\chi}_o^2\right\} \quad , \tag{13}$$

with the maximum of $p(D|\vec{\alpha},\omega,I)$ at $\hat{\vec{\alpha}} = \vec{\alpha}^{ML}$.

The integration of the prior in $\vec{\alpha}$ is straight forward:

$$\int d\vec{x}\ p(\hat{\vec{\alpha}}|\vec{x},\lambda,M_j,I) = \frac{1}{\sqrt{\det(L^T L)}} \left(\frac{\lambda}{2\pi}\right)^{\frac{N_M-N_{P_j}}{2}} \exp\left\{-\frac{\lambda}{2}\Delta\vec{c}^T S \Delta\vec{c}\right\} \quad , \tag{14}$$

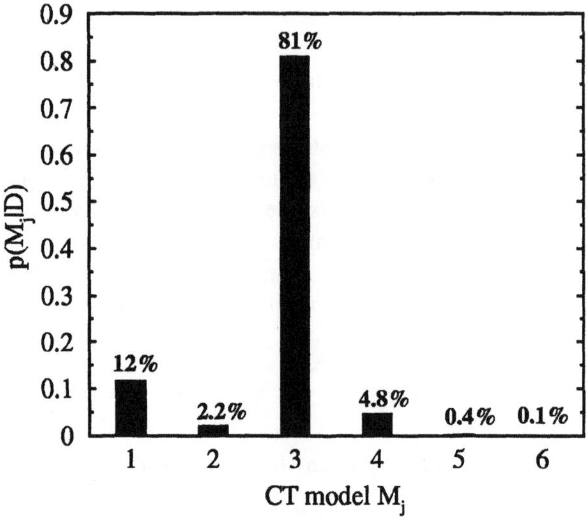

Figure 2. Probabilities for different CT models M_j.

with $S = \mathrm{I} - L\left(L^T L\right)^{-1} L^T$ and $\Delta\vec{c} = \vec{\alpha}^{ML} - \vec{c}$. Finally we need the maximum of $p(\vec{\alpha}|\vec{x}, \lambda, M_j, I)$ at $\hat{\vec{x}} = \left(L^T L\right)^{-1} L^T \Delta\vec{c}$ to state the prior in the parameter vector:

$$p(\hat{\vec{x}}|\mu, M_j, I) = \left(\frac{\mu}{2\pi}\right)^{\frac{N_{P_j}}{2}} \exp\left\{-\frac{\mu}{2}\Delta\vec{c}^T K \Delta\vec{c}\right\} \quad , \tag{15}$$

with $K = L\left(L^T L\right)^{-2} L^T$.

What is left to be done is the marginalization over the hyperparameters using Jeffreys' Prior. This gives our final result for the model comparison:

$$p(D|M_j, I) \propto \frac{\Gamma\left(\frac{N-N_M-2}{2}\right)}{\{\tilde{\chi}_o^2\}^{\frac{N-N_M-2}{2}}} \frac{\Gamma\left(\frac{N_M-N_{P_j}}{2}\right)}{\{\Delta\vec{c}^T S \Delta\vec{c}\}^{\frac{N_M-N_{P_j}}{2}}} \frac{\Gamma\left(\frac{N_{P_j}}{2}\right)}{\{\Delta\vec{c}^T K \Delta\vec{c}\}^{\frac{N_{P_j}}{2}}} \quad , \tag{16}$$

where the first term already lost all dependences on the models and is given only for reasons of completeness. As a result (see Fig. 2) we have that the collisionless high-β case is the most probable model to describe confinement in W7-AS. These results seem to indicate that collisions in the plasma are only of minor importance. Inclusion of collisions makes up the difference between models (1,2) and models (3,4). In both cases the inclusion is accompanied by an additional degree of freedom in parameter space. This additional flexibility is penalized automatically since it obviously does not lead to a better explanation of the data. The fluid models (5,6) are not supported by the data at all.

The results of Fig. 2 have been obtained with an approximation, which we call a mild one, namely a hierarchy in hyperparameters $\omega \gg \lambda \gg \mu$. Bayesian probability

M_j	1	2	3	4	5	6
$\langle\omega\rangle$			$9.4 \cdot 10^5$			
$\langle\lambda\rangle$	119	200	3065	1801	7.92	8.81
$\langle\mu\rangle$	0.19	0.32	0.72	0.86	0.66	0.26

TABLE 2. **Posterior expectation values of hyperparameters**

theory allows us to test a posteriori whether this assumption is justified. Let us consider the probability for the hyperparameters in the light of the data. Using Bayes' theorem we have

$$p(\omega, \lambda, \mu|D, M_j, I) = \frac{p(\omega, \lambda, \mu|M_j, I)p(D|\omega, \lambda, \mu, M_j, I)}{p(D|M_j, I)} \quad , \tag{17}$$

which allows us to calculate expectation values of the hyperparameters. The second term in the numerator of (17) is known from (6). The denominator is identical to our result for model comparison and is given by (16). For the first term in the numerator of (17) we note that knowledge of a particular model M_j does not imply any information on the values of the hyperparameters ω, λ, μ, that is, $p(\omega, \lambda, \mu|M_j, I)$ is logically independent of M_j. Considering further – from the prior information in I – that ω, λ, μ are scale parameters we assign Jeffreys' prior to $p(\omega, \lambda, \mu|M_j, I)$. Then we are ready to calculate

$$\langle\xi\rangle = \int d\omega\ d\lambda\ d\mu\ \xi\ p(\omega, \lambda, \mu|D, M_j, I) \tag{18}$$

where ξ stands for any of the three ω, λ, μ. The integration gives

$$\langle\omega\rangle = \frac{N - N_M - 2}{\tilde{\chi}_o^2} \quad , \tag{19}$$

$$\langle\lambda\rangle = \frac{N_M - N_{P_j}}{\Delta\vec{c}^T S \Delta\vec{c}} \quad , \tag{20}$$

$$\langle\mu\rangle = \frac{N_{P_j}}{\Delta\vec{c}^T K \Delta\vec{c}} \quad . \tag{21}$$

We can now turn to the results collected in table 2. $\langle\omega\rangle$ does of course not depend on the choice of model since it is entirely specified by the likelihood function (8). $\langle\lambda\rangle$ and $\langle\mu\rangle$ on the other hand do depend on the chosen model. We find that in all six cases the assumption $\omega \gg \lambda \gg \mu$ is excellent. This confirms that the approximation $\omega \gg \lambda \gg \mu$ which we made in arriving at (6) is very well justified for the present data set.

Finally the scaling exponents are calculated. We consider

$$\langle\Delta\vec{\alpha}\rangle = \int d\omega\ d\lambda\ d\mu\ p(\omega, \lambda, \mu|\vec{W}, M_j, I) \cdot \langle\Delta\vec{\alpha}\rangle_{\omega,\lambda,\mu} \quad . \tag{22}$$

	α_n	α_B	α_P	α_a
MC	0.499	0.781	0.412	2.183
stdev	0.025	0.047	0.024	0.081
EB	0.499	0.779	0.410	2.184
stdev	0.024	0.046	0.020	0.069

TABLE 3. Comparison of the results for the collisionless high-β model. MC: numerical integration (Monte Carlo); EB: empirical Bayes result.

where we first have to calculate $\langle\Delta\vec{\alpha}\rangle_{\omega,\lambda,\mu}$ given by

$$\langle\Delta\vec{\alpha}\rangle_{\omega,\lambda,\mu} = \frac{\int d\vec{\alpha}\ \Delta\vec{\alpha}\ p(\vec{\alpha}|\omega,\lambda,\mu,D,M_j,I)}{\int d\vec{\alpha}\ p(\vec{\alpha}|\omega,\lambda,\mu,D,M_j,I)} \tag{23}$$

After some algebra we get

$$p(\vec{\alpha}|\omega,\lambda,\mu,D,M_j,I) \propto \exp\left\{-\frac{1}{2}\left(\Delta\vec{\alpha}-\vec{\alpha}_o\right)^T\left(\omega H+\lambda S\right)\left(\Delta\vec{\alpha}-\vec{\alpha}_o\right)\right\} = G(\vec{\alpha}), \tag{24}$$

with $\Delta\vec{\alpha} = \vec{\alpha} - \vec{\alpha}^{ML}$ and $\vec{\alpha}_o = -\lambda(\omega H + \lambda S)^{-1}S\Delta\vec{c}$. From the fact that G is a Gaussian centered at $\Delta\vec{\alpha} = \vec{\alpha}_o$ we have

$$\int d\vec{\alpha}\ \left(\Delta\vec{\alpha}-\vec{\alpha}_o\right)G(\vec{\alpha}) = 0 = \int d\vec{\alpha}\ \Delta\vec{\alpha}G(\vec{\alpha}) - \vec{\alpha}_o\int d\vec{\alpha}\ G(\vec{\alpha}) \quad , \tag{25}$$

and therefore, since the factor of $\vec{\alpha}_o$ is just the normalization in (23),

$$\langle\Delta\vec{\alpha}\rangle_{\omega,\lambda,\mu} = \vec{\alpha}_o = -\lambda(\omega H+\lambda S)^{-1}S(\vec{\alpha}^{ML}-\vec{c}) \quad . \tag{26}$$

Now that we know $\langle\Delta\vec{\alpha}\rangle_{\omega,\lambda,\mu}$ we have finally to marginalize over the hyperparameters in (22). This is a very complicated integral which can only be done numerically. We employ Monte Carlo integration with importance sampling. This allows us also to check a frequently employed conceptual approximation to the strict straight forward Bayesian theory. In this approximation it is assumed that integrating over a hyperparameter is equivalent to estimating that hyperparameter from the data and then constraining it in the posterior distribution to that value [4]. This procedure, also called the empirical Bayesian estimate, is believed to perform best when many well conditioned data are available which in turn allow robust estimates of the hyperparameters. Empirical Bayes (EB) parameter estimates are included in table 3 and were obtained by substitution of (19), (20) and (21) into (26). Comparison of the parameter estimates to the exact Monte Carlo results shows that the empirical Bayes estimate performs extremely well in the present case.

3. Conclusion

In this paper we have addressed the question of dimensionally constrained energy confinement analysis for a set of W7-AS data. We distinguish between dimensionally exact and dimensionally constrained, since we have no rigorous reason at all to assume that plasma behavior may be adequately represented by a power law. In our Bayesian analysis we found that the W7-AS plasma is best described by the collisionless high-β Connor/Taylor model conditional on the assumption that the confinement function is of the power law type. Let us mention in this context that Connor and Taylor never claimed a single power law term whose exponents satisfy dimensional constraints as a function representing plasma confinement. In fact they express a general dimensionless function in a series of dimensionally exact power law terms. We are currently exploring this route.

References

1. J. Hugill and J. Sheffield *Nucl. Fusion*, **18**, p. 15, 1978.
2. J. W. Connor and J. B. Taylor *Nucl. Fusion*, **17**, p. 1047, 1977.
3. R. J. Goldston *Plasma Physics and Controlled Nuclear Fusion*, **26**, p. 87, 1984.
4. G. L. Bretthorst *J. Magn. Reson.*, **88**, p. 552, 1990.
5. P. C. Gregory and T. J. Loredo *Astrophys. J.*, **398**, p. 146, 1992.

DECONVOLUTION BASED ON EXPERIMENTALLY DETERMINED APPARATUS FUNCTIONS

V. DOSE, R. FISCHER AND W. VON DER LINDEN
Max-Planck-Institut für Plasmaphysik, EURATOM Association
POB 1533, D-85740 Garching, Germany†

Abstract.
Deconvolution of experimental measurements in e.g. electron or ion scattering can result in considerable resolution enhancement. It is usually assumed that the apparatus function which we wish to remove from the experimental signal is either known exactly or with a much higher precision than the signal. This assumption is not valid in general. In fact many situations are conceivable where measurement of the apparatus function requires the same effort as measurement of the signal. We have performed a rigorous Bayesian analysis for this general case and present applications to Rutherford backscattering from thin films.

Key words: Deconvolution, Inverse Problem, Apparatus Function, Likelihood, Image Processing

Deconvolution remains at the center of interest in the analysis of experimental data. Increasing the instrumental resolution to the desired value may not be possible, too cumbersome or expensive. The question then arises whether a sufficiently accurate measurement together with a sufficiently precise knowledge of the apparatus function suffices to arrive at the desired resolution by appropriate deconvolution procedures. The problem is old and has so far withstood a sustainable solution.

In this paper we shall choose for purposes of illustration the particular experiment of Rutherford backscattering (RBS). RBS is a surface analytical technique whose importance derives from its quantitative nature. Consider a beam of ions of mass m_0, usually either protons or He nuclei with well defined energy E_0 in the range of some MeV incident on a target of atomic mass number M_B with a thin film overlayer M_A. We shall assume - in line with the actual data - that M_A is very much larger than M_B. Ions undergoing an elastic Coulomb collision with a deflection by an angle Θ are recorded in a solid state detector. Since the projectile target interaction is coulombic, the cross section associated with the scattering is the Rutherford cross section, which is quantitatively known. All backscattered ions have energy $E_1 < E_0$. E_1 depends on E_0, m_0, $M_{A,B}$, Θ and is, as a function of

†Email: rainer.fischer@ipp.mpg.de

target mass, larger for scattering from M_A than for scattering from M_B since we have assumed $M_A >> M_B$. A typical situation is shown schematically in fig. 1.

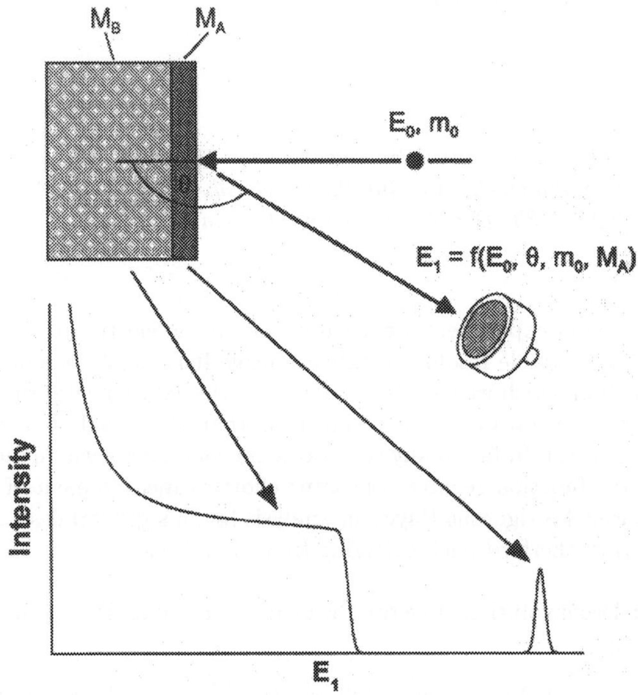

Figure 1. **A schematic of a Rutherford backscattering (RBS) experiment.**

Scattering from species A which forms only a thin film on top of the bulk substrate material B results therefore in a spectral spike slightly broadened by the apparatus transfer function while scattering from species B produces a step function. The high energy side of this step results from scattering from the topmost B-layer. Particles which have penetrated to some depth into material B before undergoing large angle scattering loose energy on their way into the scattering center and out to the detector and appear in the spectrum at progressively lower energies E_1. Note that the apparatus broadening is reflected in the rounded leading edge of the B-species spectrum.

Let us now turn to the various mechanisms contributing to the overall apparatus function. Numbers given for the various contributing effects are best estimates for the particular experimental setup which we used to obtain the data we are going to analyze here. The energy chosen for the analysis of Cu and Co films on a silicon substrate was 2.6 MeV with full width at half maximum of 13 keV. The solid state detector resolution at $E_1 \approx 1$ MeV is 15 keV wide. The electronic noise amounts to 5 keV while the contributions from detector solid angle, beam spot size and finite film thickness are estimated to be 3 keV. The resulting overall resolution is then approximately 20.7 keV. Data from a thin cobalt film on a silicon substrate

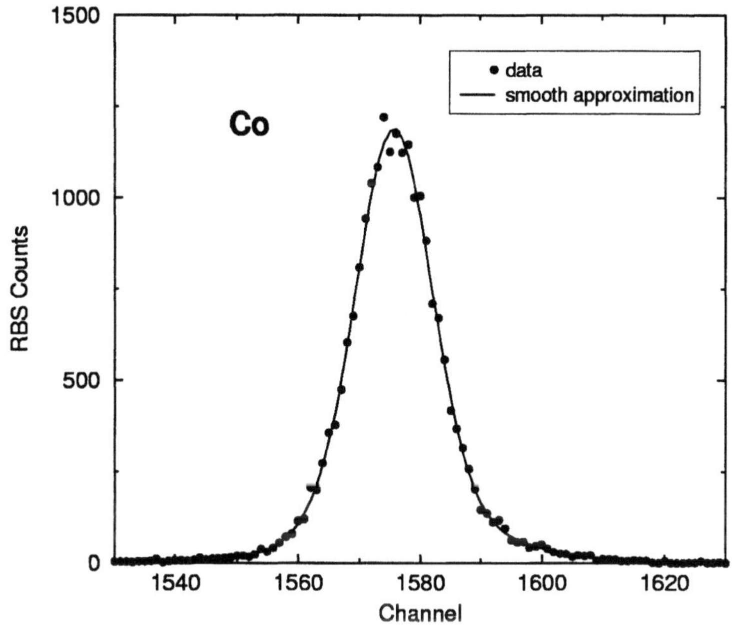

Figure 2. RBS spectrum from a thin cobalt layer on a silicon substrate using 2.6 MeV He^+ ions. Since cobalt is a single isotope element the measurement represents the apparatus function.

displayed in fig. 2 are comparable with this estimate and yield the more precise overall resolution of 19 keV (FWHM). The thin continuous line running through the data points is a parameter free smooth density estimate of the experimental data shown as full dots [1]. The target material cobalt was chosen because it is a single isotope element. The Rutherford backscattering spectrum from a monolayer of cobalt obtained with an ideal apparatus is accordingly a δ-spike, the real data in fig. 2 represent consequently the overall apparatus function.

We shall now proceed to write down the likelihood function . The data vector $\boldsymbol{d}$ (the i-th component of $\boldsymbol{d}$ corresponds to the i-th energy channel) is then modeled as

$$\boldsymbol{d} = \boldsymbol{A}\,\boldsymbol{f} + \boldsymbol{N_d}, \quad < N_{d_i}^2 >= \sigma_i^2. \tag{1}$$

where $\boldsymbol{A}$ is the apparatus broadening matrix , $\boldsymbol{f}$ the desired spectroscopic feature and $\boldsymbol{N_d}$ the vector of noise deteriorating $\boldsymbol{d}$. According to the principle of maximum entropy we then have

$$P(\boldsymbol{d}|\boldsymbol{f},\boldsymbol{A},\boldsymbol{\sigma},I) = \frac{1}{\prod_{i=1}^{N_d}\sqrt{2\pi\sigma_i^2}} \exp\left[-\frac{1}{2}\sum_i\left(\frac{d_i-(\boldsymbol{A}\,\boldsymbol{f})_i}{\sigma_i}\right)^2\right]. \tag{2}$$

Since we are dealing with a counting experiment the diagonal matrix $\boldsymbol{\sigma}$ is easily estimated from the measurement. If we assume, that the apparatus matrix $\hat{\boldsymbol{A}}$ is also known for example as the continuous curve in fig. 2, then the desired spectral distribution $\boldsymbol{f}$ results from Bayes theorem

$$P(\boldsymbol{f}|\boldsymbol{d},\hat{\boldsymbol{A}},\boldsymbol{\sigma},I) = \frac{P(\boldsymbol{d}|\boldsymbol{f},\hat{\boldsymbol{A}},\boldsymbol{\sigma},I)\,P(\boldsymbol{f}|I)}{P(\boldsymbol{d}|\hat{\boldsymbol{A}},\boldsymbol{\sigma},I)}. \tag{3}$$

We have offered two powerful procedures to obtain the answer to Eqn. (3) at the 1996 conference on maximum entropy and Bayesian methods [2,3]. One is the adaptive kernel method and the other the spline based adaptive resolution image reconstruction. Since both are well documented in the last years proceedings we shall abandon a discussion of the actual inverse problem (3) but rather consider the effects, if we replace the assumed known apparatus matrix by the real world pointwise measurement shown in fig. 2 as full dots. Let $\boldsymbol{A}$ represent this matrix and let a_{ij} be its elements. We then have to calculate $P(\boldsymbol{d}|\boldsymbol{f},\boldsymbol{A},\boldsymbol{\sigma},\boldsymbol{\Delta},I)$ by first introducing the true apparatus function $\hat{\boldsymbol{A}}$ and subsequently integrating it out,

$$P(\boldsymbol{d}|\boldsymbol{f},\boldsymbol{A},\boldsymbol{\sigma},\boldsymbol{\Delta},I) = \int d\,\hat{a}_{ij}\;P(\boldsymbol{d}|\boldsymbol{f},\hat{\boldsymbol{A}},\boldsymbol{\sigma},\boldsymbol{\Delta},I)\,P(\hat{\boldsymbol{A}}|\boldsymbol{A},\boldsymbol{\Delta},I), \tag{4}$$

where we have suppressed all logically irrelevant conditional information. $\boldsymbol{\Delta}$ is the matrix of standard deviations of the measurement of a_{ij} and again easily estimated since we deal with a counting experiment. Accordingly the prior distribution $P(\hat{\boldsymbol{A}}|\boldsymbol{A},\boldsymbol{\Delta},I)$ is

$$P(\hat{\boldsymbol{A}}|\boldsymbol{A},\boldsymbol{\Delta},I) \propto \exp\left[-\frac{1}{2}\sum_{ij}\left(\frac{\hat{a}_{ij}-a_{ij}}{\Delta_{ij}}\right)^2\right]. \tag{5}$$

Since both factors in the integral (4) are quadratic forms in the variables $\hat{a}_{ij}$ we proceed in the usual way: find the maximum and complete the square. This is somewhat massy but elementary. Assuming further that the likelihood function (2) is sufficiently localized such that all integrations can be extended to $-\infty < \hat{a}_{ij} < \infty, \forall i,j$ we arrive at

$$P(\boldsymbol{d}|\boldsymbol{f},\boldsymbol{A},\boldsymbol{\sigma},\boldsymbol{\Delta},I) = \frac{1}{\prod_{i=1}^{N_d}\sqrt{2\pi\sigma_{\text{eff},i}^2}}\, exp\left(-\frac{1}{2}\sum_{i=1}^{N_d}\frac{(d_i-(\boldsymbol{A}\,\boldsymbol{f})_i)^2}{\sigma_{\text{eff},i}^2}\right). \tag{6}$$

The most important difference between (6) and (2) is the replacement of $\boldsymbol{\sigma}$ by $\boldsymbol{\sigma}_{\text{eff}}$, again a diagonal matrix given by

$$\sigma_{\text{eff},i}^2 = \sigma_i^2 + \sum_{j=1}^{N}\Delta_{ij}^2 f_j^2. \tag{7}$$

with σ_i^2 being the contribution to the likelihood variances from the data measurement and $\sum_{j=1}^{N} \Delta_{ij}^2 f_j^2$ being the contribution from the finite precision of the measurement of the apparatus matrix $\boldsymbol{A}$.

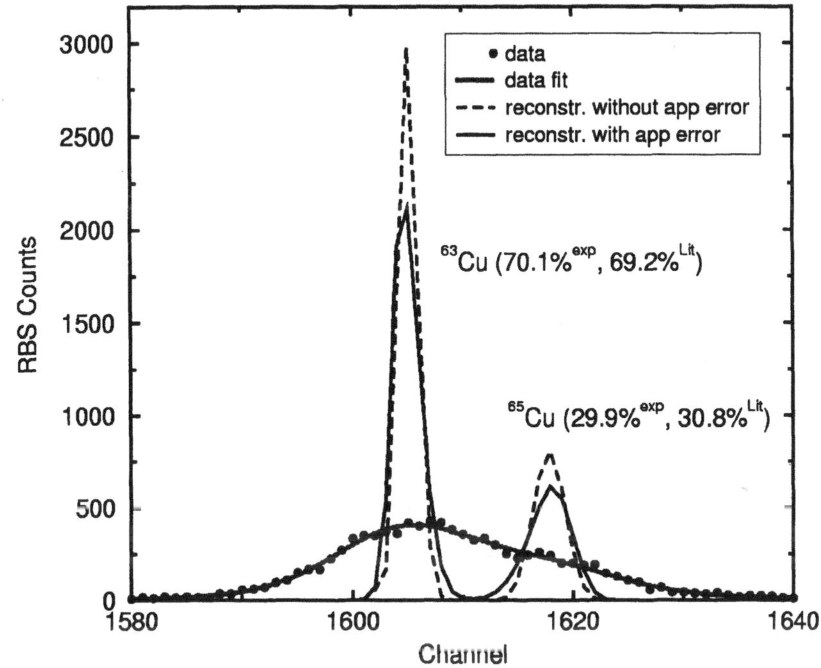

Figure 3. RBS spectrum from a thin copper layer on silicon using 2.6 MeV He^+ ions. The primary data do not separate the isotopes at $A_1 = 63$ amu and $A_2 = 65$ amu. Reconstruction using the smooth and the pointwise function of fig. 2 are given. Note the excellent agreement of the isotope abundances in comparison to tabulated data.

The data which we have processed were obtained from a thin copper film deposited again on silicon and are shown in fig. 3. Copper unlike cobalt has two different isotopes $A_1 = 63$ amu with a natural abundance of 69.2% and $A_2 = 65$ amu with a natural abundance of 30.8%. The unprocessed RBS spectrum of the copper thin film shown as full dots in fig. 3 does not resolve these two isotopes. The dashed line is obtained upon assuming that the apparatus function is identical to the smooth curve approximation of the measurement shown in fig. 2 using (3). We see a dramatic gain in resolution. The continuous curve through the data points represents the data fit. If we now replace the likelihood function (2) by the likelihood function (6) which is based on the pointwise measured cobalt spectrum in fig. 2 we obtain the continuous peaks and obviously, we have lost resolution. The difference between the results based on (2) and (6) respectively can easily be reconciled recalling that the variances of apparatus function $\boldsymbol{A}$ and spectral feature $\boldsymbol{f}$ add up under the convolution operation to the variance of the signal $\boldsymbol{d}$. Thus

$$\text{var}(\boldsymbol{d}) = \text{var}(\boldsymbol{A}) + \text{var}(\boldsymbol{f}). \tag{8}$$

The question arises which measurement accuracy is sufficient for the desired resolution. For the sake of arguments we assume that the variances of the peaks of the data $\boldsymbol{d}$, the apparatus function $\boldsymbol{A}$ and the image $\boldsymbol{f}$ are well described by their widths W_d, W_A and W_f. With $W_f = \alpha \times W_d$, where α is the inverse of the resolution enhancement, $\alpha < 1$, eqn. 8 transforms into

$$\frac{W_A}{W_d} \approx (1-\alpha^2)^{\frac{1}{2}} \approx 1-\epsilon \tag{9}$$

From conservation of the area of the images under convolution we estimate $\alpha = 1/7.5$ leading to $\epsilon = 0.009$ for the dashed line spectrum in fig. 3 and $\alpha = 1/5.4$ with $\epsilon = 0.017$ for the continuous line spectrum in fig. 3. The difference in the width of the images based on either (2) or (6) translates accordingly into a difference in width of the data of 0.8%. This is exactly the uncertainty of the width of the pointwise given apparatus function in fig. 2 if we make a quick and dirty calculation of its second moment and the error thereof. But it is not only the difference in width that matters, it is also necessarily the pointwise apparatus function which produces the smaller resolution enhancement since the smooth curve representation is always one with as low a curvature as possible when obeying the data constraints. Its width represents therefore systematically an upper limit to the width of the true apparatus function, thereby overdoing the deconvolution.

In addition, ϵ is a measure for the width accuracies necessary for the desired resolution. For an ideal measurement of a δ-peak $\frac{W_d}{W_A} = 1$ whereas for a real measurement $\frac{\Delta W_d}{W_A} = \epsilon$. For a resolution enhancement of 10 (100) the joint uncertainty of the widths of the data and the apparatus function has to be smaller than 0.5% (0.005%). This crude assessment reveals the limits of appropriate deconvolution procedures in cases where we deal with noisy data.

References

1. R. Fischer, M. Mayer, W. von der Linden, and V. Dose, "Enhancement of the energy resolution in ion-beam experiments with the maximum-entropy method," *Phys. Rev. E*, **55**, p. 6667, 1997.
2. R. Fischer, W. von der Linden, and V. Dose, "Adaptive kernels and occam's razor in inversion problems," in *MAXENT96 - Proceedings of the Maximum Entropy Conference 1996*, M. Sears, V. Nedeljkovic, N. E. Pendock, and S. Sibisi, eds., p. 21, NMB Printers, Port Elizabeth, South Africa, 1996.
3. W. von der Linden, V. Dose, and R. Fischer, "How to separate the signal from the background," in *MAXENT96 - Proceedings of the Maximum Entropy Conference 1996*, M. Sears, V. Nedeljkovic, N. E. Pendock, and S. Sibisi, eds., p. 146, NMB Printers, Port Elizabeth, South Africa, 1996.

A BAYESIAN APPROACH FOR THE DETERMINATION OF THE CHARGE DENSITY FROM ELASTIC ELECTRON SCATTERING DATA

A. Mohammad-Djafari
Laboratoire des Signaux et Systèmes (CNRS-ESE-UPS),
Plateau de Moulon, 91192 Gif-sur-Yvette, France
and
H. G. Miller
Department of Physics, University of Pretoria,
Pretoria 0002, South Africa

Abstract. The problem of the determination of the charge density from limited information about the charge form factor is an ill-posed inverse problem. A Bayesian probabilistic approach to this problem which permits to take into account both errors and prior information about the solution is presented. We will show that many classical methods can be considered as special cases of the proposed approach. We address also the problem of the basis function choice for the discretization and the uncertainty of the solution. Some numerical results for an analytical model are presented to show the performance of the proposed method.

1. Introduction

Elastic electron scattering provides a mean of determining the charge density of a nucleus, $\rho(r)$, from the experimentally determined charge form factor, $F(q)$. The connection between the charge density and the cross section is well understood and in plane wave Born approximation $F(q)$ is just the Fourier transform of $\rho(r)$ which for the case of even-even nuclei, which we shall consider, is simply given by

$$F(q) = 4\pi \int_0^{\infty} r^2 \, J_0(qr)\rho(r) \, \mathrm{d}r \tag{1}$$

where J_0 is the spherical Bessel function of zero order and q is the absolute value of the three momentum transfer. Given that the experimental measurements are performed over a limited range at a finite number of values of the momentum transfer q, a unique determination of $\rho(r)$ is not possible since the resulting inverse problem is ill posed.

One of the generally accepted procedures for determining $\rho(r)$ is to expand it in a basis and then determine the expansion coefficients from a least squares (LS) fit to the experimentally measured values of $F(q)$ [1, 2, 3, 14, 18]. The following questions then arise: how to choose a basis and how to determine the order of the expansion? Another problem with the LS methods is that increasing the number of terms in the expansion generally leads to unphysical oscillations in the charge density in spite of the fact that the charge form factor is well reproduced at the experimentally determined values of q [4, 13]. Finally, due to the fact that the problem is inherently ill posed, a small error in the data (experimental errors or measurement noise) will produce large variations in the solution which is not acceptable in practical situations.

What we are going to do is to show how a Bayesian approach can be helpful to give both correct and reasonable answers to the aforementioned questions and to propose new methods which are more stable with respect to the errors and finally to give procedures to put the correct error bars on the proposed solutions.

2. Fundamentals of the Bayesian approach

Let us start by discretizing the problem in the usual manner by expanding $\rho(r)$ in a basis $\phi_n(r)$:

$$\rho(r) = \begin{cases} \sum_{n=1}^{N} a_n \phi_n(r) & r \leq R_c \\ 0 & r > R_c \end{cases} \tag{2}$$

and substituting it in (1) yields

$$\begin{aligned} F(q) &= 4\pi \int_0^{R_c} r^2 J_0(qr) \sum_{n=1}^{N} a_n \phi_n(r) \, \mathrm{d}r \\ &= 4\pi \sum_{n=1}^{N} a_n \int_0^{R_c} r^2 J_0(qr) \phi_n(r) \, \mathrm{d}r \end{aligned} \tag{3}$$

Now, defining

$$A_{m,n} = 4\pi \int_0^{R_c} r^2 J_0(q_m r) \phi_n(r) \, \mathrm{d}r \tag{4}$$

we obtain

$$\boldsymbol{F}_c = \boldsymbol{A}\boldsymbol{a} + \boldsymbol{\epsilon} \tag{5}$$

where $\boldsymbol{a}$ is a vector containing the coefficients $\{a_n, n = 1, \cdots, N\}$, $\boldsymbol{F}_c$ is a vector containing the form factor data $\{F_c(q_m), m = 1, \cdots, M\}$ and A an $(M \times N)$ matrix containing the coefficients $A_{m,n}$ given by (4). Note also that when the vector $\boldsymbol{a}$ is determined, we can calculate $\boldsymbol{\rho} = \{\rho(r_k), k = 1, \cdots, K\}$ by

$$\boldsymbol{\rho} = \boldsymbol{\Phi}\boldsymbol{a} \tag{6}$$

where $\boldsymbol{\Phi}$ is a $K \times N$ matrix with the elements $\Phi_{kn} = \phi_n(r_k)$.

The vector $\boldsymbol{\epsilon}$ is added to take account of the errors in both measurement noise and due to discretization. We assume that the components $\{\epsilon_m, m = 1, \cdots, M\}$ are additive, zero mean (no systematic error), mutually independent (no correlation) and independent of $\boldsymbol{a}$, and they can only be characterized by their common variance σ_ϵ^2. This hypothesis is reasonable unless we know more about its characteristics.

Note that we have not yet discussed the choice of the basis functions ϕ_n and the determination of the expansion order, N. We will come back to these questions later. Let us now see how the Bayesian estimation approach works.

The main idea behind the Bayesian probabilistic approach is to represent the uncertainty or any lack of knowledge or any diffuse prior knowledge about a quantity by a probability law. For example, the knowledge (or the hypothesis) that $\{\epsilon_m, m = 1, \cdots, M\}$ are zero mean, mutually independent and that they are only characterized by their common variance σ_ϵ^2 can be described by choosing a Gaussian probability distribution for them. One may also use the Maximum Entropy Principle to enforce this choice. This means that we can write

$$p(\epsilon_m) = \frac{1}{(2\pi\sigma^2)^{1/2}} \exp\left\{-\frac{1}{2\sigma_\epsilon^2}\epsilon_m^2\right\} \tag{7}$$

or $p(\epsilon_m) = \mathcal{N}(0, \sigma_\epsilon^2)$ and

$$p(\epsilon) = \prod_{m=1}^{M} p(\epsilon_m) = \frac{1}{(2\pi\sigma^2)^{M/2}} \exp\left\{-\frac{1}{2\sigma_\epsilon^2}\|\epsilon\|^2\right\} \tag{8}$$

or simply $p(\epsilon) = \mathcal{N}(0, \sigma_\epsilon^2 \boldsymbol{I})$ where $\boldsymbol{I}$ is the $(M \times M)$ unitary matrix and σ_ϵ^2 is the common variance of ϵ_m for all m. Now, using this model (5), we can define the conditional probability law

$$p(\boldsymbol{F}_c|\boldsymbol{a}) = \frac{1}{(2\pi\sigma_\epsilon^2)^{M/2}} \exp\left\{-\frac{1}{2\sigma_\epsilon^2}\|\boldsymbol{F}_c - \boldsymbol{A}\boldsymbol{a}\|^2\right\} \tag{9}$$

or $p(\boldsymbol{F}_c|\boldsymbol{a}) = \mathcal{N}(\boldsymbol{A}\boldsymbol{a}, \sigma_\epsilon^2 \boldsymbol{I})$. It is usual to call $p(\boldsymbol{F}_c|\boldsymbol{a})$ or its logarithm, considered as a function of $\boldsymbol{a}$, the *Likelihood*.

One can stop here and define the solution of the problem (5) as the vector $\widehat{\boldsymbol{a}}$, which maximizes the likelihood (ML):

$$\widehat{\boldsymbol{a}} = \arg\max_{\boldsymbol{a}} \{p(\boldsymbol{F}_c|\boldsymbol{a})\} \tag{10}$$

or equivalently

$$\widehat{\boldsymbol{a}} = \arg\min_{\boldsymbol{a}} \{-\ln p(\boldsymbol{F}_c|\boldsymbol{a})\} \tag{11}$$

which, in the case of a Gaussian distribution (9) becomes

$$\widehat{\boldsymbol{a}} = \arg\min_{\boldsymbol{a}} \{\|\boldsymbol{F}_c - \boldsymbol{A}\boldsymbol{a}\|^2)\} \tag{12}$$

and we find here the LS solutions given by:

$$(\boldsymbol{A}^t \boldsymbol{A})\,\widehat{\boldsymbol{a}} = \boldsymbol{A}^t \boldsymbol{F}_c \tag{13}$$

The main problem with these solutions is that, very often, the matrix $\boldsymbol{A}^t\boldsymbol{A}$ is either singular or at least ill-conditionned, so that the solutions are very sensitive to the errors in the data or even on the roundoff errors during the numerical calculation.

The Bayesian approach can do better. In fact, before looking at the data, we may have some prior knowledge about $\boldsymbol{a}$. For example, we know that $\boldsymbol{a} \in \mathbb{R}^N$ and that we might prefer those vectors which have a minimal norm $\|\boldsymbol{a}\|$. To translate this prior knowledge and this preference, we may assign a Gaussian probability distribution to the vector $\boldsymbol{a}$:

$$p(\boldsymbol{a}) = \frac{1}{(2\pi\sigma_a^2)^{M/2}} \exp\left\{-\frac{1}{2\sigma_a^2}\|\boldsymbol{a}\|^2\right\} \tag{14}$$

where σ_a^2 gives an idea about the scale of the norm of the vector $\boldsymbol{a}$. Now, using the Bayes rule we can calculate the posterior probability law

$$p(\boldsymbol{a}|\boldsymbol{F}_c) = \frac{p(\boldsymbol{F}_c|\boldsymbol{a})\, p(\boldsymbol{a})}{p(\boldsymbol{F}_c)} \tag{15}$$

where the denominator

$$p(\boldsymbol{F}_c) = \int p(\boldsymbol{F}_c|\boldsymbol{a})\, p(\boldsymbol{a})\, \mathrm{d}\boldsymbol{a} \tag{16}$$

is the normalization constant (called sometimes the *Evidence*) [15].

This posterior law contains all the information we may wish about the solution. For example, we may want to know what is the probability that $\underline{\boldsymbol{a}} < \boldsymbol{a} \leq \overline{\boldsymbol{a}}$. This can be calculated by

$$P(\underline{\boldsymbol{a}} < \boldsymbol{a} \leq \overline{\boldsymbol{a}}) = \int_{\underline{\boldsymbol{a}}}^{\overline{\boldsymbol{a}}} p(\boldsymbol{a}|\boldsymbol{F}_c)\, \mathrm{d}\boldsymbol{a} \tag{17}$$

Or, we may be interested only in one of these parameters a_n and want to know what is the probability that $\underline{a}_n < a_n \leq \overline{a}_n$. This can be calculated by

$$P(\underline{a}_n < a_n \leq \overline{a}_n) = \int_{\underline{a}_n}^{\overline{a}_n} p(a_n|\boldsymbol{F}_c)\, \mathrm{d}a_n \tag{18}$$

where the marginal posterior law $p(a_n|\boldsymbol{F}_c)$ can be calculated by

$$p(a_n|\boldsymbol{F}_c) = \int \cdots \int p(\boldsymbol{a}|\boldsymbol{F}_c)\, \mathrm{d}a_1 \cdots \mathrm{d}a_{n-1}\, \mathrm{d}a_{n+1} \cdots \mathrm{d}a_N. \tag{19}$$

We can also simply define as the solution the vector $\widehat{\boldsymbol{a}}$ which corresponds to the mean value of the posterior law—called Posterior mean (PM) estimator:

$$\widehat{\boldsymbol{a}} = \int \boldsymbol{a}\, p(\boldsymbol{a}|\boldsymbol{F}_c)\, \mathrm{d}\boldsymbol{a} \tag{20}$$

or the vector $\widehat{\boldsymbol{a}}$ which maximizes this posterior distribution –called Maximum a posteriori (MAP) estimator:

$$\widehat{\boldsymbol{a}} = \arg\max_{\boldsymbol{a}} \{p(\boldsymbol{a}|\boldsymbol{F}_c)\} = \arg\min_{\boldsymbol{a}} \{-\ln p(\boldsymbol{a}|\boldsymbol{F}_c)\} \tag{21}$$

or even the vector $\widehat{\boldsymbol{a}}$ whose components $\widehat{a}_n$ correspond to the maximizer of the marginal posterior law (19) —called Marginal MAP estimator:

$$\widehat{a}_n = \arg\max_{a_n} \{p(a_n|\boldsymbol{F}_c)\} = \arg\min_{a_n} \{-\ln p(a_n|\boldsymbol{F}_c)\}. \tag{22}$$

In the following we consider only the MAP estimator (21). Using the probability distributions (9) and (14) in (15), the MAP solution is given by:

$$\widehat{\boldsymbol{a}} = \arg\min_{\boldsymbol{a}} \{||\boldsymbol{F}_c - \boldsymbol{a}||^2 + \lambda||\boldsymbol{a}||^2)\} \tag{23}$$

where $\lambda = (\sigma_\epsilon/\sigma_a)^2$ and we find the minimum norm least squares (MNLS) solution which is given explicitly by

$$\widehat{\boldsymbol{a}} = (\boldsymbol{A}^t\boldsymbol{A} + \lambda\boldsymbol{I})^{-1}\boldsymbol{A}^t\boldsymbol{F}_c. \tag{24}$$

Comparing (13) and (24) we see that, for a given N, the matrix $(\boldsymbol{A}^t\boldsymbol{A} + \lambda\boldsymbol{I})$ is always better conditioned than the matrix $(\boldsymbol{A}^t\boldsymbol{A})$ and so the solution (24) always is more stable than the solution (13).

We may also want some information about the uncertainty of this solution. For this we can use the posterior law $p(\boldsymbol{a}|\boldsymbol{F}_c)$. For example, using the likelihood (9) and the prior law (14), it is easy to show that the posterior law is Gaussian, i.e. $p(\boldsymbol{a}|\boldsymbol{F}_c) = \mathcal{N}(\widehat{\boldsymbol{a}}, \widehat{\boldsymbol{P}})$ with $\widehat{\boldsymbol{a}}$ given by (24) and $\widehat{\boldsymbol{P}} = (\boldsymbol{A}^t\boldsymbol{A} + \lambda\boldsymbol{I})^{-1}$. We can then use the diagonal elements of the posterior covariance matrix $\widehat{\boldsymbol{P}}$ to calculate the posterior variances of the estimates, i.e. $\mathrm{Var}(a_n) = \widehat{P}_{nn}$ and so put the error bars on the solution. When the posterior law is not Gaussian, we can always calculate

$$\mathrm{E}(a_n) = \int a_n\, p(a_n|\boldsymbol{F}_c)\,\mathrm{d}a_n \tag{25}$$

and

$$\mathrm{Var}(a_n) = \int (a_n - \mathrm{E}(a_n))^2\, p(a_n|\boldsymbol{F}_c)\,\mathrm{d}a_n \tag{26}$$

but in general we may not have explicite expressions for these integrals. We can however do numerical calculation either by approximating the posterior law by a Gauusian law or by a stochastic integral calculation.

One question still remains: how to determine λ and N? Three approaches are possible:

1. Assign them experimentally from the data using some knowledge on the physics of the problem. For example, the Parseval-type relation between $\rho(r)$ and $F_c(q)$:

$$\int 4\pi r^2 \rho^2(r)\,\mathrm{d}r = \frac{1}{(2\pi)^3}\int 4\pi q^2 Fc(q)\,\mathrm{d}q \tag{27}$$

   can be used to estimate σ_a^2 by:

$$\sigma_a^2 = \frac{1}{N}\sum_{n=1}^{N} a_n^2 = \frac{1}{M}\sum_{m=1}^{M} F_{c_m}^2 \tag{28}$$

   and having an estimate of the noise variance σ_ϵ^2 we can determine λ.

2. Consider λ and N as two extra parameters (hyperparameters) to estimate jointly with the unknown parameter $\boldsymbol{a}$. We can then assign a prior law for them. For example Jeffrey's priors $p(\lambda) = \frac{1}{\lambda}$ for λ and a uniform $p(N) = 1/N_{\max}$ for N. (Other choices are possible, for example a Gamma prior λ which eliminates $\lambda = 0$ and $\lambda = \infty$ and a binomial prior for N which eliminates $N = 0$ and $N = N_{\max}$.)
   Finally, we can estimate them jointly with $\boldsymbol{a}$ by

$$(\widehat{\boldsymbol{a}}, \widehat{\lambda}, \widehat{N}) = \arg \max_{(\boldsymbol{a},\lambda,N)} \{p(\boldsymbol{a}, \lambda, N|\boldsymbol{F}_c)\} \tag{29}$$

   where

$$p(\boldsymbol{a}, \lambda, N|\boldsymbol{F}_c) \propto p(\boldsymbol{F}_c|\boldsymbol{a}, \lambda, N)\, p(\boldsymbol{a}|\lambda, N)\, p(\lambda)\, p(N). \tag{30}$$

   We must however be careful to verify that this joint criterion has at least a local optimum.
3. Consider λ and N as two extra parameters as in the preceeding case but not on the same level. This means that we can try to estimate them first by

$$(\widehat{\lambda}, \widehat{N}) = \arg \max_{(\lambda,N)} \{p(\lambda, N|\boldsymbol{F}_c)\} \tag{31}$$

   where

$$p(\lambda, N|\boldsymbol{F}_c) = \int p(\boldsymbol{a}, \lambda, N|\boldsymbol{F}_c)\, \mathrm{d}\boldsymbol{a} \tag{32}$$

   and then use them in (21). Note, however that finding an analytical expression for $p(\lambda, N|\boldsymbol{F}_c)$ is not always possible and its numerical calculation may be very costly.
4. Consider λ and N as two nuisance parameters, integrate them out and estimate $\boldsymbol{a}$ directly by

$$\widehat{\boldsymbol{a}} = \arg\max_{\boldsymbol{a}} \{p(\boldsymbol{a}|\boldsymbol{F}_c)\} = \arg\max_{\boldsymbol{a}} \left\{ \sum_{n=1}^{N} \int p(\boldsymbol{a}, \lambda, n|\boldsymbol{F}_c)\, \mathrm{d}\lambda \right\} \tag{33}$$

(For more details on these methods, their relative characteristics, their practical implementations and their relatives performances see [20, 17, 19, 16]).

Still one question remains: the choice of the basis-functions.

3. Choice of the basis functions

Two approaches are used to select the basis functions. We call them the *operator based parametric* approach and the *nonparametric* approach and we will discuss both in detail in the following sections. We propose then a new third approach which tries to eliminate the limitations and to keep the advantages of the preceeding approaches. We call this third approach ***physically based parametric***.

3.1. Operator based parametric approach

The first approach is to choose special purpose basis functions based on the properties of the operator linking the data to the unknowns. For example in our case, due to the fact that the kernel of the integral operator of the direct problem is a Bessel function, we may also use the Bessel functions as the basis functions for $\rho(r)$

$$\rho(r) = \begin{cases} \sum_{n=1}^{N} a_n J_0(q_n r) & r \leq R_c \\ 0 & r > R_c \end{cases} \tag{34}$$

This will permit us, using the orthogonality relation

$$\int_0^{R_c} r^2 \, J_l(q_n r)\, J_l(q_m r)\, \mathrm{d}r = \frac{R_c^3}{2}\, J_{l+1}^2(q_n R_c)\, \delta_{n,m}, \tag{35}$$

to find an explicit expression for the charge form factor as a function of the coefficients a_n :

$$F(q) = \frac{4\pi R_c^2}{q} \sum_{n=1}^{N} a_n \frac{(-1)^n}{(qR_c)^2 - (n\pi)^2} \sin(qR). \tag{36}$$

With this choice, note also that, if the form factor $F(q)$ was known exactly at $q_n = \frac{n\pi}{R_c}$ then the coefficients a_n could be calculated analytically by

$$a_n = \frac{F(q_n)}{2\pi R_c^3 \left[J_1(q_n R)\right]^2}. \tag{37}$$

In general, however, the cross section is measured at momentum transfers different from $q_n = \frac{n\pi}{R_c}$.

Now, assume that we are given M measurements at arbitrary momentum transfers $\boldsymbol{q} = \{q_1, q_2, \ldots, q_M\}$ and we wish to determine the N expansion coefficients $\boldsymbol{a} = \{a_1, a_2, \ldots, a_N\}$. In this case Eq. (36) leads to

$$\boldsymbol{F}_c = \boldsymbol{A}\boldsymbol{a} + \boldsymbol{\epsilon} \tag{38}$$

as in (5).

The main advantage of this approach is the fact that $\boldsymbol{a}$ is a small dimension vector and so is the matrix $\boldsymbol{A}$ and we have an explicit analytical expression for calculating its elements.

But at least one main disadvantage to such a choice is that our prior knowledge on $\boldsymbol{a}$ may be limited. For example, if we know that $\rho(r)$ is a positive function we cannot easily incorporate this information in the parameters, $\boldsymbol{a}$.

3.2. Nonparametric approach

The second approach is to choose the basis functions as general as possible and independently of the direct problem operator, for example, either:

$$\phi_n(r) = \delta(r - n\Delta) \tag{39}$$

or

$$\phi_n(r) = \begin{cases} 1 & \text{if } (n-1)\Delta < r \leq n\Delta \\ 0 & \textit{elsewhere} \end{cases} \tag{40}$$

with Δ chosen appropriately small (maximum needed resolution) to be certain we are able to approximate any function $\rho(r)$ as precisely as desired. But, this means that N will probably be large. This is a disadvantage, but this can be compensated, as we will see further, by the fact that the coefficients a_n now have a direct physical meaning: the samples of $\rho(r)$ in the first case and the mean values of $\rho(r)$ in the intervals $(n-1)\Delta < r \leq n\Delta$ in the second case. This means, for example, that the prior knowledge such as the smoothness or the positivity of the function $\rho(r)$ can be transmitted to the coefficients a_n easily.

Let us choose (39) and go further into the details. Replacing (2) with the basis-functions (39) in (1) we obtain:

$$F(q) = \sum_{n=1}^{N} a_n \int_0^{R_c} \mathrm{d}r\, 4\pi r^2\, J_0(qr)\delta(r - n\Delta) = 4\pi \sum_{n=1}^{N} a_n \Delta\, (n\Delta)^2\, J_0(n\Delta q) \tag{41}$$

Denoting by

$$A_{m,n} = 4\pi\, \Delta\, (n\Delta)^2\, J_0(n\Delta q_m) \tag{42}$$

we obtain $\boldsymbol{F}_c = \boldsymbol{A}\boldsymbol{a} + \boldsymbol{\epsilon}$ as in (38). If we use (40) in place of (39), the only change will be in the expression of the Matrix elements $A_{m,n}$ which become

$$A_{m,n} = 4\pi \int_{(n-1)\Delta}^{n\Delta} r^2\, J_0(q_m r)\, \mathrm{d}r. \tag{43}$$

To make a distinction between this approach and the preceding one, let us denote $\boldsymbol{a}$ by $\boldsymbol{\rho}$ and $\boldsymbol{A}$ by $\boldsymbol{B}$:

$$\boldsymbol{F}_c = \boldsymbol{B}\boldsymbol{\rho} + \boldsymbol{\epsilon} \tag{44}$$

Let us now compare (44) and (38): $\boldsymbol{a}$ in (38) is a vector of small dimension while $\boldsymbol{\rho}$ in (44) is a vector of much larger dimension.

Here we can also define either the LS solution:

$$\widehat{\boldsymbol{\rho}} = \arg\min_{\boldsymbol{\rho}} \{\|\boldsymbol{F}_c - \boldsymbol{B}\boldsymbol{\rho}\|^2\} = (\boldsymbol{B}^t\boldsymbol{B})^{-1}\boldsymbol{B}^t\boldsymbol{F}_c \tag{45}$$

or the MNLS solution:

$$\widehat{\boldsymbol{\rho}} = \arg\min_{\boldsymbol{\rho}} \{\|\boldsymbol{F}_c - \boldsymbol{B}\boldsymbol{\rho}\|^2 + \lambda\|\boldsymbol{\rho}\|^2\} = (\boldsymbol{B}^t\boldsymbol{B} + \lambda\boldsymbol{I})^{-1}\boldsymbol{B}^t\boldsymbol{F}_c \tag{46}$$

but neither of these solutions may be satisfactory.

In (44) it is possible to incorporate the smoothness and the positivity of the function $\rho(r)$ into a more appropriate prior distribution for the components ρ_n. For example, to enforce the smoothness of $\rho(r)$ we can assign

$$p(\boldsymbol{\rho}) = p(\rho_1) \prod_{n=2}^{N} p(\rho_n|\rho_{n-1}) = \mathcal{N}(\rho_{n-1}, \sigma_0^2) \tag{47}$$

with

$$p(\rho_1) = \mathcal{N}(\rho_0, \sigma_0^2) \propto \exp\left[-\frac{1}{2\sigma_0^2}(\rho_1 - \rho_0)^2\right] \tag{48}$$

$$p(\rho_n|\rho) = p(\rho_n|\rho_{n-1}) = \mathcal{N}(\rho_{n-1}, \sigma_0^2) \propto \exp\left[-\frac{1}{2\sigma_0^2}(\rho_n - \rho_{n-1})^2\right] \tag{49}$$

which leads to

$$p(\boldsymbol{\rho}) \propto \exp\left[-\frac{1}{2\sigma_0^2}\sum_{n=1}^{N}(\rho_n - \rho_{n-1})^2\right] \tag{50}$$

Using this prior distribution in (15), the MAP estimator becomes

$$\widehat{\boldsymbol{\rho}} = \arg\max_{\boldsymbol{\rho}}\{p(\boldsymbol{\rho}|\boldsymbol{F}_c)\} = \arg\min_{\boldsymbol{\rho}}\{J(\boldsymbol{\rho})\} \tag{51}$$

with

$$J(\rho) = \|\boldsymbol{F}_c - \boldsymbol{B}\boldsymbol{\rho}\|^2 + \lambda\sum_{n=1}^{N}(\rho_n - \rho_{n-1})^2 \tag{52}$$

Defining the matrix

$$\boldsymbol{D} = \begin{pmatrix} 1 & -1 & & & \\ & 1 & -1 & & \\ & & \ddots & \ddots & \\ & & & & \\ & & & 1 & -1 \end{pmatrix} \tag{53}$$

it is easy to show that

$$J(\rho) = \|\boldsymbol{F}_c - \boldsymbol{B}\boldsymbol{\rho}\|^2 + \lambda\,\|(\boldsymbol{D}\boldsymbol{\rho}\|^2 + \lambda\,(\rho_1 - \rho_0)^2 \tag{54}$$

Let us temporarily assume that $\rho_1 = \rho_0$. We then have an explicit solution for the minimizer of (54) which is given by:

$$\widehat{\boldsymbol{\rho}} = (\boldsymbol{B}^t\boldsymbol{B} + \lambda\boldsymbol{D}^t\boldsymbol{D})^{-1}\boldsymbol{B}^t\boldsymbol{F}_c \tag{55}$$

Comparing this solution with the MNLS solution (46) gives us the possibility to see the difference in which the term $\boldsymbol{D}^t\boldsymbol{D}$ is used in place of $\boldsymbol{I}$. Indeed, due to the fact that $\boldsymbol{D}$ corresponds to a first order derivative, we may designate the MNLS solution as the zero order regularized solution in contrast to the first order regularized solution. It is possible to extend this to more general regularized solutions by an appropriate choice of the matrix $\boldsymbol{D}$.

Now, let us go back to (54). ρ_0 is now a new extra hyperparameter which may play a great role in the solution of our inverse problem where the data do not contain information about the DC level of the function $\rho(\boldsymbol{r})$.

One way to enforce the positivity of the solution is to choose a prior distribution such as :

$$p(\boldsymbol{\rho}) = \sum_{n=1}^{N} p(\rho_n) = \sum_{n=1}^{N} \frac{\beta^\alpha}{\Gamma(\alpha)}\,\rho_n^{\alpha-1}\exp[-\beta\rho_n] \tag{56}$$

which can also be written as:

$$p(\rho) \propto \exp\left[-\sum_{n=1}^{N}(1-\alpha)\ln\rho_n + \beta\rho_n\right] \tag{57}$$

and which is called an ***Entropic prior*** in [20]. Using this prior law in (15), the MAP estimator becomes

$$\widehat{\rho} = \arg\max_{\rho}\{p(\rho|F_c)\} = \arg\min_{\rho}\{J(\rho)\} \tag{58}$$

with

$$J(\rho) = \|F_c - B\rho\|^2 + \lambda_1\sum_{n=1}^{N}\ln\rho_n + \lambda_2\sum_{n=1}^{N}\rho_n \tag{59}$$

where λ_1 and λ_2 are related to α, β and σ_ϵ^2. Other choices are possible [20].

To enforce both positivity and the smoothness we propose here to choose

$$p(\rho) \propto \exp\left[-\lambda_1\sum_{n=1}^{N}(\rho_n-\rho_{n-1})^2 - \lambda_2\sum_{n=1}^{N}\ln\rho_n - \lambda_3\sum_{n=1}^{N}\rho_n\right] \tag{60}$$

which leads to

$$\widehat{\rho} = \arg\min_{\rho>0}\{J(\rho)\} \tag{61}$$

with

$$J(\rho) = \|F_c - B\rho\|^2 + \sum_{n=1}^{N}\lambda_1(\rho_n-\rho_{n-1})^2 + \lambda_2\ln\rho_n + \lambda_3\rho_n \tag{62}$$

3.3. Physically based parametric approach

In this approach we choose special purpose basis functions based on the physics of the problem. For example, in our case, since the charge density is a single-valued function defined in a finite domain, the Fourier-Bessel (FB) basis functions which satisfy both the orthogonality and the concentration property conditions, can be used for expansion:

$$\rho(r) = \begin{cases} \sum_{n=1}^{N} a_n J_0(q_n r) & r \le R_c \\ 0 & r > R_c \end{cases} \tag{63}$$

where $q_n = \frac{n\pi}{R_c}$. Excepted the original motivation, this choice is exactly the same as in the first approach and all the relations developed and discussed there can be used.

We may choose other basis functions which are more appropriate to translate our prior knowledge on the desired solution. For example, in our case, we know a priori that, the solution is smooth, positive and a decreasing function. Then we can choose the following function :

$$\rho(r) = \begin{cases} \sum_{n=1}^{N} a_n \exp(-q_n r^2) & r \le R_c \\ 0 & r > R_c \end{cases} \tag{64}$$

Using this expansion in eq. (1) we find $\boldsymbol{F}_c = \boldsymbol{A}\boldsymbol{a} + \boldsymbol{\epsilon}$ as in (5) or in (38), where

$$A_{m,n} = 4\pi \int_0^{R_c} \mathrm{d}r\, r^2\, J_0(q_m r) \exp(-q_n r^2). \tag{65}$$

With this choice we keep the main advantage of the first approach which is the small dimension of the vector $\boldsymbol{a}$ and the main advantages of the second approach which is the translation of our prior knowledge of the positivity of the function $\rho(r)$. This is due to the fact that if we impose the positivity constraint on the coefficients a_n we insure that the solution remains always positive.

In the next section we will illustrate the performance of these different solutions for the estimation of the charge density from elastic electron scattering data.

4. Issues on the uncertainty of the solution

In any scientific problem solving, a proposed solution should be given in any way with a measure of its uncertainty. In Bayesian approach, the posterior probability gives us naturally the necessary tool. To see this, let us come back to our problem and make a summary. We have a set of data $\boldsymbol{F}$ and we want to estimate $\rho(r)$ or more precisely $\boldsymbol{\rho}$ for some locations r_i. Let assume that we have chosen a constant discretization step and so, we want to estimate a vector $\boldsymbol{\rho} = \{\rho(r_i)\, i = 1, \cdots, K\}$.

We presented two approaches: parametric and nonparametric. In the first approach, we have

$$\boldsymbol{F} = \boldsymbol{A}\boldsymbol{a} + \boldsymbol{\epsilon} \tag{66}$$

$$\boldsymbol{\rho} = \boldsymbol{\Phi}\boldsymbol{a} \tag{67}$$

and in the second

$$\boldsymbol{F} = \boldsymbol{B}\boldsymbol{\rho} + \boldsymbol{\epsilon} \tag{68}$$

In both cases, we are interested in $\boldsymbol{\rho}$. In the first approach, we assigned $p(\boldsymbol{a})$ and $p(\boldsymbol{F}|\boldsymbol{a})$, calculated the posterior $p(\boldsymbol{a}|\boldsymbol{F})$, defined a solution $\widehat{\boldsymbol{a}}$ for the parameters $\boldsymbol{a}$, and finally, a solution $\widehat{\boldsymbol{\rho}} = \boldsymbol{\Phi}\widehat{\boldsymbol{a}}$ for $\boldsymbol{\rho}$. In the second, we assigned directly $p(\boldsymbol{\rho})$ and $p(\boldsymbol{F}|\boldsymbol{\rho})$, calculated the posterior $p(\boldsymbol{\rho}|\boldsymbol{F})$, and finaly defined a solution $\widehat{\boldsymbol{\rho}}$. In both cases, we can use the posterior laws to quantify the uncertainty of the solutions. There are, at least, three approaches:

- Simply generate samples from the posterior law $p(\boldsymbol{\rho}|\boldsymbol{F})$ using for example a monte carlo method, and show all these samples to see the distribution of the proposed solution.
- Calculate the posterior mean and the posterior variance of the solution at each point either analytically (when possible) or numerically using for example the samples generated by a monte carlo method.
- Calculate the posterior mean and covariance of the solution either analytically (when possible) or approximate it numerically by any quadrature algorithm.

To illustrate this, let us consider the cases where all the probability laws are Gaussian. Then, all the calculations can be done analytically. The following table summarizes all the steps for the calculation of the solutions and their posterior covariances in the abovementioned two cases:

Table 1: A comparison between parametric and nonparametric approaches.

Nonparametric	Parametric
$\boldsymbol{F} = \boldsymbol{B}\boldsymbol{\rho} + \boldsymbol{\epsilon}$	$\boldsymbol{F} = \boldsymbol{A}\boldsymbol{a} + \boldsymbol{\epsilon}$ $\boldsymbol{\rho} = \boldsymbol{\Phi}\boldsymbol{a}$
$p(\boldsymbol{\rho}) = \mathcal{N}(\boldsymbol{\rho}_0, \sigma_\rho^2 \boldsymbol{P}_0)$ $p(\boldsymbol{F} \mid \boldsymbol{\rho}) = \mathcal{N}(\boldsymbol{B}\boldsymbol{\rho}, \sigma_\epsilon^2 \boldsymbol{I})$	$p(\boldsymbol{a}) = \mathcal{N}(\boldsymbol{a}_0, \sigma_a^2 \boldsymbol{I})$ $p(\boldsymbol{F} \mid \boldsymbol{a}) = \mathcal{N}(\boldsymbol{A}\boldsymbol{a}, \sigma_\epsilon^2 \boldsymbol{I})$
$p(\boldsymbol{\rho} \mid \boldsymbol{F}) = \mathcal{N}(\widehat{\boldsymbol{\rho}}, \boldsymbol{P}_\rho)$ $\widehat{\boldsymbol{\rho}} = [\boldsymbol{B}^t\boldsymbol{B} + \lambda \boldsymbol{P}_0^{-1}]^{-1}\boldsymbol{B}^t(\boldsymbol{F} - \boldsymbol{B}\boldsymbol{\rho}_0)$ $\boldsymbol{P}_\rho = [\boldsymbol{B}^t\boldsymbol{B} + \lambda \boldsymbol{P}_0^{-1}]^{-1}$ with $\lambda = \sigma_\epsilon^2/\sigma_\rho^2$	$p(\boldsymbol{a} \mid \boldsymbol{F}) = \mathcal{N}(\widehat{\boldsymbol{a}}, \boldsymbol{P}_a)$ $\widehat{\boldsymbol{a}} = [\boldsymbol{A}^t\boldsymbol{A} + \lambda \boldsymbol{I}]^{-1}\boldsymbol{A}^t(\boldsymbol{F} - \boldsymbol{A}\boldsymbol{a}_0)$ $\boldsymbol{P}_a = [\boldsymbol{A}^t\boldsymbol{A} + \lambda \boldsymbol{P}_0^{-1}]^{-1}$ with $\lambda = \sigma_\epsilon^2/\sigma_a^2$ $p(\boldsymbol{\rho} \mid \boldsymbol{F}) = \mathcal{N}(\boldsymbol{\Phi}\widehat{\boldsymbol{a}}, \boldsymbol{\Phi}\boldsymbol{P}_a\boldsymbol{\Phi}^t)$
Special cases	Special cases
$\sigma_\epsilon \longrightarrow 0$: $\begin{cases} \widehat{\boldsymbol{\rho}} = [\boldsymbol{B}^t\boldsymbol{B}]^{-1}\boldsymbol{B}^t\boldsymbol{F} \\ \boldsymbol{P}_\rho = 0 \end{cases}$	$\sigma_\epsilon \longrightarrow 0$: $\begin{cases} \widehat{\boldsymbol{\rho}} = \boldsymbol{\Phi}[\boldsymbol{A}^t\boldsymbol{A}]^{-1}\boldsymbol{A}^t\boldsymbol{F} \\ \boldsymbol{P}_\rho = 0 \end{cases}$
$\sigma_\rho \longrightarrow 0$: $\begin{cases} \widehat{\boldsymbol{\rho}} = \boldsymbol{\rho}_0 \\ \boldsymbol{P}_\rho = 0 \end{cases}$	$\sigma_a \longrightarrow 0$: $\begin{cases} \widehat{\boldsymbol{\rho}} = \boldsymbol{\Phi}\boldsymbol{a}_0 \\ \boldsymbol{P}_\rho = 0 \end{cases}$
$\sigma_\epsilon \longrightarrow \infty$: $\begin{cases} \widehat{\boldsymbol{\rho}} = \boldsymbol{\rho}_0 \\ \boldsymbol{P}_\rho = \sigma_\rho^2 \boldsymbol{P}_0 \end{cases}$	$\sigma_\epsilon \longrightarrow \infty$: $\begin{cases} \widehat{\boldsymbol{\rho}} = \boldsymbol{\Phi}\boldsymbol{a}_0 \\ \boldsymbol{P}_\rho = \sigma_a^2 \boldsymbol{\Phi}\boldsymbol{\Phi}^t \end{cases}$
$\sigma_\rho \longrightarrow \infty$: $\begin{cases} \widehat{\boldsymbol{\rho}} = [\boldsymbol{B}^t\boldsymbol{B}]^{-1}\boldsymbol{B}^t\boldsymbol{F} \\ \boldsymbol{P}_\rho = \sigma_\epsilon^2 [\boldsymbol{B}^t\boldsymbol{B}]^{-1} \end{cases}$	$\sigma_a \longrightarrow \infty$: $\begin{cases} \widehat{\boldsymbol{\rho}} = \boldsymbol{\Phi}[\boldsymbol{A}^t\boldsymbol{A}]^{-1}\boldsymbol{A}^t\boldsymbol{F} \\ \boldsymbol{P}_\rho = \sigma_\epsilon^2 \boldsymbol{\Phi}[\boldsymbol{A}^t\boldsymbol{A}]^{-1}\boldsymbol{\Phi}^t \end{cases}$

When the posterior covariance matrix $\boldsymbol{P}_\rho$ is calculated, we can use it to give some information about the uncertainty of the solution. For example, we can use its diagonal elements to calculate $\sigma_k = \sqrt{P_{kk}}$ and use it to error bars on the solution.

5. Numerical experiments

In order to demonstrate the preceding considerations we make use of the following analytical model. For a charge density given by a symetrized Fermi distribution [5]

$$\rho(r) = \alpha \, \frac{\cosh(R/d)}{\cosh(R/d) + \cosh(r/d)} \tag{69}$$

an analytical expression for the coresponding charge form factor can easily be obtained [6, 7]:

$$F(q) = -\frac{4\pi^2 \alpha d}{q} \, \frac{\cosh(R/d)}{\sinh(R/d)} \left[\frac{R\cos(qR)}{\sinh(\pi q d)} - \frac{\pi d \sin(qR)\cosh(\pi q d)}{\sinh^2(\pi q d)} \right]. \tag{70}$$

Only two of the parameters α, R and d are independent since the charge density must fulfill the normalization condition

$$4\pi \int r^2 \, \rho(r) \, \mathrm{d}r = Z. \tag{71}$$

Figure 1 shows the theoretical charge density $\rho(r)$ of ^{12}C (Z=6) obtained from (69) for $r \in [0, 0.8]$ with $R = 1.1 \ \mathrm{A}^{\frac{1}{3}}$ and $d = 0.626$ fm and the theoretical charge form factor $F_c(q)$ obtained by (70) for $q \in [0, 8]$ fm^{-1} and the nine simulated experimental data:

$$q = [0.001, .5, 1.0, 2.0, 3.0, 4.0, 5.0, 6.0, 7.0] \, \mathrm{fm}^{-1}$$

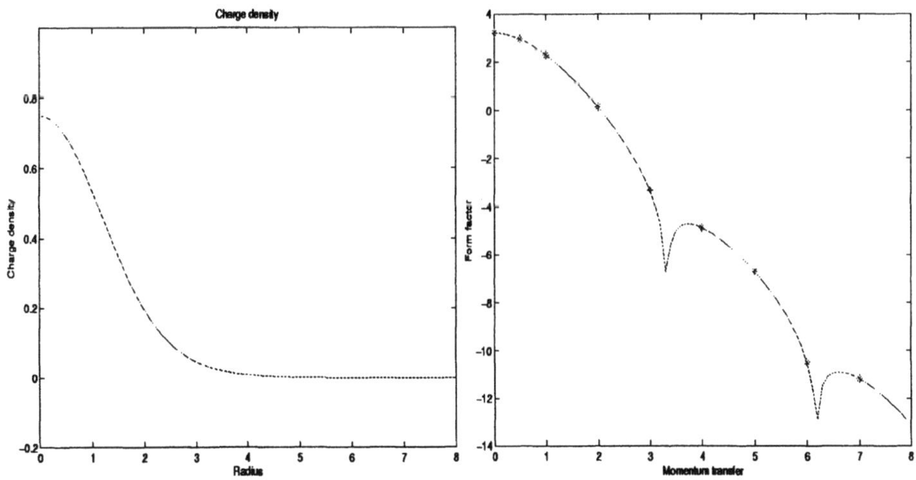

Figure 1. Theoretical charge density $\rho(r)$ [left], charge form factor $\log ||F_c(q)||$ and the data [stars] used for numerical experiments [right].

5.1. Experiments with operator based parametric models

We use these data in the parametric model (34) with $R_c = 8$ fm and estimate the coefficients $\boldsymbol{a}$ by

$$\begin{aligned} \text{LS}: &\quad \widehat{\boldsymbol{a}} = (\boldsymbol{A}^t\boldsymbol{A})^{-1}\boldsymbol{A}^t\boldsymbol{F}_c \\ \text{MNLS}: &\quad \widehat{\boldsymbol{a}} = (\boldsymbol{A}^t\boldsymbol{A} + \lambda\boldsymbol{I})^{-1}\boldsymbol{A}^t\boldsymbol{F}_c \\ \text{MAP}: &\quad \widehat{\boldsymbol{a}} = (\boldsymbol{A}^t\boldsymbol{A} + \lambda\boldsymbol{D}^t\boldsymbol{D})^{-1}\boldsymbol{A}^t\boldsymbol{F}_c \end{aligned}$$

Then using these coefficients we calculate $\rho(r)$ by (34) and $F_c(q)$ by (36).

Figure 2 and Figure 3 show the reconstructed charge densities $\widehat{\boldsymbol{\rho}}$ and the corresponding charge form factors $\widehat{\boldsymbol{F}}_c$ obtained by LS and by MNLS for $N = 5$ and $N = 10$.

Figure 4 shows the reconstructed charge densities by LS and by MNLS for different expansion order N from 5 to 10. Note that the LS solutions are very sensitive and vary greatly with N, but the MNLS solution stays more stable with respect to N.

Figure 5 shows the reconstructed charge densities and the corresponding charge form factors obtained by MNLS and MAP for $N = 30$.

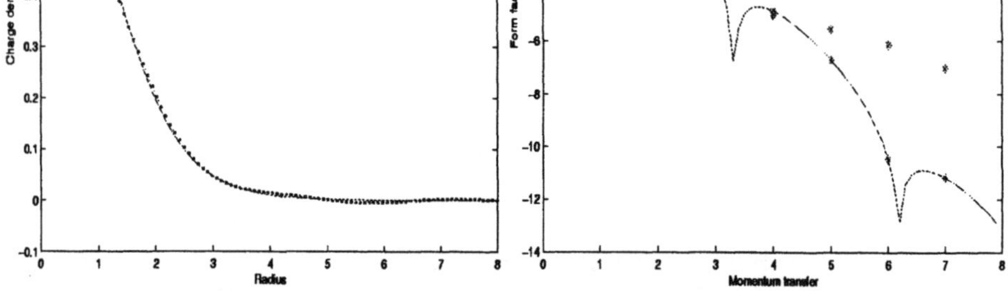

Figure 2. Parametric reconstruction of $\widehat{\rho}(r)$ obtained by LS (point) and by MNLS (dotted) for $N = 5$ [left] and the corresponding reconstructed charge form factors $\log\|\widehat{F}_c(q)\|$ [right]. Two solutions are practically indistinguishable and both not very satisfactory due to a large bias of the solution for small radius r. Note also that both solutions fit well the data.

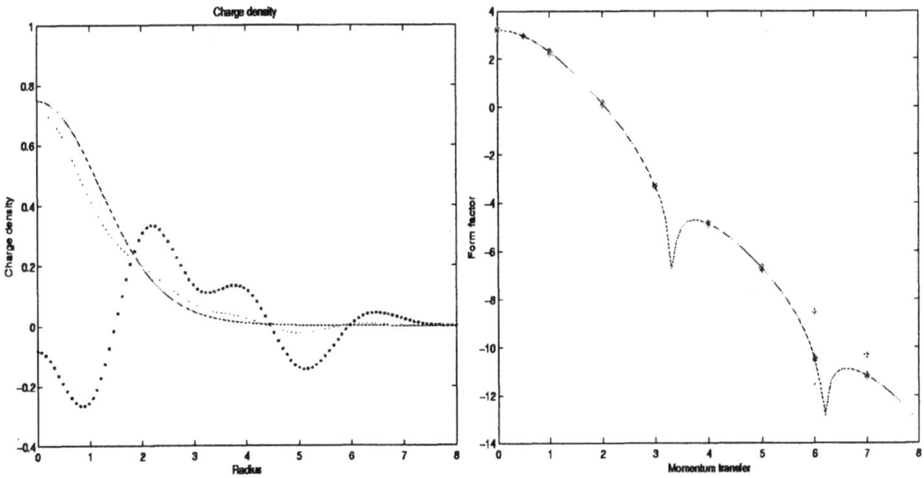

Figure 3. Parametric reconstruction of $\widehat{\rho}(r)$ obtained by LS (point) and by MNLS (dotted) for $N = 10$ [left] and the corresponding reconstructed charge form factors $\log ||F_c(q)||$ [right]. Note that the LS solution fits very well the data but is very unstable but the MNLS solution, which does not fit perfectly the data, is at least more stable.

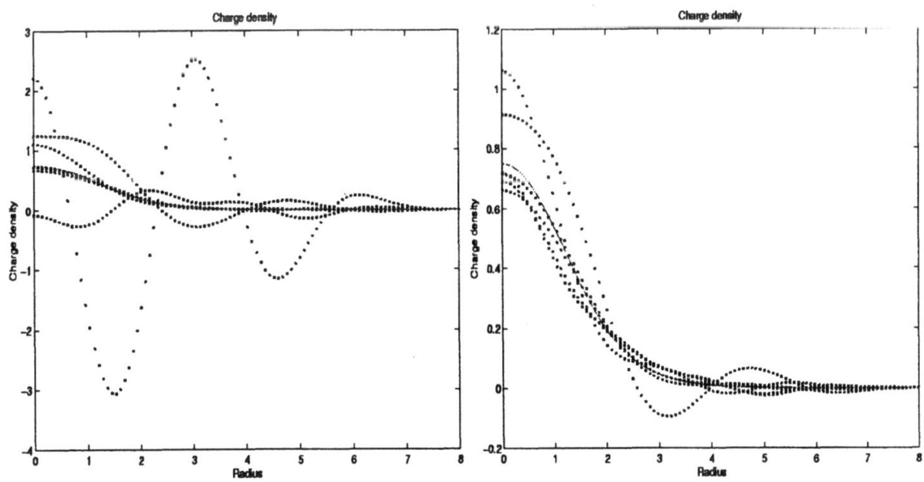

Figure 4. Parametric reconstruction of $\widehat{\rho}(r)$ obtained by LS (left) and by MNLS (right) for different values of $N := 5 : 1 : 10$. Note that the LS solutions vary greatly with N, but the MNLS solutions stay more stable with respect to N.

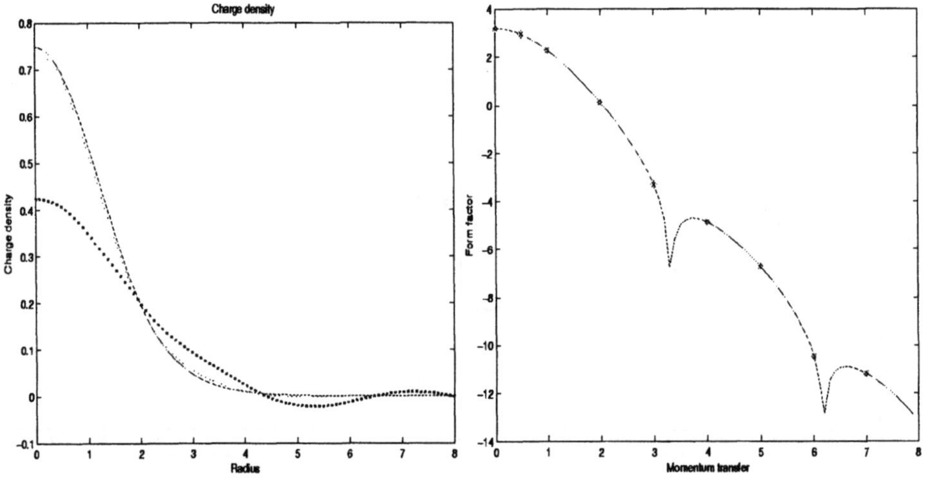

Figure 5. Parametric reconstruction of $\widehat{\rho}(r)$ obtained by MNLS (point) and by MAP (dotted) for $N = 30$ [left] and the corresponding reconstructed charge form factors $\log ||\widehat{F_c}(q)||$ [right]. In this case the LS solution is completely unrealistic and is not presented. The MNLS solution has a large bias for small raduis. The MAP solution is very satisfactory. Note also that both solutions satisfy the data constraint practically in the same way.

5.2. Experiments with non parametric models

The same data are then used with the non parametric model (44) with $N = 100$ and $R_c = 8$ fm and $\widehat{\boldsymbol{\rho}}$ is calculated by

$$\begin{aligned} \text{LS :} \quad & \widehat{\boldsymbol{\rho}} = (\boldsymbol{B}^t \boldsymbol{B})^{-1} \boldsymbol{B}^t \boldsymbol{F}_c \\ \text{MNLS :} \quad & \widehat{\boldsymbol{\rho}} = (\boldsymbol{B}^t \boldsymbol{B} + \lambda \boldsymbol{I})^{-1} \boldsymbol{B}^t \boldsymbol{F}_c \\ \text{MAP1 :} \quad & \widehat{\boldsymbol{\rho}} = (\boldsymbol{B}^t \boldsymbol{B} + \lambda \boldsymbol{D}^t \boldsymbol{D})^{-1} \boldsymbol{B}^t \boldsymbol{F}_c \end{aligned}$$

Figure 6 shows the estimated $\widehat{\boldsymbol{\rho}}$ and the corresponding $\widehat{\boldsymbol{F}}_c$ by MNLS and by MAP.

Figure 7 shows two solutions obtained by a parametric and a nonparametric method and their associated error bars.

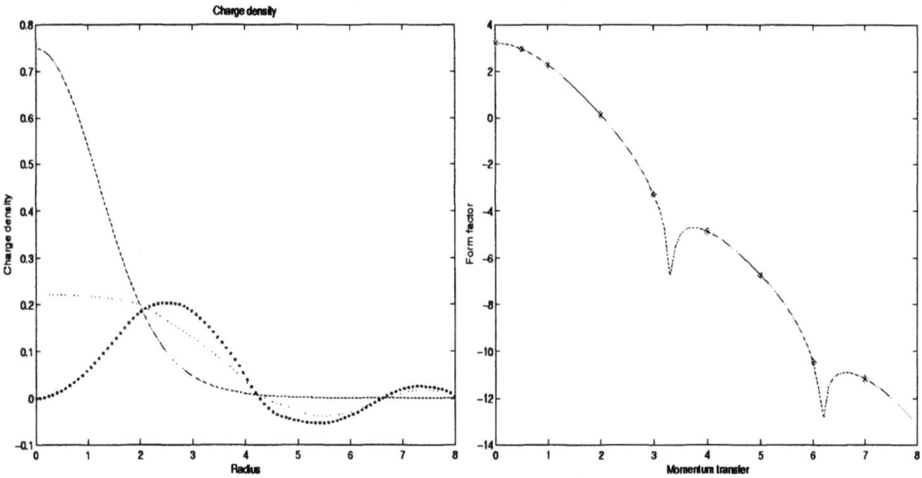

Figure 6. Non parametric reconstruction of $\widehat{\rho}(r)$ obtained by MNLS (point) and MAP1 (dotted) for $N = 100$ [left] and the corresponding reconstructed charge form factors $\log ||\widehat{F}_c(q)||$ [right].

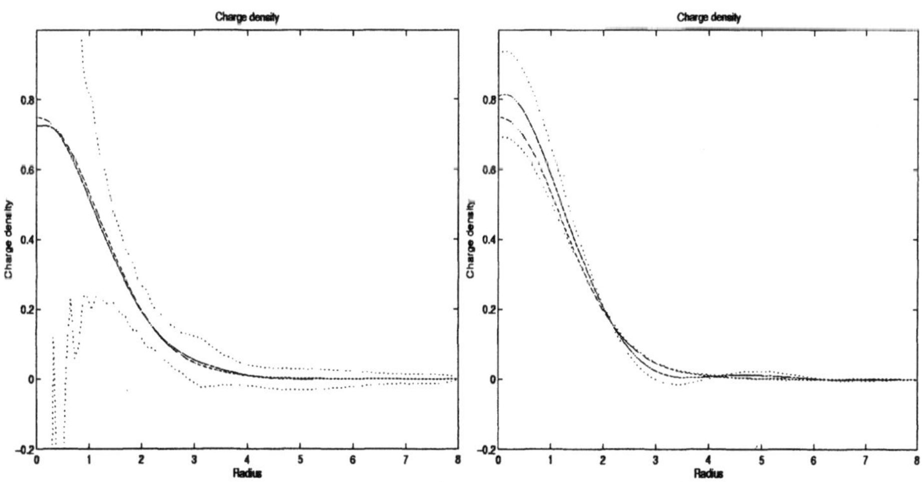

Figure 8. Uncertainty in parametric and non parametric methods:
Left: Parametric reconstruction of $\widehat{\rho}(r)$ obtained by MAP1
Right: Nonparametric reconstruction of $\widehat{\rho}(r)$ obtained by MAP1

6. Conclusion

We considered the problem of the determination of the charge density from a limited number of charge form factor measures as an ill-posed inverse problem. We proposed a Bayesian probabilistic approach to this problem and showed how many classical methods can be considered as special cases of the proposed approach. We addressed also the problem of the basis function choice for the discretization and the uncertainty of the solution. We illustrated the performances of the proposed methods by some numerical results.

References

1. J. L. Friar and J. W. Negele, Nucl. Phys. A **212**, 93 (1973).
2. B. Dreher *et al.*, Nucl. Phys. A **235**, 219 (1974).
3. J. Heisenberg and H. P. Blok, Ann. Rev. Nucl. Part. Sc. **33**, 569 (1983).
4. D. S. Watkins, *Fundamentals of Matrix Computations* (Wiley, New York, 1991).
5. M. E. Grypeos, G. A. Lalazissis, S. E. Massen, and C. P. Panos, J. Phys. G **17**, 1093 (1991).
6. R. E. Kozak, Am. J. Phys. **59**, 74 (1991).
7. R. Anni, G. Co', and P. Pellegrino, preprint (1994).
8. C. R. Rao and S. K. Mitra, *Generalized Inverse of Matrices and its Applications* (Wiley, New York, 1971).
9. J. Baker-Jarvis, J. Math. Phys. **30**, 302 (1989).
10. J. Baker-Jarvis, M. Racine, and J. Alameddine, J. Math. Phys. **30**, 1459 (1989).
11. N. Canosa, H. G. Miller, A. Plastino and R. Rossignoli, Physica **A220**, 611 (1995).
12. S. F. Gull and G.J. Daniell, Nature **272**, 686 (1978).
13. H.G. Miller, Y. Tzeng, G.D. Yen, N. Canosa, R. Rossignoli and A. Plastino, to be published.
14. Buck and Macaulay, "Linear inversion by the method of maximum entropy," in *Maximum Entropy and Bayesian Methods 89*, (J. Skilling, ed.), Kluwer Academic Publishers, 1990.
15. J. Skilling, "Classical maximum entropy" in *Maximum Entropy and Bayesian Methods 89*, (J. Skilling, ed.), pp. 45–52, Kluwer Academic Publishers, 1989.
16. A. Mohammad-Djafari, "A full Bayesian approach for inverse problems," in *Maximum Entropy and Bayesian Methods 95*, (K. Hanson and R. Silver, ed.), Kluwer Academic Publishers, 1996.
17. D.J.C. MacKay, "Hyperparameters: Optimize or integrate out?" in *Maximum Entropy and Bayesian Methods 93*, (G. Heidbreder, ed.), pp. 43–59, Kluwer Academic Publishers, 1996.
18. V.A. Macaulay and B. Buck, "A fresh look at model selection in inverse scattering," in *Maximum Entropy and Bayesian Methods 94*, (J. Skilling and S. Sibisi ed.), Kluwer Academic Publishers, 1996.
19. A. Mohammad-Djafari and J. Idier, "A scale invariant Bayesian method to solve linear inverse problems", pp. 121–134. in *Maximum Entropy and Bayesian Methods 94*, (G. Heidbreder, ed.), Kluwer Academic Publishers, 1996.
20. A. Mohammad-Djafari and J. Idier, "Maximum entropy prior laws of images and estimation of their parameters," pp. 285–293. in *Maximum Entropy and Bayesian Methods 90*, (T. Grandy, ed.), Kluwer Academic Publishers, 1991.

INTEGRATED DEFORMABLE BOUNDARY FINDING USING BAYESIAN STRATEGIES

AMIT CHAKRABORTY AND JAMES S. DUNCAN
Dept. of Electrical Engineering and Diagnostic Radiology
Yale University, New Haven, CT 06520 †

Abstract. Precise segmentation and analysis of underlying structures in an image is crucial for a variety of image analysis and computer vision applications including robot vision, pattern recognition and biomedical image processing such as image guided surgery, registration of multimodality images, cardiac motion tracking, etc. However, a robust identification and measurement of such structure is not always achievable by using a single technique that depends on a single image feature. Thus, it is necessary to make use of various image features, such as gradients, curvatures, homogeneity of intensity values, textures, etc. as well as model-based information (such as shape). Integration provides a way to make use of the rich information provided by the various information sources, whereby consistent information from the different sources are reinforced while noise and errors are attenuated. Integration is achieved in this work by using region information in addition to gradient information within the deformable boundary finding framework. The integration problem is framed in a Bayesian framework using shape and region priors to influence a gradient-based deformable boundary finding.

1. Introduction

As pointed out in the abstract, precise segmentation of pertinent structures while being very important, is difficult especially when using a single technique that depends on a single image-derived source of information. The basic premise of this paper is that we believe that to be able to do this reliably, one needs to make use of integrated methods that make use of both boundary and region information.

Most segmentation methods can be divided primarily into region-based and boundary-based approaches. Region-based methods [1–3], rely on the homogeneity of spatially localized features such as grey level intensity, texture and other pixel statistics. Homogeneity does not necessarily mean identical pixel values within a particular region, rather it means that the variation within a region is of a smaller extent than that between regions. The advantage of such methods is that since

†The first author is currently with Siemens Corporate Research, 755 College Road East, Princeton, NJ 08540. Email: (chakrab,duncan)@noodle.med.yale.edu

they rely directly on the grey level image, they are less susceptible to noise than methods that use derivative information. Also, if the high frequency information in an image is either missing or is unreliable, the segmented images remain relatively unaffected. However, the problem with typical region-based segmentation methods, such as the split and merge techniques [4], is that the resulting segmentation depends considerably on the choice of seed points and the region's shape is too dependent on the choice of the actual algorithm used. In addition, such methods often result in an over-segmented image. Rule based systems [5] do better, but are extremely application-specific. Other region-based methods either use probabilistic techniques [1,6–8,2] or use non-linear diffusion methods [3,9] (see [2] or [3] for the exact mathematical relationship between these methods). The main idea behind these methods is to do what can be termed as *edge-preserved smoothing*. However, isolating objects from the resulting image still requires considerable effort as they also suffer from the problems of poor localization and over-segmentation (related to the problem of choosing the appropriate scale). In contrast to region-based methods, boundary methods primarily use gradient information to locate object boundaries. Low level boundary methods that perform edge detection [10,11] typically result in false or broken edges. Thus, we are more interested in deformable whole boundary methods [12–17] which rely on the gradient features at a subset of the spatial positions of an image (near an object boundary).

While both the region and boundary methods have their advantages and disadvantages, their problems are not necessarily identical, i.e. they are not affected in the same way by the various problems. While the presence of noise limits the performance of any image processing algorithm, the region-based methods are less affected than gradient-based boundary finding because the gradient is very noise sensitive. Also, if the high frequency information in the image either is missing or is unreliable, boundary finding is more error-prone compared to region-based segmentation. Shape variations, on the other hand, can be better handled using a deformable boundary finding framework when we consider such variations to be generally around an average shape and such information can easily be incorporated as priors [12]. Further, since conventional boundary finding relies on changes in the grey level, rather than their actual values, it is less sensitive to changes in the grey scale distributions such as shading artifacts over images. Also, gradient-based methods in general do a better job of edge localization. This brings us to the realization that integrated methods are likely to perform better than either of the methods alone by being able to combine the complementary strengths of these individual methods, as pointed out in [18,19].

This paper is an effort towards our goal of developing a fully bi-directional framework for integrating boundary finding and region-based segmentation. This would lead to a system where the two modules would operate in parallel so that at each step the outputs of each model get updated using information from the outputs of both the models from the previous iteration.

2. Region-Based Segmentation

The main idea here is to classify a particular image into a number of regions or classes. Thus for each pixel in the image it is necessary to decide or estimate which class it belongs to. There are a variety of approaches to region-based segmentation. In the discussion below, we briefly describe the approach we have used which is adapted from the efforts of [20] and [21]. It has the capability of classifying images into distinctive regions based on either their grey level values or for a more general case, based on their texture properties. We consider textured images more general because grey level images can be considered to be the most simple form of texture images [22].

Let S denote the $M \times M$ image lattice, i.e. $S = \{(i,j), 1 \leq i,j \leq M\}$. Let $\{L_s, s \in S\}$ and $\{Y_s, s \in S\}$ denote the labels and the zero mean array obtained from the image data respectively. Note that the labels can belong to only a certain number of region classes. This number is prespecified. Let N_s denote the neighborhood of a site s (a first order neighborhood consists of only four neighbors, eight in the case of a second order system and so on). Now, if one further assumes that all the nearest neighbors of s also have the same label as s, one can write the following expression for the conditional density of the intensity at the pixel site s [20,22]:

$$P(Y_s = y_s | Y_r = y_r, r \in N_s, L_s = l) = \frac{exp(-U(Y_s = y_s | Y_r = y_r, r \in N_s, L_s = l))}{Z(l | y_r, r \in N_s)} \quad (1)$$

where $Z(l|y_r, r \in N_s)$ is the partition function of the conditional Gibbs distribution, and

$$U(Y_s = y_s | Y_r = y_r, r \in N_s, L_s = l) = \frac{1}{2\sigma_l^2}\left(y_s^2 - 2\sum_{r \in N_s} \Theta^l_{s,r} y_s y_r\right) \quad (2)$$

In (2), σ_l and Θ^l are the GMRF model parameters of the lth region class. Without going further into the details, one can see that for a first order GMRF model, if $\hat{\mu}$ is the mean and $\hat{\sigma}^2$ is the variance of the sub-image under consideration, then the feature vector for each region is denoted by (see [22,20] for further details):

$$F = (\theta_1, \theta_2, \theta_3, \theta_4, \hat{\mu}, \hat{\sigma}^2) \quad (3)$$

where the θ's correspond to the influence from the four neighbors that make up the first order neighborhood. These parameters for each region are estimated by a least squares estimate (see [22] for details) method using a window around a user specified point, representative of that particular region. Once the intensity image Y^* has been modeled, the next task is to determine the classification. This is achieved by maximizing the posterior (the Maximum *a posteriori* (MAP) method) distribution of the region labels given the intensity image:

$$P(L|Y^*) = \frac{P(Y^*|L)P(L)}{P(Y^*)} \quad (4)$$

where L corresponds to the classified image with L_s describing the label at the sth pixel. The label field L is modeled as a first or second order MRF, which says that $P(L_s|L_{S/s}) = P(L_s|L_r, r \in \hat{N}_s)$ where $L_{S/s}$ is the whole label field excluding the site s, and $\hat{N}_s$ denotes the appropriate neighborhood of site s. It acts as a prior that emphasizes the property that neighboring pixels of the classified image share the same label (see [21] for details). Maximizing (4) gives an optimal Bayesian estimate. In our results, we used the coordinate-wise descent method of [21], similar to the iterated conditional mode (ICM) algorithm [23]. Here, one starts with an initial labeling, L^0 and then iterates sequentially (or in parallel) through *each pixel, replacing the current label at that pixel with the label that maximizes* $P(L_s|Y^*, L_{S/s})$, where $L_{S/s}$ represents the label set for the whole image except the site s, and the other symbols have the same connotations as before. Without going into the details (see [20,21]) it can be shown that this is equivalent to maximizing $\forall s \in S$

$$\arg\max_{L_s} P(L_s|Y^*, L_{S/s}) = P(L_s|Y^*, L_r, r \in \hat{N}_s) \tag{5}$$

which is reduced to determining

$$L_s^* = \arg\max_{L_s} P(y_s^*|Y_{S/s}^*, L_s, L_r, r \in \hat{N}_s) P(L_s|L_r, r \in \hat{N}_s) \tag{6}$$

for each of the sites s using the ICM algorithm.

The functional in (6) is optimized at each step where the first term, the conditional probability term is given by (2) and the second term is the prior term given by the MRF assumption. To summarize the algorithm, at each iteration, each pixel is updated to maximize (6) using the data at site s and the neighborhood, and the current estimate of the labels of the neighborhood. This continues so long as the number of changes is above a certain fraction of the image size.

3. Boundary Parameterization

As already mentioned, we are primarily interested in finding a class of objects with smooth boundaries that are continuously deformable. Simple edge detectors are of limited or no use under such circumstances as the output of edge detectors does not necessarily correspond to object boundaries except for very high quality images. Hence we are concerned with a whole boundary method so that a global structure can be imposed on the problem. A variety of deformable contour methods have been proposed (see the introduction for a brief review). In this paper, we use the approach due to [12] which uses a Fourier parameterization for the boundary. It expresses a curve in terms of an orthonormal basis [24,25]. It makes the estimation problem easier as the cross correlation matrix becomes diagonal. We note that for most practical situations we constrain ourselves to a limited number of harmonics. Thus,

$$v(t) = \begin{pmatrix} x(t) \\ y(t) \end{pmatrix} = \begin{bmatrix} a_0 \\ c_0 \end{bmatrix} + \sum_{k=1}^{\infty} \begin{bmatrix} a_k & b_k \\ c_k & d_k \end{bmatrix} \begin{bmatrix} cos(kt) \\ sin(kt) \end{bmatrix} \tag{7}$$

where, $v(t)$ is the contour vector consisting of the x and y coordinates and a_k, b_k, c_k and d_k are the corresponding Fourier coefficients.
The curve is thus represented by

$$\vec{p} = (a_0, c_0, a_1, b_1, c_1, d_1, ...) \tag{8}$$

The estimation of these parameters to find the boundary is posed as an optimization problem, where, the objective function measures the strength of the boundary given the set of parameters from the image. We shall come to this issue in the following section.

4. Integration

The basic premise of this work is that the integration of region and boundary information will result in an improved algorithm for solving the segmentation problem. In this section, we first discuss the motivation behind our approach and then describe it mathematically.

As input to the integration algorithm, we have both the actual image I and the region classified image I_r, which is obtained from I after passing it through the region segmentation step discussed before. This information is introduced as an added prior into the gradient-based deformable boundary finding framework. In its simplest form, this region term enforces the boundary to enclose a single region in I_r. As we shall later see, the assumption of a single region is not strictly necessary, but for the sake of simplicity, we will continue with it and modifications to it will be discussed later. The traditional boundary finding problem does not use the original image directly, but uses instead the gradient image I_g. As in the approach of [12], we shall use the magnitude of the gradient vector at each pixel location. We have also used the gradient direction along with the magnitude (see [26]). The gradient image can be obtained by convolving I with the derivative of a Gaussian. We take the magnitude of the corresponding vector image to get I_g. Hence, the input to the integration system is the gradient image I_g and the region classified image I_r.

We shall pose the boundary estimation problem using gradient and region homogeneity information in the maximum *a posteriori* framework. Our aim is to maximize $P(\vec{p}|I_g, I_r)$, where as described in the previous section, $\vec{p}$ is the vector of parameters used to parameterize the contour. It can be shown that (see [27] for details) this optimization can be posed in the following form:

$$\arg\max_{\vec{p}} M(\vec{p}, I_g, I_r) = \arg\max_{\vec{p}} [M_{prior}(\vec{p}) + M_{gradient}(I_g, \vec{p}) + M_{region}(I_r, \vec{p})] \tag{9}$$

We now explain each of the terms in (9).

Prior shape term: The first term in equation (9) corresponds to the prior shape term. This can be obtained from previous experience or from expert knowledge about the natural structures that exist in the image. When it is non-uniform, it biases the model towards a particular range of shapes. Use of prior boundary information can especially be important in clinical applications such as delineating the corpus callosum or the endocardium of the heart, where the nature of the shape does not change a great deal from individual to individual even though the exact reconstruction is different. Prior boundary information under such circumstances can be obtained from previously outlined boundaries. These are first parameterized and then the mean and variance of these parameters are calculated to obtain a multivariate Gaussian prior, which could then be used to constrain the optimization at a later stage for further instances of similar structures (see [12] for more details). Since there might be other objects in the image, we would always need an initial estimate of the boundary to start the optimization process. The prior shape, if available, can be used for this purpose as well. The prior information could be of particular importance if the boundary is ill-defined.

Gradient term: The second term in equation (9) is the likelihood term, which depends on the gradient image. It is a measure of the likelihood of the contour of the described object being the true boundary, once the parameters defining the boundary are given. This is expressed in terms of a contour integral, where the integral is computed over $C_{\vec{p}}$, the curve described by the boundary $(x(\vec{p},t), y(\vec{p},t))$. At each point on the contour, the strength of the boundary is evaluated by the magnitude of the gradient at that particular point, given by the gradient image. Thus, the likelihood of the whole contour being the true boundary becomes proportional to the sum of the magnitude of the gradients at all the points that lie on the boundary. If we assume that the noise can be approximated by a zero mean Gaussian, and further assume that the pixels on the boundary are independent, then we may express the above term in the probability expression as the following line integral ([12]) where σ_n^2 is the variance of the underlying noise process (k_1 being a constant):

$$M_{gradient}(I_g,\vec{p}) = \frac{k_1}{\sigma_{n_1}^2} \int_{C_{\vec{p}}} I_g(x(\vec{p},t), y(\vec{p},t))dt \qquad (10)$$

The derivation of the above equation is straightforward and can be found in [12].

Region term: For every image, the region-based segmentation step is carried out only once using the method described in section 2. The last term (of equation (9)) incorporates this region information into the boundary finding framework. As we have already mentioned, we would like the boundary to enclose within it a homogeneous or slowly varying (w.r.t. grey level or texture) region. In other words, we expect the interior of the boundary to be filled with the region of a single class. This realizes the expectation that the variations within an object are assumed to be smaller than those between objects. We emphasize here that this similarity measure may be based on image intensity, texture properties or any other image attribute that can be handled by the region process. However, this method can also deal with situations like outlining an annulus, where the central part might belong to a separate class.

A very simple way to do this would be to associate positive values with the class associated with the target region and negative value otherwise. Then we take an area integral of all the points that lie within the boundary at any instance during the optimization process. Clearly, this would reach a maximum when the boundary is so placed such that it includes all of the points with a positive value and excludes all of the points that have a negative value from the surrounding regions (see [27] for details). Thus,

$$M_{region}(I_r, \vec{p}) \propto \iint_{A_{\vec{p}}} I_r(x,y)dA \tag{11}$$

Simplified Integration using Green's theorem: The objective function involving all the three above terms can be expressed as

$$\begin{aligned}\arg\max_{\vec{p}} M(\vec{p}, I_g, I_r) &= \arg\max_{\vec{p}} [M_{prior}(\vec{p}) + M_{gradient}(I_g, \vec{p}) + M_{region}(I_r, \vec{p})] \\ &\equiv \arg\max_{\vec{p}} \left[M_{prior}(\vec{p}) + K_1 \int_{C_{\vec{p}}} I_g(x(\vec{p},t), y(\vec{p},t))dt + K_2 \iint_{A_{\vec{p}}} I_r(x,y)dA \right]\end{aligned} \tag{12}$$

where K_1 and K_2 are the weighting constants which signifies the relative importance of the two terms in the above equation. Of the last two terms in (12), one is a line integral and the other is an area integral. In general, computing a line integral is much less expensive compared to an area integral. Thus we would save a lot of computation, especially when we carry out an iterative optimization procedure, if we could convert the area integral to a line integral. We already compute a line integral in the second term which is present in the original boundary finding method. The above conversion can be obtained using Green's theorem [28] as follows. Let,

$$M_r(x,y) = \int_0^x I_r(z,y)dz; \quad N_r(x,y) = -\int_0^y I_r(x,z)dz \tag{13}$$

Hence, using Green's Theorem,

$$\iint_{A_{\vec{p}}} I_r(x,y)dA = \frac{1}{2}\int_{C_{\vec{p}}} [N_r(x,y)\frac{\partial x}{\partial t} + M_r(x,y)\frac{\partial y}{\partial t}]dt \tag{14}$$

Thus, finally we have,

$$\begin{aligned}\arg\max_{\vec{p}} M(\vec{p}, I_g, I_r) = \arg\max_{\vec{p}} [M_{prior}(\vec{p}) + \int_{C_{\vec{p}}} (K_1 I_g(x,y) \\ + K_2\{N_r(x,y)\frac{\partial x}{\partial t} + M_r(x,y)\frac{\partial y}{\partial t}\})dt]\end{aligned} \tag{15}$$

In this section we have presented a deformable boundary finding procedure that introduces a prior term which incorporates information that we obtain from region-based segmentation. Further, using Green's theorem, we reduced the whole problem to computing line integrals only rather than both line and area integrals. Since, M_r and N_r are evaluated only once for an image, and is not repeated at every iteration, the computational speedup due to the use of the Green's theorem formalism is from *O(target area)* to *O(target perimeter)*. For most objects this difference is quite significant because the number of pixels within the target object is generally much larger than the number of sample points on the contour.

5. Results

Experiments were carried out both with synthetic and natural images to verify the performance of the above method. The output of all these experiments are object boundaries. Synthetic images were used to quantitatively evaluate the performance of the integrated method against the stand-alone gradient-based method. The synthetic image has one target object in the center surrounded by a background as shown in Figure 1(a).The target object has both convex and concave parts to it. Further, it also has some high curvature points that make the boundary finding process non-trivial. To make it even more difficult, we have smoothed the image using a Gaussian kernel so that the edges become fuzzy. On top of that, white noise was added which would again affect the boundaries the most as they were smoothed. The result is a considerably lower SNR at the boundaries that at any other location in the image. Thus, this procedure represents many of the difficulties associated with structure segmentation in biomedical images. No prior information was used for these images. The region-based segmentation was carried out using only the mean and the variances of the two regions under consideration. For the example using synthetic image, we used 8 harmonics. In Figure 1, we show the result for a signal to noise ratio given by SNR=2.0. The initial boundary is considerably displaced and disfigured when compared to the actual boundary. We see that the combined method produces the best result.

Next, we compare the performance of the three methods under varying amounts of noise, for different initializations and also for their convergence speeds (see Figure 2). For each noise level, the three methods were executed using the same initial boundary placement. The three methods were allowed to run for the same number of iterations or until the change between successive iterations fell below a certain predetermined level. The Y-axis gives a measure of the distance between the approximated contour and the true one. (This is a scaled version of the square of the distance between the contours.) In this and all the following plots, 'Gradient' represents the gradient-based method, 'Region' corresponds to the boundary found based only on the region classified image, and 'Combined' refers to the proposed integrated method. As we might expect, the performance of all the three methods worsen as we increase the amount of noise. However, as can be seen, the combined method is the least sensitive to noise. Also, it performs much better compared to the traditional gradient-based method. Since the gradient-based method relies upon the first derivative, it is more susceptible to noise than the region-based method, which uses homogeneity within the image. On the other hand, the combined method seems to give uniformly better results.

Finally, we apply the algorithm to real clinical images. Figure 3a shows a short axis Magnetic Resonance (MR) image of a dog heart. The aim is to outline the endocardium. A simple region-based segmentation using only the means and the variances was done and the output is shown in Figure 3b. Again we used 8 harmonics to describe the boundary. The initial boundary is shown in Figure 3c.

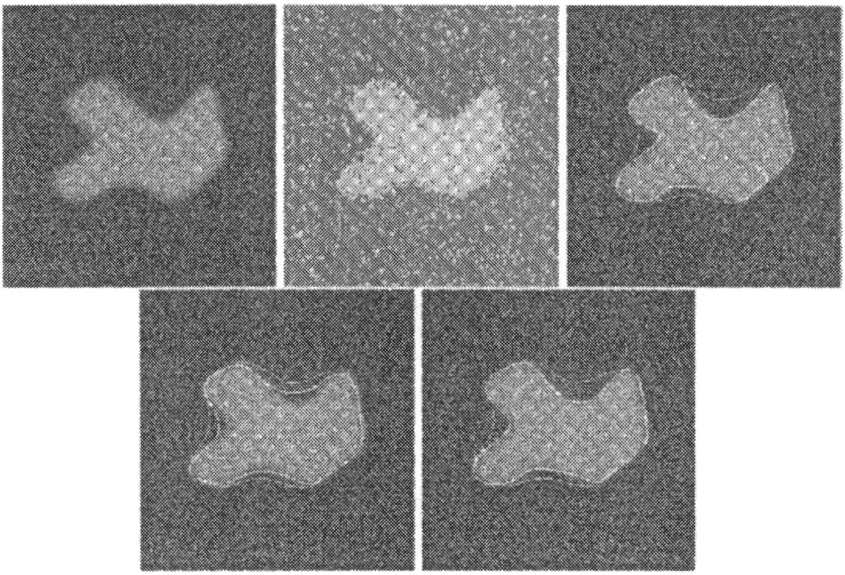

Figure 1. Performance of the three methods for a noisy image (SNR=2.0). In the last three images, the more gray contours represent the initial boundaries and the brighter ones the final derived boundaries. (a)Top,Left: Original image; (b)Top,Middle: Region classified image; (c)Top,Right: Output of Boundary finding using gradient information only; (d)Bottom,Left: Output of Boundary finding using region information only; (e)Bottom: Output of Boundary finding using the integrated approach.

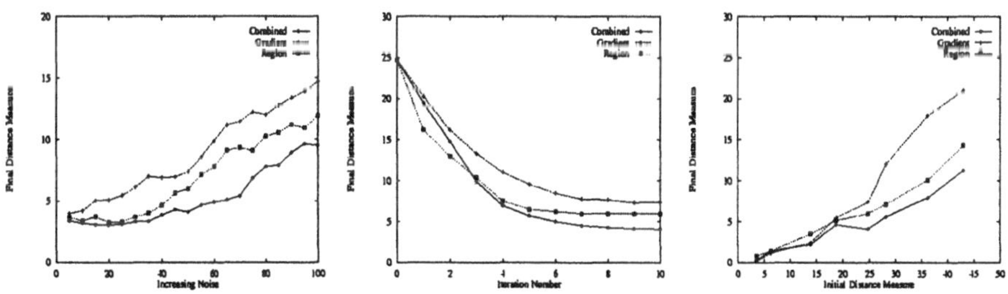

Figure 2. Performance plots (a)Left: Performance under noise using the three methods. The SNR varies from 4.0 to 1.0; (b)Middle: Convergence speed using the three methods (SNR=2.0); (c)Right: Performance under different Initializations (SNR=2.0)

Figure 3d shows the probable boundary as outlined by an expert. We display this only for the sake of comparison. Once again no prior information was used. In all the experiments, we used the same initialization and the same number of harmonics to describe the boundaries for both the methods. As we can see, the image quality is very poor and the edges seem to be very fuzzy. Thus the gradient information is weak. If we apply gradient-based boundary finding, due to a lack of strong edge information, the boundary will diverge after a few iterations, as shown in Figure 3e. On the other hand, as we can see the region information in itself is also not

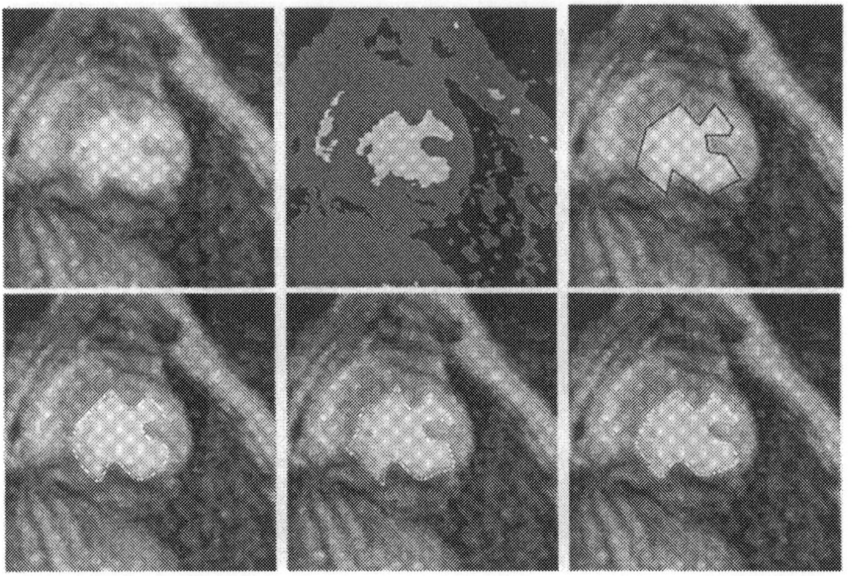

Figure 3. Example with an MR heart image. (a)Top,Left: Original image; (b)Top,Middle: Region classified image; (c)Top,Right: Original image with the initialization as marked by a user; (d)Bottom,Left: Original image with the contour of the endocardium drawn by an expert; (e)Bottom,Middle: Output of Boundary finding using gradient information only; (f)Bottom,Right: Output of Boundary finding using an integrated (gradient + region) approach using the same initialization.

very good. Figure 3f shows the results of the integrated method, which though not perfect, is much better compared to the other method. The main reason for this improved performance is that neither region-based segmentation nor gradient-based boundary finding will actually fail as there is some information in both the gradient and the region classified image. By themselves, they have the limitations previously described and thus neither method alone produces desirable results. Once we combine them, the output improves due to the information fusion, which relieves some of the limitations found when using the algorithms separately.

6. Conclusions

We have presented in this paper a new technique for integrating gradient and region information within the deformable boundary finding framework. We note that the region homogeneity constraint does not necessarily mean regions of constant value. All it requires is that the intra-region variation is smaller than the inter-region variation. Further, it is also not constrained just to the use of intensity value. Other region features like texture can also be used within this framework. The method is posed within a Bayesian framework of maximization of the *a posteriori* probability. Appropriate use of Green's theorem makes the algorithm computationally attractive. As the examples show, the integrated approach is more robust to both increased amounts of noise, as well as increasingly displaced initialization of the initial boundary.

Application of this method on real medical images results in noticeable improvement as has been shown. We have used it for clinical research purposes for outlining the endocardial and epicardial boundaries of the heart and the results are much better than what we had achieved earlier using the purely boundary-based method of [12].

However, there remain areas of potential improvement. We note that some of the probabilistic assumptions may not hold if the parameterized boundary is too far away from the true boundary at any instant. However, the resulting objective function that is generated from the probabilistic assumptions can also be interpreted as finding the boundary that best matches (agrees) with the gradient and region image. This interpretation continues to be valid. Thus in some sense, the method transcends its probabilistic limitations albeit in an ad hoc way. This can be made mathematically more complete. Also, the method can be extended to find multiple objects simultaneously. Generating priors, maybe from a multiscale image representation, could also be an area of potential investigation.

Acknowledgements This work was supported in part by the National Institutes of Health NIH-NHLBI, NIMH-P01MH and NIH-NINDS under Grants R01-HL-44803, 445351-01 and 1-R01NS35193-1, respectively; and the National Science Foundation under Grant IRI-9530768.

References

1. D. Geman and S. Geman, "Stochastic relaxation, Gibbs distribution and Bayesian restoration of images," *IEEE Transactions on Pattern Analysis and Machine Intelligence*, **6**, pp. 721–741, 1984.
2. D. Geiger and A. Yuille, "A common framework for image segmentation," *International Journal of Computer Vision*, **6**, pp. 227–243, 1991.
3. P. Perona and J. Malik, "Scale-space and edge detection using anisotropic diffussion," *IEEE Transactions on Pattern Analysis and Machine Intelligence*, **12**, pp. 629–639, 1990.
4. J. Kittler and J. Illingworth, "On threshold selection using clustering criteria," *IEEE Transactions on System, Man and Cybernetics*, **15**, pp. 652–655, 1985.
5. P. Burt, T. Hong, and A. Rosenfeld, "Segmentation and estimation of region properties through co-operative hierarchical computation," *IEEE Transactions on System, Man and Cybernetics*, **11**, pp. 802–809, 1981.
6. D. Mumford and J. Shah, "Boundary detection by minimizing functionals," *Proceedings, IEEE Conf. Computer Vision and Pattern Recognition*, p. 22, 1985.
7. Y. Leclerc, "Constructing simple stable descriptions for image partitioning," *International Journal of Computer Vision*, **3**, pp. 73–102, 1989.
8. A. Blake and A. Zisserman, *Visual Reconstruction*, MIT Press, Cambridge, MA, 1987.
9. L. Alvarez, P. Lions, and J. Morel, "Image selective smoothing and edge detection by nonlinear diffusion," *SIAM Journal of Numerical Analysis*, **29**, pp. 845–866, 1992.
10. J. Canny, "A computational approach to edge-detection," *IEEE Transactions on Pattern Analysis and Machine Intelligence*, **8**, pp. 679–698, 1986.
11. D. Marr and E. Hildreth, "Theory of edge detection," *Proc. Royal Society, London*, **B207**, pp. 187–217, 1980.
12. L. Staib and J. Duncan, "Boundary finding with parametrically deformable models," *IEEE Transactions on Pattern Analysis and Machine Intelligence*, **14**, pp. 161–175, 1992.

13. M. Kass, A. Witkin, and D. Terzopoulus, "Snakes: Active contour models," *International Journal of Computer Vision*, **1**, pp. 312–331, 1988.
14. A. Amini, T. Weymouth, and R. Jain, "Using dynamic programming for solving variational problems in vision," *IEEE Transactions on Pattern Analysis and Machine Intelligence*, **12**, pp. 855–867, September 1990.
15. T. Cootes, C. Taylor, D. Cooper, and J. Graham, "Active shape models - Their training and application," *Computer Vision, Graphics and Image Processing: Image Understanding*, **61**, pp. 38–59, 1995.
16. V. Caselles, R. Kimmel, and G. Sapiro, "Geodesic active contours," *International Journal of Computer Vision*, to appear.
17. R. Malladi, J. Sethian, and B. Vemuri, "Shape modeling with front propagation: A level set approach," *IEEE Transactions on Pattern Analysis and Machine Intelligence*, **17**, pp. 158–175, 1995.
18. T. Pavlidis and Y. Liow, "Integrating region growing and edge detection," *IEEE Transactions on Pattern Analysis and Machine Intelligence*, **12**, pp. 225–233, 1990.
19. C. Chu and J. Agarwal, "The integration of image segmentation maps using region and edge information," *IEEE Transactions on Pattern Analysis and Machine Intelligence*, **15**, pp. 1241–1252, 1993.
20. B. Manjunath and R. Chellappa, "Unsupervised texture segmentation using markov random field models," *IEEE Transactions on Pattern Analysis and Machine Intelligence*, **13**, pp. 478–482, 1991.
21. R. Leahy, T. Herbet, and R. Lee, "Application of Markov random fields in medical imaging," *Information Processing in Medical Imaging*, pp. 1–14, 1989.
22. B. M. T. Simchony and R. Chellappa, "Stochastic and deterministic networks for texture segmentation," *IEEE Transactions on Acoustics, Speech and Signal Processing*, **38**, pp. 1039–1049, 1990.
23. J. Besag, "On the statistical analysis of dirty pictures," *Journal of the Royal Statistical Society*, **48**, pp. 259–302, 1986.
24. F. Kuhl and C. Giardina, "Elliptic Fourier features of a closed contour," *Computer Graphics and Image Processing*, **18**, pp. 236–258, 1982.
25. C. S. Lin and C. L. Hwang, "New forms of shape invariants from elliptic Fourier descriptors," *Pattern Recognition*, **20**, (5), pp. 535–545, 1987.
26. M. Worring, A. Smeulders, L. Staib, and J. Duncan, "Parametrized feasible boundaries in gradient vector fields," *Information Processing in Medical Imaging*, pp. 48–61, 1993.
27. A. Chakraborty, L. Staib, and J. Duncan, "Deformable boundary finding in medical images by integrating gradient and region information," *IEEE Transactions on Medical Imaging*, pp. 859–870, Dec 1996.
28. P. Baxandall and H. Liebeck, *Vector Calculus*, Oxford University Press, 1986.

SHAPE RECONSTRUCTION IN X-RAY TOMOGRAPHY FROM A SMALL NUMBER OF PROJECTIONS USING DEFORMABLE MODELS

Ali Mohammad–Djafari[1] and Ken Sauer[2]

[1] Laboratoire des Signaux et Systèmes (CNRS–SUPELEC–UPS)
École Supérieure d'Électricité
Plateau de Moulon, 91192 Gif–sur–Yvette Cedex, France

[2] Department of Electrical Engineering
University of Notre Dame
Notre Dame, IN 46556, USA

Abstract. X-ray tomographic image reconstruction consists of determining an object function from its projections. In many applications such as nondestructive testing, we look for a fault region (air) in a homogeneous, known background (metal). The image reconstruction problem then becomes the determination of the shape of the default region. Two approaches can be used: modeling the image as a binary Markov random field and estimating the pixels of the image, or modeling the shape of the fault and estimating it directly from the projections. In this work we model the fault shape by a deformable polygonal disc or a deformable polyhedral volume and propose a new method for directly estimating the coordinates of its vertices from a very limited number of its projections. The basic idea is not new, but in other competing methods, in general, the fault shape is modeled by a small number of parameters (polygonal shapes with very small number of vertices, snakes and deformable templates) and these parameters are estimated either by least squares or by maximum likelihood methods. We propose modeling the shape of the fault region by a polygon with a large number of vertices, allowing modeling of nearly any shape and estimation of its vertices' coordinates directly from the projections by defining the solution as the minimizer of an appropriate regularized criterion. This formulation can also be interpreted as a maximum a posteriori (MAP) estimate in a Bayesian estimation framework. To optimize this criterion we use either a simulated annealing or a special purpose deterministic algorithm based on iterated conditional modes (ICM). The simulated results are very encouraging, especially when the number and the angles of projections are very limited.

key words: Computed tomography, Shape reconstruction, Nondestructive testing, Bayesian MAP estimation

1. Introduction

Tomographic image reconstruction in non destructive testing (NDT) is recent and consists in determining an object $f(x, y)$ from its projections $p(r, \phi)$:

$$p(r,\phi) = \iint f(x,y)\delta(r - x\cos\phi - y\sin\phi)\,\mathrm{d}x\,\mathrm{d}y \tag{1}$$

In many image reconstruction applications, especially in NDT, we know that $f(x,y)$ has a constant value c_1 inside a region P (fault region) and another constant value c_2 outside that region (safe or background region), e.g. metal & air.

$$f(x,y) = \begin{cases} c_1 & \text{if} \quad (x,y) \in P, \\ c_2 & \text{elsewhere} \end{cases} \tag{2}$$

The image reconstruction problem becomes then the determination of the shape of the fault region P. In this work, without loss of generality, we assume that $c_1 = 1$ and $c_2 = 0$ and model the shape of the object by its contour.

There has been many works in image reconstruction dealing with this problem. To emphasis the originality and the place of this work, we give here a summary of the different approaches for this problem:

• The first approach consists in discretizing the equation (1) to obtain:

$$\boldsymbol{p} = \boldsymbol{H}\boldsymbol{f} + \boldsymbol{n} \tag{3}$$

where, $\boldsymbol{f}$ is the discretized values of the object $f(x,y)$ (the pixel values of the image), $\boldsymbol{p}$ is values of the projection data $p(r,\phi)$, $\boldsymbol{n}$ is a vector to represent the modeling and measurement errors (noise) and $\boldsymbol{H}$ the discretized Radon operator. Then the solution is defined as the minimizer of a compound criterion

$$J(\boldsymbol{f}) = Q(\boldsymbol{p} - \boldsymbol{H}\boldsymbol{f}) + \lambda\Omega(\boldsymbol{f}), \tag{4}$$

where λ is the regularization parameter and Q and Ω have to be chosen appropriately to reflect our prior knowledge on the noise and on the image. This is the classical regularization approach of general image reconstruction problem. One can also interpret $J(\boldsymbol{f})$ as the maximum a posteriori (MAP) criterion in the Bayesian estimation framework where $\exp[-Q(\boldsymbol{f})]$ represents the likelihood term and $\exp[-\Omega(\boldsymbol{f})]$ the prior probability law.

This approach has been used with success in many applications (e.g. [1, 2, 3, 4]) but the cost of its calculation is huge due to the great dimension of $\boldsymbol{f}$. Many works have been done on choosing appropriate regularization functionals or equivalently appropriate prior probability laws for $\boldsymbol{f}$ to enforce some special properties of the image such as smoothness, positivity or piecewise smoothness [5, 6, 7, 8, 9]. Among these, one can mention mainly two types of functions for $\Omega(\boldsymbol{f})$:

Entropic laws:

$$\Omega(\boldsymbol{f}) = \sum_{j=1}^{N} \phi(f_j) \quad \text{with} \quad \phi(x) = \{|x|^p, -x\log x, \log x, \cdots\} \tag{5}$$

Markovian laws:

$$\Omega(\boldsymbol{f}) = \sum_{j=1}^{N} \sum_{i \in \mathcal{N}_j} \phi(f_j, f_i) \tag{6}$$

with convex potential functions:

$$\phi(x,y) = \left\{|x-y|^p, -|x-y|\log\frac{x}{y}, \log\cosh|x-y|, \cdots\right\} \tag{7}$$

or nonconvex potential functions:

$$\phi(x,y) = \left\{\min\{|x-y|^2, 1\}, \frac{-1}{1+|x-y|^2}, \cdots\right\} \tag{8}$$

See for example [6] for the entropic laws, [7, 5] for scale invariant markovian laws with convex potential functions and [8, 10, 9, 11] for markovian laws with non convex potential functions and other specific choices.

• The second approach consists in modeling directly the closed contour of the object as the zero-crossing of a smooth function $u(x, y)$:

$$\partial D = \{(x,y) : u(x,y) = 0\} \quad \text{and} \quad f(x,y) = \left\{ \begin{array}{ll} c_1 & \text{if } u(x,y) > 0, \\ c_2 & \text{if } u(x,y) < 0 \end{array} \right. , \tag{9}$$

and in defining a time evolution for u (consequently the corresponding contour ∂D and function f) such that

$$\partial D(t) = \{(x,y) : u(x,y,t) = 0\} \tag{10}$$

and such that as time $t \mapsto \infty$ we arrive at a function $u(x, y)$ such that the associated $f(x, y)$ is a solution to the inverse problem in a least square (LS) sense. This means that the evolution of u and consequently the corresponding contour ∂D and object f is such that the LS criterion

$$J(f) = ||p - R(f)||^2 \tag{11}$$

decreases during the evolution. In this approach the function u is assimilated to a surface (of heat front wave) and the evolution of the contours ∂D is constrained to be perpendicular to the surface. This means that the variation of the interior region $(\delta x, \delta y)$ is such that

$$(\delta x, \delta y) = \alpha(x,y,t)\frac{\nabla u}{|\nabla u|} \tag{12}$$

This is the Level-Set approach originally developed by Osher and Sethian [12] for problems involving the motion of curves and surfaces and then adapted, used and referred as ***snakes*** or ***active contour models*** by many authors in computer vision [13, 14] and recently for inverse problems [15]. See also [16, 17] for the application of the deformable templates in tomographic image reconstruction.

This approach also needs pixel or voxel representation of the image and its calculation cost is huge, specially in 3D imaging systems. We are presently working on this approach trying to extend it for minimizing a regularized criterion in place of the the LS criterion (11) and implementing it in 2D and 3D tomographic image reconstruction.

• The third approach starts by giving a parametric model for the object and then tries to estimate these parameters using least squares (LS) or maximum likelihood (ML) methods. In general, in this approach one chooses a parametric model such as superposition of circular or elliptical homogeneous regions to be able to relate analytically the projections to these parameters. For example, for a superposition of elliptical homogeneous regions we have:

$$f(x,y)=\sum_{k=1}^{K} d_k f_k(x-\alpha_k, y-\beta_k) \tag{13}$$

$$\text{with } f_k(x,y)=\begin{cases} 1 & \text{if } \quad (x/a_k)^2+(y/b_k)^2<1, \\ 0 & \text{elsewhere} \end{cases} \tag{14}$$

where $\boldsymbol{\theta}=\{d_k,\alpha_k,\beta_k,a_k,b_k,k=1,\cdots,K\}$ is a vector of parameters defining the parametric model of the image (density values, coordinates of the centers and the two diameters of the ellipses). It is then easy to calculate analytically the projections and the relation between the data and the unknown parameters becomes:

$$p(r,\phi)=h(r,\phi;\boldsymbol{\theta})+n(r,\phi) \tag{15}$$

where $h(r,\phi;\boldsymbol{\theta})$ has an analytic expression in $\boldsymbol{\theta}$. The LS or the ML estimate when the noise is assumed to be zero mean, white and Gaussian, is then given by:

$$\widehat{\boldsymbol{\theta}}=\arg\min_{\boldsymbol{\theta}}\left\{\|p(r,\phi)-h(r,\phi;\boldsymbol{\theta})\|^2\right\} \tag{16}$$

This approach has also been used with success in image reconstruction [18, 19, 20, 21]. But, the range of applicability of these methods is limited to the cases where the parametric models are actually appropriate.

• The fourth approach which is more appropriate to our problem of shape reconstruction, consists in modeling directly the contour of the object by a function, say $g(\theta)$ in cylindrical coordinates (ρ,θ) such as:

$$D=\left\{(x,y): x^2+y^2=g^2(\theta)\right\}. \tag{17}$$

The next step is then to relate the projections $p(r,\phi)$ to $g(\theta)$ which, in this case is:

$$p(r,\phi)=\int_0^{2\pi}\int_0^{g(\theta)} \delta(r-\rho\cos(\phi-\theta))\rho\,\mathrm{d}\rho\,\mathrm{d}\theta. \tag{18}$$

and finally to discretize this relation to obtain:

$$\boldsymbol{p}=\boldsymbol{h}(\boldsymbol{g})+\boldsymbol{n} \tag{19}$$

where $\boldsymbol{g}$ represents the discretized values of $g(\theta)$ defining the contour of the object and $\boldsymbol{h}(\boldsymbol{g})$ represents the discretized version of the nonlinear operator (18) relating projection data $\boldsymbol{p}$ and $\boldsymbol{g}$. Then, one defines the solution as the argument which minimizes

$$J(\boldsymbol{g}) = ||\boldsymbol{p} - \boldsymbol{h}(\boldsymbol{g})||^2 + \lambda\Omega(\boldsymbol{g}), \tag{20}$$

where $\Omega(\boldsymbol{g})$ has to be chosen appropriately to reflect some regularity property of the object's contour.

In this case also one can consider $J(\boldsymbol{g})$ as the MAP criterion with $Q(\boldsymbol{g}) = ||\boldsymbol{p} - \boldsymbol{h}(\boldsymbol{g})||^2$ as the likelihood term and $\Omega(\boldsymbol{g})$ as the prior one. This approach has been used in image restoration [22], but it seems to be new in image reconstruction applications and the proposed method in this work is in this category. The originality of our work is to model the contour of the object by a piecewise linear function which means that the object is modeled as a polygonal region whose vertices are estimated directly from the projection data.

2. Proposed method

In this paper we propose to model the contour of the object (fault region) as a periodic piecewise linear function or equivalently to model the shape of the object as a polygonal region with a great number N of vertices to be able to approximate any shape. Then we propose to estimate directly the coordinates $\{(x_j, y_j), j = 1, \cdots, N\}$ of the vertices of this polygonal region from the projection data.

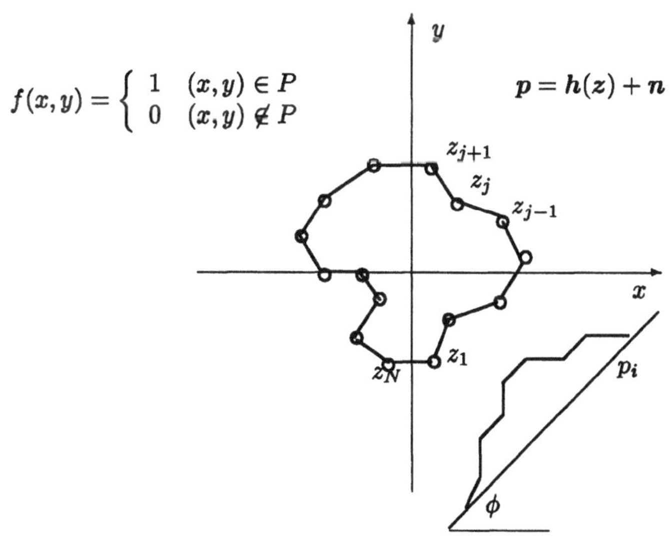

Figure 1: Proposed shape reconstruction modeling.

The idea of modeling the shape of the object as a polygonal region is not new and some works have been done in image reconstruction applications. See for example Milanfar, Karl & Willsky [23, 24, 25], but in general, in these works, either a hypothesis of convexity of the polygonal region has been used which is very restrictive in real applications or the number of vertices of the polygon is restricted to a very small number. In our work we do not make either of these hypotheses; the polygon can be convex or not and we choose N sufficiently great to to be able to approximate appropriately any shape.

The solution is then defined as the minimizer of the following criterion

$$J(\boldsymbol{z}) = ||\boldsymbol{p} - \boldsymbol{h}(\boldsymbol{z})||^2 + \lambda\Omega(\boldsymbol{z}), \tag{21}$$

where $\boldsymbol{z} = \boldsymbol{x} + i\boldsymbol{y}$ is a complex vector whose real and imaginary parts represent the x and the y coordinates of the polygon vertices, $\boldsymbol{h}(\boldsymbol{z})$ represents the direct operator which calculates the projections for any given $\boldsymbol{z}$ and $\Omega(\boldsymbol{z})$ is chosen to be a function which reflects the regularity of the object contour. One can for example choose the following function:

$$\Omega(\boldsymbol{z}) = \left(\sum_{j=1}^{N} |z_{j+1} - z_j| - 2\pi R_0\right)^2 \tag{22}$$

which favors a shape whose contour length is near a prior known value $2\pi R_0$. In this work we used the following function:

$$\Omega(\boldsymbol{z}) = \sum_{j=1}^{N} |z_{j-1} - 2z_j + z_{j+1}|^2 = 4\sum_{j=1}^{N} |z_j - (z_{j-1} - z_{j+1})/2|^2 , \tag{23}$$

which favors a shape whose local curvature is limited. Note that $|z_j - (z_{j-1} - z_{j+1})/2|$ is just the Euclidian distance between the point z_j and the midpoint of the line segment passing through z_{j-1} and z_{j+1} and so this choice favors a shape whose local curvature is limited. We can also give a probabilistic interpretation to this choice. In fact we can consider z_j as random variables with the following Markovian law:

$$p(z_j|\boldsymbol{z}) = p(z_j|z_{j-1}, z_{j+1}) \propto \exp\left[-\frac{1}{2\sigma^2} |z_j - (z_{j-1} - z_{j+1})/2|^2\right]. \tag{24}$$

Other functions are possible and are studied in this work.

In both cases, the criterion $J(\boldsymbol{z})$ is multi-modal essentially due to the fact that $\boldsymbol{h}(\boldsymbol{z})$ is a nonlinear function of $\boldsymbol{z}$. Calculating the optimal solution corresponding to the global minimum of (21) needs then carefully designed algorithms. For this we propose the two following strategies:

- The first is to use a global optimization technique such as simulated annealing. This technique has given satisfactory results as it can be seen from the simulations in the next section. However, this algorithm needs a great number of iterations overall calculations is not very important due to the fact that, in this algorithm, at each iteration only one of the vertices z_j is changed. So, at each step we need to calculate the variation of the criterion due to this change which can be done with a reasonable cost.

• The second is to find an initial solution in the attractive region of the global optimum and to use a local descent type algorithm such as the ICM (Iterated conditional modes) of Besag [26, 27, 28] to find the solution.

The main problem here is how to find this initial solution. For this, we used a moment based method proposed by Milanfar, Karl & Willsky [23, 24, 25] which is accurate enough to obtain an initial solution which is not very far from the optimum. The basic idea of this method is to relate the moments of the projections to the moments of a class of polygonal regions obtained by an affine transformation of a regular polygonal region, and so to estimate a polygonal region whose corners are on an ellipse and whose moments up to the second order matches those of the projections.

However, there is no theoretical proof that this initial solution will be in the attractive region of the global optimum. The next section will show some results comparing the performances of these two methods as well as a comparison with some other classical methods.

3. Simulation results

To measure the performances of the proposed method and keeping the objective of using this method for NDT applications where the number of projections are very limited, we simulated two cases where the objects have polygonal shapes with $N = 40$ corners (hand-made) and calculated their projections for only 5 directions

$$\phi = \{-45, -22.5, 0, 22.5, 45\} \quad \text{degrees.}$$

Then, we added some noise (white, Gaussian and centered) on them to simulate the measurement errors. The signal to noise ratio (SNR) was chosen 20dB. We define SNR as follows:

$$SNR = 10 \log \frac{\sum_i (p_i - \bar{p})^2}{\sum_i (n_i - \bar{n})^2}$$

where p_i and n_i are respectively data and noise samples and $\bar{p}$ and $\bar{n}$ their respective mean values.

Finally, from these data we estimated the solutions by the proposed method using either the simulated annealing (SA) or the iterated conditional modes (ICM) algorithms.

Figure 2 shows these two objects and their relative simulated projections data.

In Figures 3 and 4, we give the reconstruction results obtained by the proposed method using either the SA algorithm (Figure 3) or the ICM algorithm (Figure 4). In these figures we show the original objects, the initial solutions, the intermediate solutions during the iterations and the final reconstructed objects obtained after 200 iterations.

Note that, the SA algorithm is theoretically independent of initialization while the ICM is not. However, in these figures, for the purpose of the comparison, we show the results obtained by the two algorithms with the same initialization.

To show that the method is not very sensitive to the prior knowledge of the vertices number, we give in Figure 5, the reconstruction results of the object 2 in 4 cases: $N = 10$, $N = 20$ and $N = 30$ and $N = 40$. As we can remark all the reconstructed results seem satisfactory.

In Figure 6 we show a comparison between the results obtained by the proposed method and those obtained either by a classical backprojection method or by some other methods in the first approach using (3) and (4) with different regularization functionals $\Omega(\boldsymbol{f})$, more specifically:

- *Gaussian Markov models* (5) with the potential function $\phi(x,y) = |x - y|^2$ which can also be considered as a quadratic regularization method; and
- *Compound Markov models* with nonconvex potential functions $\phi(x,y) = \min\{|x - y|^2, 1\}$ which is a truncated quadratic potential function.

In the first case the criterion to optimize is convex and we used a conjugate gradient (CG) algorithm to find the optimized solution. In the second case the criterion is not convex and we used a Graduated nonconvexity (GNC) based optimization algorithm developed in [10, 9, 11] to find the solution.

Note that, these results are given here to show the relative performances of these methods in a very difficult situation where we have only five projections. In fact, in more comfortable situations (more projections uniformly distributed around the object and high SNR) all these methods, even the simplest one such as the classical backprojection will give similar and satisfactory results. Here, we compare the results obtained from the same restricted set of data.

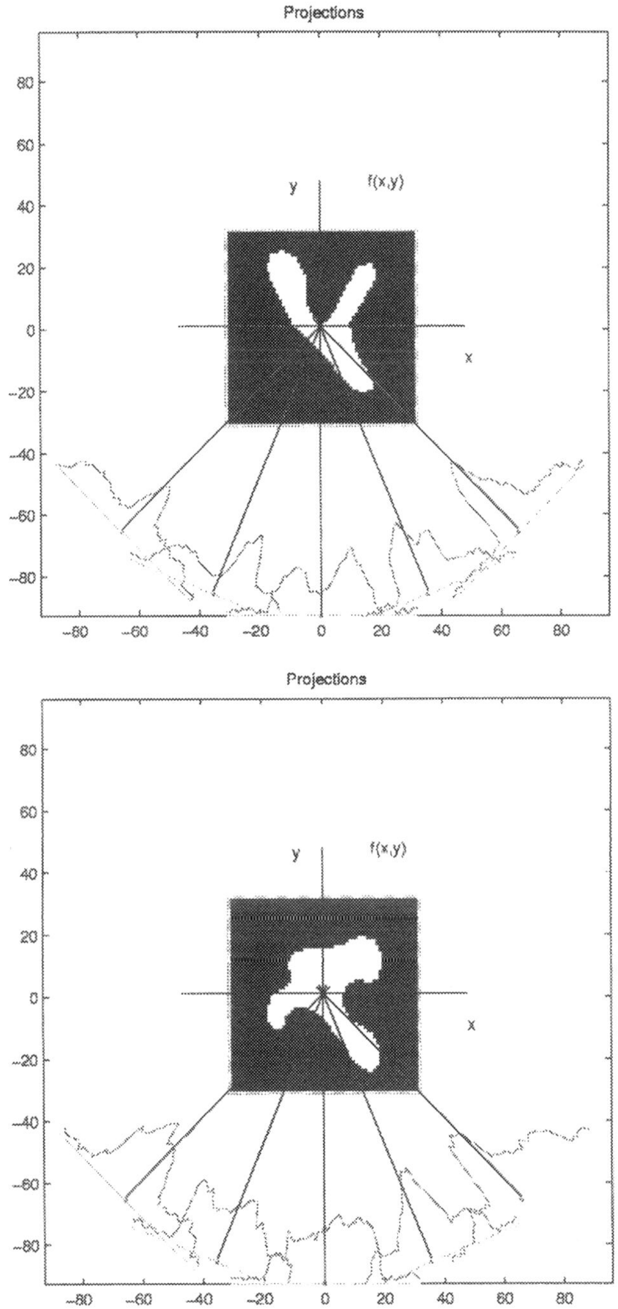

Figure 2: Original images and simulated projection data.

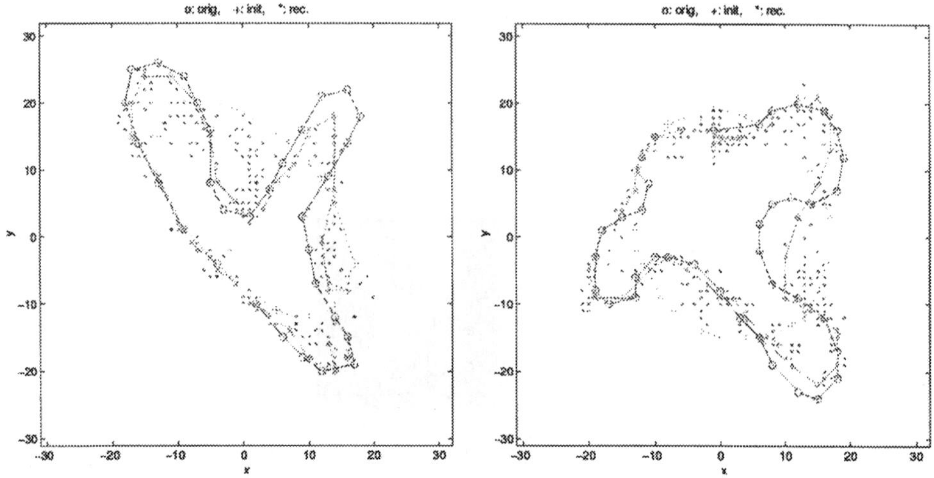

Figure 3: Reconstruction results using simulated annealing.
o) Original objects, +) Initializations, .) Evolution of the solutions during the iterations and
⋆) Final reconstructed objects.

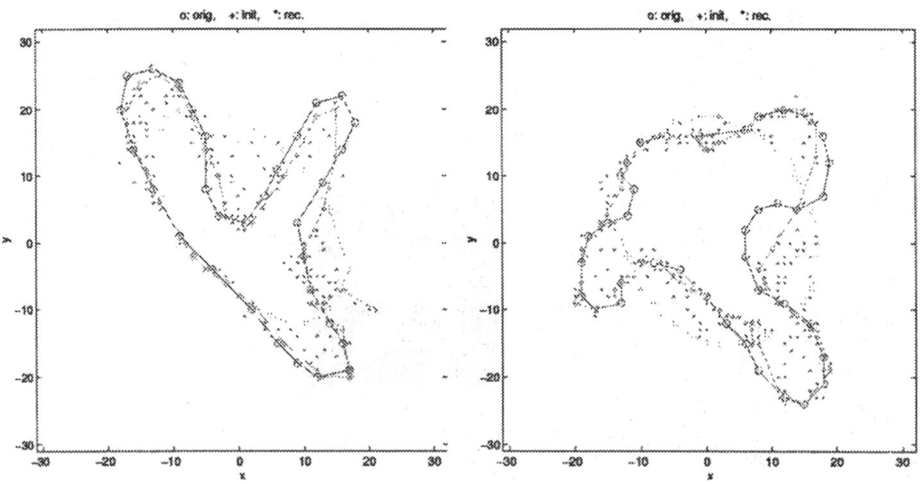

Figure 4: Reconstruction results using a moment-based initialization and a local descent (ICM) minimizer.
o) Original objects, +) Initializations, .) Evolution of the solutions during the iterations and
⋆) Final reconstructed objects.

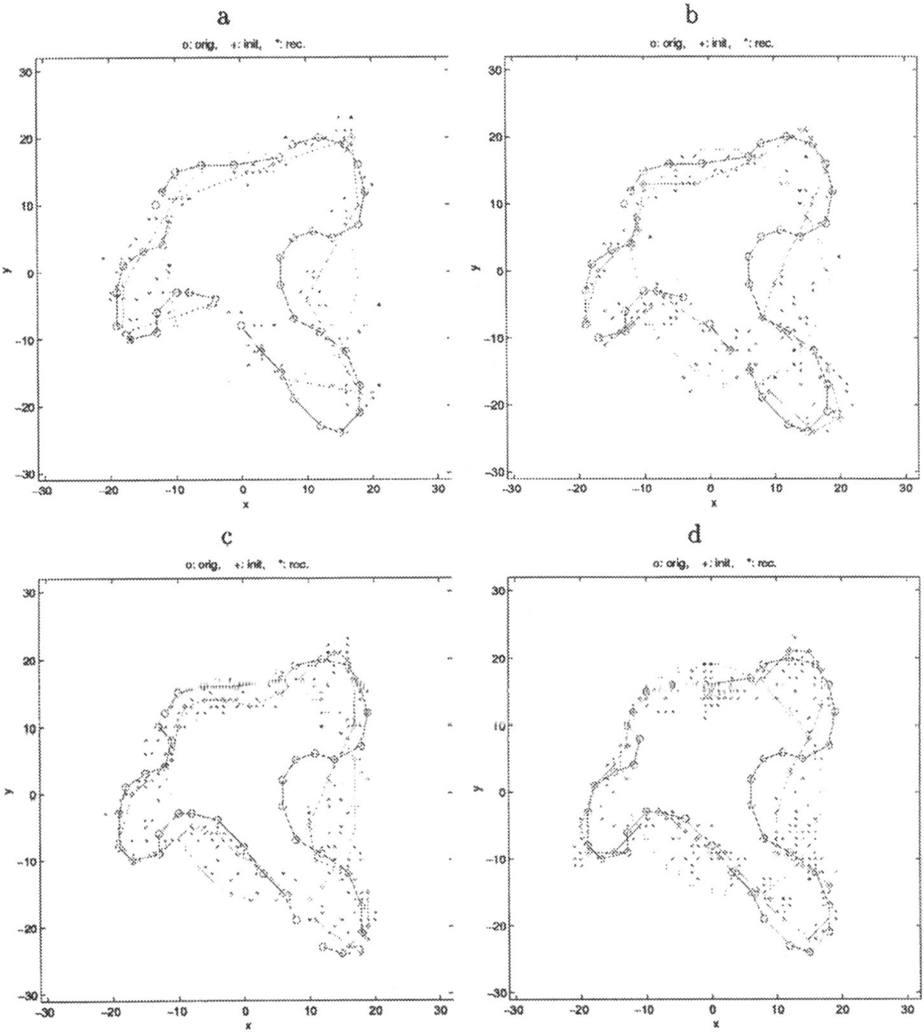

Figure 5: Reconstruction results with different number of vertices N of the same object. The number of vertices of the original object is $N = 40$.
a) $N = 10$, b) $N = 20$ c) $N = 30$ and d) $N = 40$.

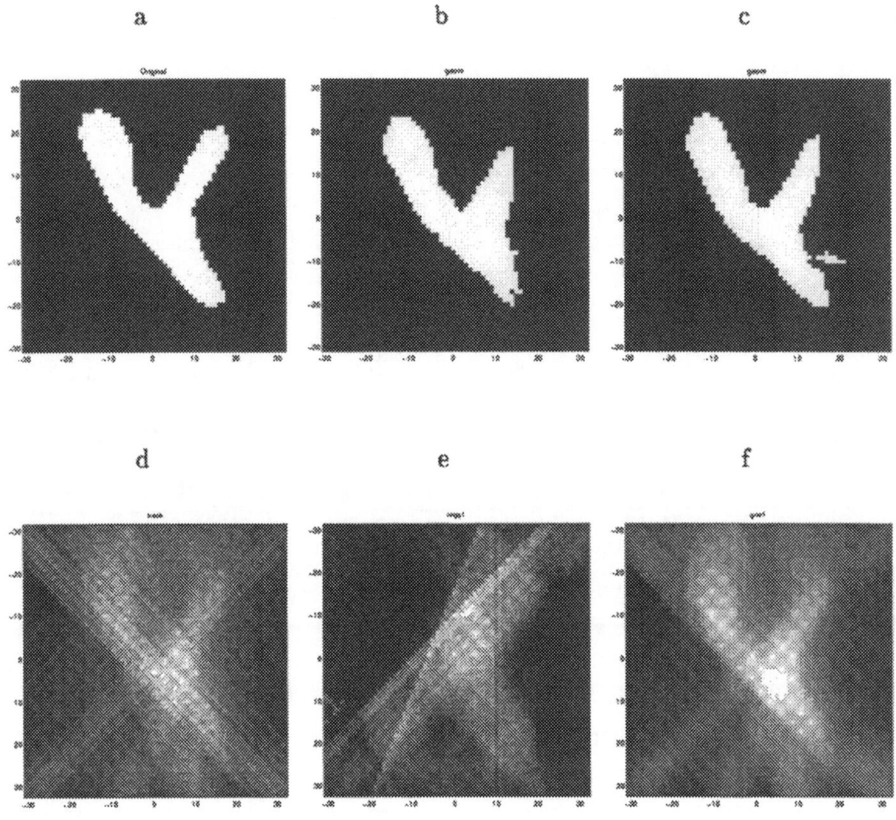

Figure 6: A comparison with backprojection and some other classical methods:
a) Original objects,
b) Results obtained by the proposed method using the SA optimization algorithm,
c) Results obtained by the proposed method using the ICM algorithm,
d) Backprojection,
e) Gaussian Markov modeling MAP estimation and
f) Compound Markov modeling and GNC optimization algorithm.

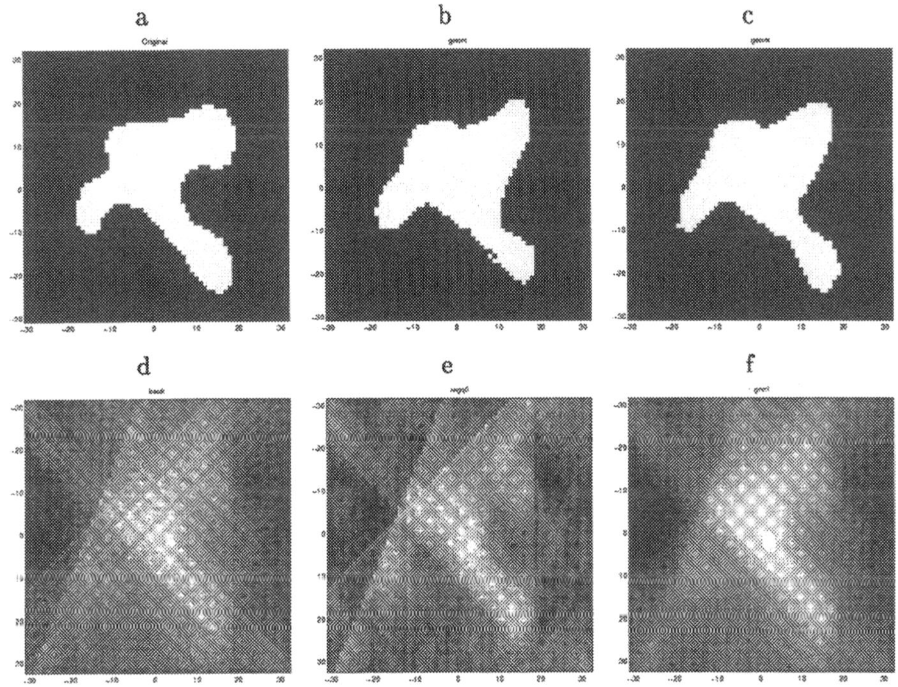

Figure 7: A comparison with backprojection and some other classical methods:
a) Original objects,
b) Results obtained by the proposed method using the SA optimization algorithm,
c) Results obtained by the proposed method using the ICM algorithm,
d) Backprojection,
e) Gaussian Markov modeling MAP estimation and
f) Compound Markov modeling and GNC optimization algorithm.

4. Conclusions

A new method for tomographic image reconstruction of a compact binary object from a small number of its projections is proposed. The basic idea of the proposed method is to model the compact binary object as a polygonal region whose vertices' coordinates are estimated directly from the projections using the Bayesian MAP estimation framework or equivalently by optimizing a regularized criterion.

Unfortunately, this criterion is not unimodal. To find the optimized solution two algorithms have been proposed:
– a global optimization algorithm based on simulated annealing (SA) and
– a local descent-based method based on the Iterated Conditional Modes (ICM) algorithm proposed originally by Besag, with a good initialization obtained by using a moment based method.

The first algorithm seems to give entire satisfaction. The second can also give satisfaction, but it may also be plugged in a local minimum. In both algorithms the main calculation cost is due to the calculus of the variation of the criterion when one of the vertices' coordinates is changed. We have written an efficient program to do this [29, 30].

An extension of this work to 3D image reconstruction with a small number of conic projections is in preparation [31, 32]. The final objective of the proposed method is for non destructive testing (NDT) image reconstruction applications where we can use not only X-rays but also ultrasound or Eddy currents or a combination of them [33, 34, 11] to localize and to characterize more accurately any anomalies (air bulbs) in metallic structures.

References

1. G. Herman, H. Tuy, H. Langenberg, and P. Sabatier, *Basic Methods of Tomography and Inverse Problems.* Adams Hilger, 1987.
2. A. Kak and M. Slaney, *Principles of Computerized Tomographic Imaging.* New York, NY: IEEE Press, 1987.
3. S. Geman and D. McClure, "Statistical methods for tomographic image reconstruction," in *Proc. of the 46-th Session of the ISI, Bulletin of the ISI,* vol. 52, pp. 22–26, 1987.
4. G. Demoment, "Image reconstruction and restoration : Overview of common estimation structure and problems," *IEEE Transactions on Acoustics Speech and Signal Processing,* vol. ASSP-37, pp. 2024–2036, December 1989.
5. S. Brette, J. Idier, and A. Mohammad-Djafari, "Scale invariant Markov models for linear inverse problems," in *Proc. of the Section on Bayesian Statistical Sciences,* (Alicante, Spain), pp. 266–270, American Statistical Association, 1994.
6. A. Mohammad-Djafari and J. Idier, *A scale invariant Bayesian method to solve linear inverse problems,* pp. 121–134. MaxEnt93, Santa Barbara, U.S.A.: Kluwer Academic Publishers, G. Heidbreder ed., 1996.
7. C. Bouman and K. Sauer, "A generalized Gaussian image model for edge-

preserving MAP estimation," *IEEE Transactions on Image Processing*, vol. IP-2, pp. 296–310, July 1993.

8. L. Bedini, I. Gerace, and A. Tonazzini, "A deterministic algorithm for reconstructing images with interacting discontinuities," *Computer Vision and Graphics and Image Processing*, vol. 56, pp. 109–123, March 1994. AMD.
9. M. Nikolova, J. Idier, and A. Mohammad-Djafari, "Inversion of large-support ill-posed linear operators using a piecewise Gaussian MRF," tech. rep., GPI–LSS, to appear in IEEE Transactions on Image Processing, Gif-sur-Yvette, France, 1997.
10. M. Nikolova, A. Mohammad-Djafari, and J. Idier, "Inversion of large-support ill-conditioned linear operators using a Markov model with a line process," in *Proceedings of IEEE ICASSP*, vol. V, (Adelaide, Australia), pp. 357–360, 1994.
11. M. Nikolova and A. Mohammad-Djafari, "Eddy current tomography using a binary Markov model," *Signal Processing*, vol. 49, pp. 119–132, May 1996.
12. S. Osher and J. Sethian, "Fronts propagating with curvature-dependent speed: Algorithms based on hamilton-jacobi formulations," *Journal of Computational Physics*, vol. 79, pp. 12–49, 1988.
13. F. Catté, P. Lions, J. Morel, and T. Coll, "Image selective smoothing and edge detection by nonlinear diffusion," *SIAM J. Num. Anal.*, vol. 29, pp. 182–193, 1992.
14. R. Malladi, J. Sethian, and B. Vemuri, "Shape modelling with front propagation: A level-set approch," *IEEE Transactions on Pattern Analysis and Machine Intelligence*, vol. 17, 1995.
15. F. Santosa, "A level-set approch for inverse problems involving obstacles," *Control, Optimisation and Calculus of Variations*, vol. 1, pp. 17–33, 1996.
16. K. M. Hanson, G. S. Cunningham, and R. J. McKee, "Uncertainties in tomographic reconstructions based on deformable models," in *Image Processing*, K. M. Hanson, ed., *Proc. SPIE*, vol. 3034, pp. 276–286, 1997.
17. K. M. Hanson, G. S. Cunningham, and R. J. McKee, "Uncertainty assessment for reconstructions based on deformable models," *Int. J. Imaging Systems and Technology*, vol. 8, pp. 506–512, 1997.
18. L. A. Shepp and Y. Vardi, "Maximum likelihood reconstruction for emission tomography," *IEEE Trans. on Medical Imaging*, vol. MI-1, pp. 113–122, 1982.
19. D. Rossi and A. Wilsky, "Reconstruction from projections based on detection and estimation of objects," *ieeeASSP*, vol. ASSP-32, no. 4, pp. 886–906, 1984.
20. J. Prince and A. Wilsky, "Reconstructing convex sets from support line measurements," *ieeePAMI*, vol. 12, no. 3, pp. 377–389, 1990.
21. J. Prince and A. Wilsky, "Convex set reconstruction using prior shape information," *CVGIP*, vol. 53, no. 5, pp. 413–427, 1991.
22. N. Friedland and A. Rosenfeld, "Compact object recognition using energy-function-based optimization," *IEEE Transactions on Pattern Analysis and Machine Intelligence*, vol. 14, no. 7, pp. 770–777, 1992.
23. P. Milanfar, W. Karl, and A. Wilsky, "A moment-based variational approach to tomographic reconstruction," *IEEE Transactions on Image Processing*,

vol. 25, no. 9, pp. 772–781, 1994.
24. P. Milanfar, W. Karl, and A. Wilsky, "Reconstructing binary polygonal objects from projections: A statistical view," *CGVIP*, vol. 56, no. 5, pp. 371–391, 1994.
25. P. Milanfar, G. Vergese, W. Karl, and A. Wilsky, "Reconstructing polygons from moments with connections to array processing," *IEEE Transactions on Signal Processing*, vol. 43, no. 2, pp. 432–443, 1995.
26. J. E. Besag, "On the statistical analysis of dirty pictures (with discussion)," *Journal of the Royal Statistical Society B*, vol. 48, no. 3, pp. 259–302, 1986.
27. J. E. Besag, "Digital image processing : Towards Bayesian image analysis," *Journal of Applied Statistics*, vol. 16, no. 3, pp. 395–407, 1989.
28. J. E. Besag and P. Green, "Spatial statistics and Bayesian computation," *J. R. Statist. Soc. B*, vol. 55, pp. 25–37, 1993.
29. A. Mohammad-Djafari, "Image reconstruction of a compact object from a few number of projections," in *IASTED, Int. Conf. on Signal and Image Processing (SIP'96)*, (Orlando, Florida, USA), pp. 325–329, 1996.
30. A. Mohammad-Djafari, "A Bayesian approach to shape reconstruction of a compact object from a few number of projections," in *Maximum Entropy and Bayesian Methods*, (MaxEnt96, Aug. 96, Berg-en-Dal, South Africa), Kluwer Academic Publishers, 1996.
31. A. Mohammad-Djafari and Y. Khayi, "Array processing techniques and shape reconstruction in tomography," in *Proceedings of IEEE ICASSP*, vol. II, (Munich, Germany), pp. 3785–3789, 1997.
32. A. Mohammad-Djafari, K. Sauer, Y. Khayi, and E. Cano, "Reconstruction of the shape of a compact object from a few number of projections," in *IEEE Int. Conf. on Image Processing (ICIP)*, vol. 1, (IEEE ICIP 97, Santa Barbara, CA, USA), pp. 165–169, 1997.
33. S. Gautier, G. Le Besnerais, A. Mohammad-Djafari, and B. Lavayssière, *Data fusion in the field of non destructive testing*. Santa Fe, U.S.A.: Kluwer Academic Publishers, K. Hanson ed., 1995.
34. D. Prémel and A. Mohammad-Djafari, "Eddy current tomography in cylindrical geometry," *IEEE Transactions on Magnetics*, vol. M-31, pp. 2000–2003, May 1995.

AN EMPIRICAL MODEL OF BRAIN SHAPE

J. C. GEE
Department of Neurology and GRASP Laboratory
University of Pennsylvania, Philadelphia, PA 19104

AND

L. LE BRIQUER
GRASP Laboratory
University of Pennsylvania, Philadelphia, PA 19104

Abstract. A method is presented to systematically encode brain shape variation, observed from actual image samples, in the form of empirical distributions that can be applied to guide the Bayesian analysis of future image studies. Unlike eigendecompositions based on intrinsic features of a physical model, our modal basis for describing anatomic variation is derived directly from spatial mappings which bring previous brain samples into alignment with a reference configuration. The resultant representation ensures parsimony, yet captures information about the variation across the entire volumetric extent of the brain samples, and facilitates analyses that are governed by the measured statistics of anatomic variability rather than by the physics of some assumed model.

Key words: Bayesian image analysis, shape models, cerebral anatomy

1. Introduction

An understanding of the natural variation in human neuroanatomy is of fundamental clinical and scientific interest. In the interpretation of medical images, for example, the localization of a brain structure is confounded by the complex ways in which its shape can vary from one individual to another. For certain disorders that affect organ shape, symptomatic individuals can be identified only after the naturally occurring variations are distinguished from the disease-related changes in organ shape. To perform the requisite measurements, however, requires a level of precision that will in many instances exceed what is possible through interactive analysis. Morever, as the amount of data involved in medical applications continues to grow, the viability of manually-based approaches will be further undermined.

In this work, a computational approach is presented for brain shape characterization, where the morphological variability in a sample population is automatically inferred with respect to a canonical anatomic configuration. The assumption is that at a certain spatial scale normal instances of the same anatomy share a common topological scheme and that differences between individuals appear only as variations in the local shape details. By modeling individuals as warped versions of a canonical representation of anatomy, a population can be described by the distribution of high-dimensional spatial transformations which align the members with the canonical anatomy.

This approach of characterizing organ shape variation using highly accurate and detailed spatial mappings onto a common reference configuration for the organ of interest obviates methodological difficulties that arise in morphological or shape analyses which do not normalize adequately or which require manual labeling of the organ. Equally important, the proposed method is fully volumetric, so that the shape statistics of any substructure as well as the spatial correlations between structures can be determined. While this allows complex variations to be modeled, our shape representation is made tractable by scaling its complexity with the actual amount of variation measured in a population.

2. Inferring the Spatial Mapping

Our warping algorithm used to perform the spatial normalization is a Bayesian generalization of the elastic matching technique commonly applied in deformable modeling [1]. Conceptually, elastic matching works as follows: Given two image volumes, designated as the "reference" and "target", respectively, in figure 1, the reference volume is deformed so that its appearance corresponds to that of the target volume. The alignment is achieved by numerically modeling the reference volume as an elastic continuum to which virtual loads are applied so that it assumes the shape of the target. By definition, matching is accomplished when similar regions are brought into correspondence, so our loads are specified in terms of an image-similarity function. In effect, the solution to the elastostatics problem determines the deformation $\boldsymbol{u}(\boldsymbol{x})$ which minimizes the cost function π:

$$\pi(\boldsymbol{u}) = \int_{\text{Reference}} \{\textit{elastic strain}(\boldsymbol{u}) - \textit{similarity}(\text{Reference}(\boldsymbol{x}), \text{Target}(\boldsymbol{x}+\boldsymbol{u}))\}\, d\boldsymbol{x},$$

where an optimal balance is sought between the internal strain energy, measuring the amount of elastic distortion induced in the deformed volume, and the quality of the registration, as determined by the forementioned similarity functional. In our implementation [2], the deformations or mappings are represented as high-dimensional vector fields, specifying the 3-D displacement at each point of one volume into the position of its corresponding point in the second volume.

Since any brain region or structure identified in the canonical anatomy consists of a set of points, the deformation that maps these points into their corresponding position in the image of a subject's brain yields immediate information about the location and shape of that structure in the subject—this is the operational basis for atlas[1]-based approaches toward the automated localization of anatomy. Thus, the variation within the set of warps from the canonical configuration to

[1]In its most basic form, an atlas consists of a labeled representation of anatomy.

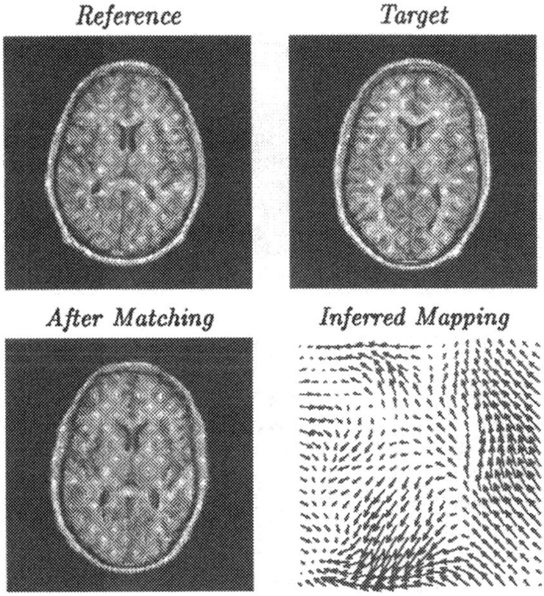

Figure 1. Aligning a pair of magnetic resonance image sections from different subjects by warping one image (Reference) so that its appearance (bottom left) matches that of the other (Target) after the elastic deformation (bottom right).

a sample population can be directly related to the anatomic variability among the individuals in that sample, as exemplified in figure 2. The goal of this work is to encode these shape statistics and to then utilize this prior knowledge about anatomy in the analysis of future data sets.

3. Empirical Models of Anatomic Variation

Assume spatial mappings from a referential anatomy to a group of k subjects are available in the form of vector displacement fields $\boldsymbol{u}^{(i)}, i = 1, \ldots, k$. The variance of the mappings,

$$var(k) = 1/k[\boldsymbol{u}^{(1)} \ldots \boldsymbol{u}^{(k)}][[\boldsymbol{u}^{(1)}]^{\mathrm{T}} \ldots [\boldsymbol{u}^{(k)}]^{\mathrm{T}}]^{\mathrm{T}} - \bar{\boldsymbol{u}}^{(k)}[\bar{\boldsymbol{u}}^{(k)}]^{\mathrm{T}},$$

computed in the Cartesian space will be singular because the number of observed mappings will be small in comparison with the dimensionality of the mappings: in practice, the number of displacement unknowns is usually on the order of 10^4 but can be even more numerous in order to accommodate the complex morphological variations that occur naturally in human anatomy. Because of the limited sample size that will be encountered in most practical instances, we focus on constructing prior models based on the first and second order statistics of the shape measurements.

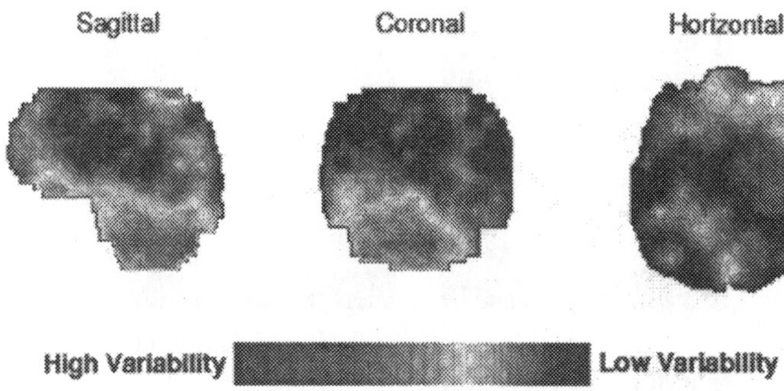

Figure 2. Anatomic variation in a group of 8 subjects, measured by warping a referential anatomy into spatial alignment with each individual in the group and then calculating the variation within the set of inferred warps. The variability is depicted using three orthogonal sections through the reference space, where the intensity value at a point corresponds to the maximum eigenvalue of the 3-D displacement variance of that point from the reference to the subject population.

3.1. AN ORTHONORMAL BASIS SPANNING THE OBSERVED MAPPINGS

In order to derive from the observed variance an information matrix on which our Gaussian shape prior can be based, the statistics must be represented in a subspace of the original Cartesian space. In particular, the subspace should grow in dimension as additional observations are introduced, and its basis must permit recursive updating. To satisfy these requirements, an orthonormal basis spanning only the space of the observed mappings can be constructed using the Gram-Schmidt process as follows:

Initial observation $\boldsymbol{u}^{(1)}$:

- $\boldsymbol{u}^{(1)} = w_1^{(1)}\boldsymbol{v}_1, \quad \text{where} \quad \boldsymbol{v}_1 = \frac{\boldsymbol{u}^{(1)}}{\|\boldsymbol{u}^{(1)}\|}, \quad w_1^{(1)} = <\boldsymbol{u}^{(1)}, \boldsymbol{v}_1> = \|\boldsymbol{u}^{(1)}\|;$

After $k-1$ observations, the orthonormal basis consists of $\{\boldsymbol{v}_1, \ldots, \boldsymbol{v}_r\}$:

- The new observation $\boldsymbol{u}^{(k)}$ is projected on the basis $\{\boldsymbol{v}_1, \ldots, \boldsymbol{v}_r\}$: $w_{\text{proj}}^{(k)} = [w_1^{(k)} \ldots w_r^{(k)}]^{\mathrm{T}} = [\boldsymbol{v}_1{}^{\mathrm{T}} \ldots \boldsymbol{v}_r{}^{\mathrm{T}}]^{\mathrm{T}}\boldsymbol{u}^{(k)}$;
- If the orthogonal component, $\boldsymbol{u}_\perp^{(k)} = \boldsymbol{u}^{(k)} - \boldsymbol{u}_{\text{proj}}^{(k)}$, is the null vector or is negligibly small, the basis is unchanged;
- Otherwise, the basis is augmented with the vector $\boldsymbol{v}_{r+1} = \boldsymbol{u}_\perp^{(k)}/\|\boldsymbol{u}_\perp^{(k)}\|$: $\boldsymbol{u}^{(k)} = [\boldsymbol{v}_1 \ldots \boldsymbol{v}_{r+1}][w_{\text{proj}}^{(k)}\ w_{r+1}^{(k)}]^{\mathrm{T}}, \quad \text{where} \quad w_{r+1}^{(k)} = \|\boldsymbol{u}_\perp^{(k)}\|$.

The initial basis vector $\boldsymbol{v}_1$ is just the normalized version of the first observation and, like the remaining members of the basis, takes the form of a spatial mapping.

3.2. RECURSIVE COMPUTATION OF THE MEAN AND VARIANCE

As the mappings are processed to build the new basis with which they are then re-expressed, their mean and variance are simultaneously computed following the standard recursive scheme—note that the quantities are expressed in the new basis representation:

Initial observation $\boldsymbol{u}^{(1)}$:

– $\boldsymbol{s}^{(1)} = w^{(1)}, \quad \boldsymbol{ss}^{(1)} = w^{(1)} {w^{(1)}}^{\mathrm{T}}, \quad$ where $\quad \boldsymbol{u}^{(1)} = w^{(1)} \boldsymbol{v}_1$;

After $k-1$ observations, $\boldsymbol{s}^{(k-1)}$ is an r-dimensional vector, and $\boldsymbol{ss}^{(k-1)}$ is an r-dimensional matrix:

– If the orthonormal basis is unchanged after processing the new observation $\boldsymbol{u}^{(k)}$, then
$\boldsymbol{s}^{(k)} = \boldsymbol{s}^{(k-1)} + w_{\mathrm{proj}}^{(k)}, \quad \boldsymbol{ss}^{(k)} = \boldsymbol{ss}^{(k-1)} + w_{\mathrm{proj}}^{(k)} {w_{\mathrm{proj}}^{(k)}}^{\mathrm{T}}$;

– Otherwise, $\boldsymbol{s}^{(k)} = \begin{bmatrix} \boldsymbol{s}^{(k-1)} \\ 0 \end{bmatrix} + \begin{bmatrix} w_{\mathrm{proj}}^{(k)} \\ w_{r+1}^{(k)} \end{bmatrix}$,

$$\boldsymbol{ss}^{(k)} = \begin{bmatrix} \boldsymbol{ss}^{(k-1)} & 0 \\ 0 & 0 \end{bmatrix} + \begin{bmatrix} w_{\mathrm{proj}}^{(k)} \\ w_{r+1}^{(k)} \end{bmatrix} \begin{bmatrix} w_{\mathrm{proj}}^{(k)} \\ w_{r+1}^{(k)} \end{bmatrix}^{\mathrm{T}};$$

– Mean $\bar{\boldsymbol{w}}^{(k)} = \frac{1}{k} \boldsymbol{s}^{(k)}$;
Variance $var_V^{(k)} = \frac{1}{k} \boldsymbol{ss}^{(k)} - \bar{\boldsymbol{w}}^{(k)} {\bar{\boldsymbol{w}}^{(k)}}^{\mathrm{T}}$.

3.3. EIGENDECOMPOSITION OF VARIANCE

Our motivation for choosing a new basis of reduced dimension was to circumvent the semidefinite nature of the variance matrix that results when the original Cartesian space is used. Under the new basis, it is possible that $var_V^{(k)}$ remains positive semidefinite, but the singularity will be limited to only a single null eigenvalue. Consider the eigendecomposition:

$$var_V^{(k)} = \mathcal{G}^{(k)} \Lambda^{(k)} {\mathcal{G}^{(k)}}^{-1},$$

where $\mathcal{G}^{(k)} = [\boldsymbol{g}_1^{(k)} \dots \boldsymbol{g}_r^{(k)}]$ are the eigenvectors of $var_V^{(k)}$ expressed in the orthonormal basis $V = \{\boldsymbol{v}_1, \dots, \boldsymbol{v}_r\}$, and $\Lambda^{(k)}$ denotes the diagonal matrix of eigenvalues. If there exists a null eigenvalue in $\Lambda^{(k)}$, it is removed; the corresponding eigenvector is similarly deleted from $\mathcal{G}^{(k)}$. In Cartesian space, $\{\boldsymbol{g}_1^{(k)} \dots \boldsymbol{g}_r^{(k)}\}$ forms a new orthonormal basis of the subspace spanning the observed mappings:

$$[\boldsymbol{e}_1^{(k)} \dots \boldsymbol{e}_r^{(k)}] = [\boldsymbol{v}_1 \dots \boldsymbol{v}_r][\boldsymbol{g}_1^{(k)} \dots \boldsymbol{g}_r^{(k)}],$$

and its subset $\{\boldsymbol{e}_1^{(k)} \dots \boldsymbol{e}_{r-1}^{(k)}\}$ spans the space of the displacement variance:

$$var_{\mathcal{E}}^{(k)} = [\boldsymbol{e}_1^{(k)} \dots \boldsymbol{e}_{r-1}^{(k)}] \Lambda^{(k)} [{\boldsymbol{e}_1^{(k)}}^{\mathrm{T}} \dots {\boldsymbol{e}_{r-1}^{(k)}}^{\mathrm{T}}]^{\mathrm{T}}.$$

The mean $\bar{w}^{(k)}$, expressed in the basis $\{g_1^{(k)} \dots g_r^{(k)}\}$, becomes

$$\bar{z}^{(k)} = [g_1^{(k)\mathrm{T}} \dots g_r^{(k)\mathrm{T}}]^{\mathrm{T}} \bar{w}^{(k)},$$

or, in the Cartesian space, is derived as follows:

$$\bar{u}^{(k)} = [e_1^{(k)} \dots e_r^{(k)}] \bar{z}^{(k)}.$$

The mean $\bar{z}$ and variance Λ specify completely a Gaussian distribution for the observed anatomic variation in the subject group, with respect to the assumed referential anatomy:

$$\begin{aligned} p(z) &\propto \exp - \left\{ \sum_{i=1}^{r-1} \frac{1}{2\lambda_i} (z_i - \bar{z}_i)^2 \right\} \\ &\propto \exp - \left\{ \frac{1}{2} \eta^{\mathrm{T}} \Lambda^{-1} \eta \right\}, \end{aligned} \tag{1}$$

where $\eta = z - \bar{z}$, and $u = \bar{u} + [e_1 \dots e_{r-1}]\eta$.

Figure 3 shows the first two principal modes of the eigendecomposition of the displacements obtained for the subject group referenced in figure 2.

4. Bayesian Image Matching: Application to Anatomical Localization

For anatomical localization, the knowledge about anatomic variation learned from a sample population and encoded as described above is used to design a new matching stage which subserves the final high-dimensional estimation (based on the usual continuity constraints) that is necessary to achieve the best anatomic correspondence. Specifically, the new matching procedure searches a space $\Omega_{\mathcal{E}}$, centered at the mean of the previous observations and spanned by the principal modes $\{e_1, \dots, e_{r-1}\}$ of the variance in those observations. Since the empirical statistics about anatomic variability specify precisely the relevant prior probability model for the unknown spatial mapping, the procedure is naturally formulated within a Bayesian estimation-theoretic framework. In the following, the posterior mode is sought as the point estimate—a stochastic method for estimating the mean can be found in [1].

Following [1], a Gibbs model is constructed for the likelihood function $p(I_s, I_t|u)$, or the conditional probability of observing the image pair $\{I_s, I_t\}$ given any particular value of the unknown mapping u between them:

$$p(I_s, I_t \mid u) \propto \exp - \left\{ \int_s \mathit{similarity}(I_s(x), I_t(x+u))\, dx \right\}. \tag{2}$$

Because of its nonlinear potential function, the Gibbs distribution in (2) is replaced with a Gaussian approximation at each iteration of our algorithm for determining the posterior mode.

To solve for z in the space $\Omega_{\mathcal{E}}$, the Gaussian likelihood at the current iteration is rewritten in the basis $\{e_1 \ldots e_{r-1}\}$:

$$\begin{aligned} p(I_s, I_t \mid u) &\approx \exp - \left\{ \frac{1}{2} u^{\mathrm{T}} A u - u^{\mathrm{T}} b + k \right\} \\ &\approx \exp - \left\{ \frac{1}{2} [u - \bar{u}]^{\mathrm{T}} K [u - \bar{u}] + [u - \bar{u}]^{\mathrm{T}} (K \bar{u} - b) + \right. \\ & \qquad \left. \frac{1}{2} \bar{u}^{\mathrm{T}} K \bar{u} - \bar{u}^{\mathrm{T}} b + k \right\} \\ &\approx \exp - \left\{ \frac{1}{2} \eta^{\mathrm{T}} K \eta - \eta^{\mathrm{T}} f \right\}, \end{aligned}$$

where the constant terms have been ignored in the last expression, and

$$K = [e_1{}^{\mathrm{T}} \ldots e_{r-1}{}^{\mathrm{T}}]^{\mathrm{T}} A \, [e_1 \ldots e_{r-1}], \quad f = [e_1{}^{\mathrm{T}} \ldots e_{r-1}{}^{\mathrm{T}}]^{\mathrm{T}} (b - K\bar{u}). \quad (3)$$

Given the prior specified in (1), the negative logarithm of the posterior distribution,

$$\pi(\eta) = \frac{1}{2} \eta^{\mathrm{T}} [K + \Lambda^{-1}] \eta - \eta^{\mathrm{T}} f,$$

has minimum value at the solution of the linear system

$$[K + \Lambda^{-1}] \eta = f.$$

Our iterative estimation of the maximum *a posteriori* solution to the image matching problem in which the admissible mappings are drawn from an empirical prior distribution of the anatomic variation can be summarized as follows:

for $(k = 0;\ k < N;\ k{+}{+})$ {   Loop over N iterations.
 if $(k = 0)$ {
    $\eta = 0;\ u^{(0)} = \bar{u} = [e_1 \ldots e_r]\bar{z};$ Displacement estimate is initialized to the prior mean.
 } else {
    Calculate Gaussian approximation to the likelihood at $u^{(k-1)}$: A, b;
    Express Gaussian with respect to η using (3): K, f;
    Solve $[K + \Lambda^{-1}]\eta = f$;
    Set $u^{(k)} = \bar{u} + [e_1 \ldots e_{r-1}]\eta$;
 }

5. Discussion

In contrast to the available techniques for shape characterization in medicine [3–5], the proposed method is fully volumetric, its analysis extends throughout the three-dimensional extent of the organ; in the case of the brain, the shape statistics of any region as well as the spatial correlations between regions can be determined. The method subsumes the advantages offered by the growing number of morphometric approaches which utilize spatial mappings onto a canonical configuration to extract and compare shape measurements. A fundamental distinction, however, is that the current method is foremost a means to encode the observed shape variation of an organ and to apply this information in analysis. The complexity of the representation scales with the actual amount of variation measured in a population, and is refined in recursive fashion, thus allowing a large number of observations to be sequentially processed, each with time proportional to the complexity (size of the basis) of the model under refinement. Evaluation of the method is ongoing, but preliminary studies focused on cerebral anatomy indicate that our approach is capable of capturing subtle shape variations in a population [6,7]. In future work, we will also explore the relationship of our method to approaches for learning generic image models that have been proposed in the machine vision literature [8–12].

Acknowledgements

This work was supported by the U.S.P.H.S. under grant RO1-NS-33662. The first author is grateful to Ruzena Bajcsy for her unfailing support and encouragement of this work.

References

1. J. C. Gee and R. K. Bajcsy, "Elastic matching: Continuum mechanical and probabilistic analysis," in *Brain Warping*, A. Toga, ed., Academic Press, San Diego, In preparation.
2. J. C. Gee, D. R. Haynor, M. Reivich, and R. Bajcsy, "Finite element approach to warping of brain images," in *Medical Imaging 1994: Image Processing*, M. H. Loew, ed., SPIE, Bellingham, 1994.
3. T. F. Cootes, D. H. Cooper, C. J. Taylor, and J. Graham, "Trainable method of parametric shape description," *Image and Vision Computing*, **10**, (5), pp. 289–294, 1992.
4. A. Hill, T. F. Cootes, and C. J. Taylor, "A generic system for image interpretation using flexible templates," in *Proc. British Machine Vision Conference*, pp. 276–285, 1992.
5. J. Martin, A. Pentland, S. Sclaroff, and R. Kikinis, "Characterization of neuropathological shape deformations," *IEEE Trans. Pattern Anal. Machine Intell.*, To appear.
6. L. Le Briquer and J. C. Gee, "Design of a statistical model of brain shape," in *Information Processing in Medical Imaging*, J. S. Duncan and G. Gindi, eds., pp. 477–482, Springer-Verlag, Heidelberg, 1997.
7. J. C. Gee, "Atlas warping for brain morphometry," in *Medical Imaging 1998: Image Processing*, K. M. Hanson, ed., SPIE, Bellingham, To appear, 1998.
8. D. Heeger and J. Bergen, "Pyramid-based texture analysis/synthesis," in *ACM SIGGRAPH*, 1995.
9. S. C. Zhu, Y. Wu, and D. Mumford, "Filters, random fields, and maximum entropy (FRAME) – Towards a unified theory for texture modeling," *Int. J. Comput. Vision*, To appear.
10. K. Popat and R. W. Picard, "Cluster-based probability model and its application to image and texture processing," *IEEE Trans. Image Process.*, **6**, (2), pp. 268–284, 1997.
11. J. DeBonet and P. Viola, "A non-parametric multi-scale statistical model for natural images," in *Adv. in Neural Info. Processing*, To appear, 1997.
12. E. P. Simoncelli, "Statistical models for images: compression, restoration and synthesis," in *31st Asilomar Conf. Signals, Systems, and Computers*, IEEE, New York, 1997.

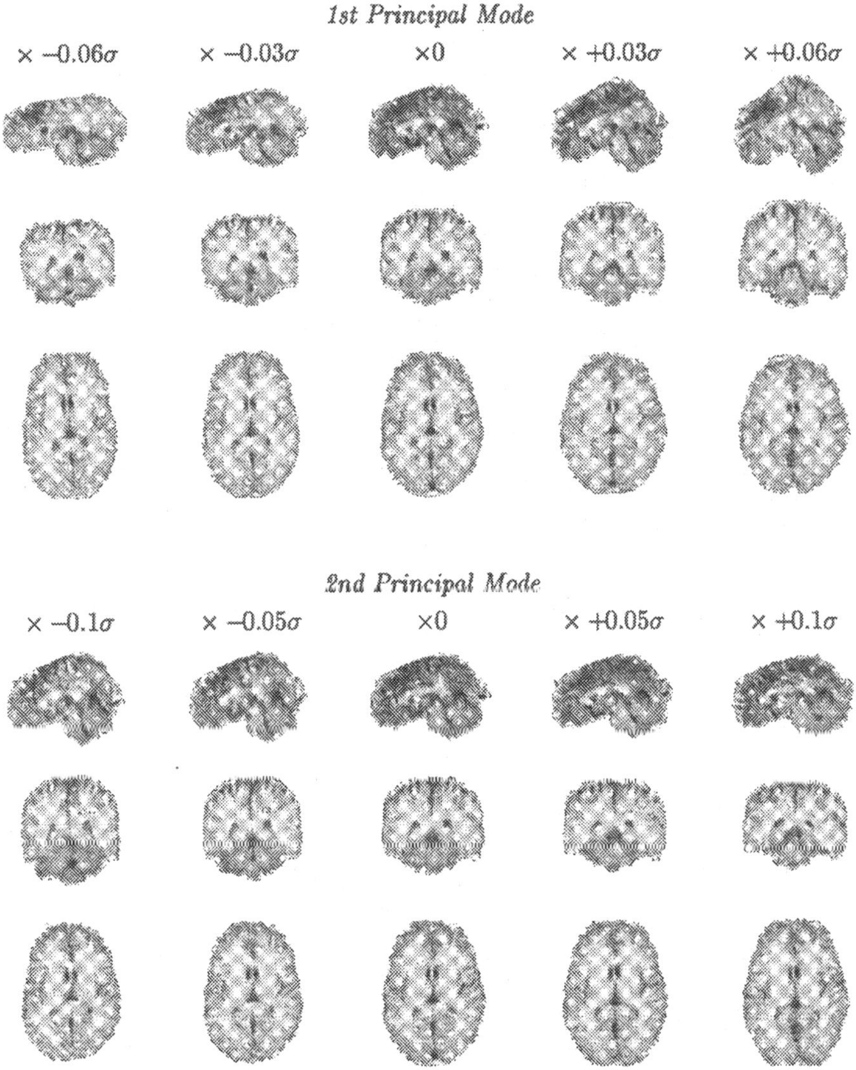

Figure 3. Eigendecompositions of the anatomic variability in a group of subjects with respect to a referential anatomy. The morphological variation captured by each principal mode is visualized by deforming the referential anatomy using the mappings specified by the mode and its magnitude-scaled versions. Each deformed instance is depicted using the same sagittal (top), coronal (middle), and horizontal (bottom) sections through the reference space.

DIFFICULTIES APPLYING RECENT BLIND SOURCE SEPARATION TECHNIQUES TO EEG AND MEG

K. H. KNUTH
Dynamic Brain Imaging Laboratory
Department of Neuroscience
Albert Einstein College of Medicine
Bronx NY 10461, USA
kknuth@balrog.aecom.yu.edu

Abstract. High temporal resolution measurements of human brain activity can be performed by recording the electric potentials on the scalp surface (electroencephalography, EEG), or by recording the magnetic fields near the surface of the head (magnetoencephalography, MEG). The analysis of the data is problematic due to the fact that multiple neural generators may be simultaneously active and the potentials and magnetic fields from these sources are superimposed on the detectors. It is highly desirable to un-mix the data into signals representing the behaviors of the original individual generators. This general problem is called blind source separation and several recent techniques utilizing maximum entropy, minimum mutual information, and maximum likelihood estimation have been applied. These techniques have had much success in separating signals such as natural sounds or speech, but appear to be ineffective when applied to EEG or MEG signals. Many of these techniques implicitly assume that the source distributions have a large kurtosis, whereas an analysis of EEG/MEG signals reveals that the distributions are multimodal. This suggests that more effective separation techniques could be designed for EEG and MEG signals.

1 Mixing and Separation of EEG Signals

Propagation and mixing of signals from simultaneous active sources occurs in many physical situations. Probably the most familiar situation is the "cocktail party problem" where there are many speakers, or sources of acoustic signals, and the listener detects mixtures of these signals. While the listener has only two ears, we will consider cases that are much less difficult where there are as many detectors as sources. Often we are searching for a signal embedded in noise of some sort. In cases where much information is known about the signal, such as temporal behavior or frequency content, there are useful standard techniques that can assist in isolating the signal. However, when little specific information is known about the temporal or frequency structure or when signal and noise overlap significantly in the time or frequency domain separation becomes difficult.

This is a serious problem in the interpretation of electrophysiological data. During a given experiment there are many simultaneous sources of electrical activity in the brain and the currents generated by all of these sources propagate throughout the head and are linearly superimposed on one another. Thus any detector of electromagnetic activity in the brain will pick up a linear combination of all of the active sources.

In the experimental situation there are m detectors, which are used to record the electric potentials or magnetic fields generated by synchronous neural activity in the brain. These detectors are either electrodes, which are placed on the scalp for electroencephalography (EEG) or superconducting quantum interference devices (SQUIDs) coupled to coils placed near the head for magnetoencephalography (MEG). If during the recording there are n active neural generators then mixtures of these n signals are received by the m detectors. This linear instantaneous mixing process can be represented in terms of a matrix multiplication between an $m \times n$ mixing matrix and the n-dimensional source vector $\mathbf{s}(t)$. The m-dimensional signal, $\mathbf{x}(t)$, can be represented by

$$\mathbf{x}(t) = \mathbf{A}\,\mathbf{s}(t)\,. \tag{1}$$

The most obvious way to separate these signals is to find the matrix $\mathbf{A}^{-1}$; however, there is actually more freedom present in the problem. The order of the sources is not important and so any permutation of the sources is also considered a solution. In addition, since both $\mathbf{A}$ and $\mathbf{s}(t)$ are unknown, the scaling of the sources is indeterminate and we therefore seek an unmixing matrix, $\mathbf{W}$ of the form:

$$\mathbf{W} = \Sigma\Pi\mathbf{A}^{-1}, \tag{2}$$

where Σ is a diagonal scaling matrix, and Π is a permutation matrix. The resulting unmixed signals can then be expressed as

$$\mathbf{u}(t) = \mathbf{W}\,\mathbf{x}(t)\,. \tag{3}$$

2 Summary of Blind Source Separation Techniques

In the last several years there has been much work on the problem of blind source separation, which has resulted in many diverse approaches. Most of these approaches use higher-order statistics, minimum mutual information, and maximum entropy in their solutions. In addition to the maximum entropy solution, which is relevant to this discussion, we will mention a higher-order extension to the more popular principal component analysis.

Principal component analysis (PCA) is a technique that has been used to attempt to separate EEG signals into their principal, or decorrelated, components. It is based on the fact the cross-cumulants of statistically independent signals are identically zero. In PCA one diagonalizes the covariance matrix, which results in the decorrelation of the signals. Because this technique deals only with the covariance, it takes into account only the second-order statistics and disregards information represented by the higher-order statistics. In addition, by using only second-order statistics to describe the signals one implicitly assumes that the sources are Gaussian, which is often inappropriate. Algebraic techniques utilizing the information present in the higher-order statistics exist. The most direct technique relies on the diagonalization of the fourth-order tensor derived from the fourth-order cross-cumulants (Cardoso 1995). Other similar techniques exists where one minimizes functions expressed in terms of higher-order moments or cumulants of the input signals, which represent series expansions of the mutual information (Comon 1994; Yang & Amari 1997).

The term independent component analysis (ICA) refers to finding the statistically independent sources responsible for the data. The approach developed by Bell and Sejnowski (1995) uses a model where the mixed input is presented to a neural network that uses a sigmoidal nonlinearity to transform the input into the output. The weights are then varied so that the joint entropy of the output is maximized. In maximizing the joint entropy of the output one expects to minimize the mutual information between the outputs thus ensuring their statistical independence.

The maximum likelihood approach uses a latent variable model of the sources and finds a mixing matrix that maximizes the likelihood of the model given the data. It has been shown (MacKay 1996; Pearlmutter & Parra 1996; Cardoso 1997) that the maximum likelihood technique is equivalent to ICA where the models of the probability distributions of the source amplitudes are represented by the sigmoidal nonlinearites in Bell and Sejnowski's neural network. The maximum likelihood approach is essentially equivalent to the Bayesian approach, except that in the former the assumption is implicitly made that the mixing matrix prior is a constant.

3 Derivation of the Bell-Sejnowski Algorithm

Here we present a derivation of the Bell-Sejnowski algorithm starting with Bayes Theorem (a similar derivation can be found in MacKay 1996). The main purpose of this derivation is to elucidate the assumptions that result in the algorithm. We use a model where the data, $\mathbf{x}(t)$, is generated by a linear transformation of the source behavior, $\mathbf{s}(t)$, as described in eqn. (1). The first simplifying assumption is that the mixing matrix, $\mathbf{A}$, is square, i.e. the number of detectors, m, equals the number of sources, n. This greatly simplifies the formulation of the problem since the inverse of a non-singular square matrix is well defined. To relax this assumption would most likely require the use of the pseudoinverse of the matrix $\mathbf{A}$ (Penrose 1955). In an experimental setting, we usually have a situation where the number of detectors is much greater than the number of the sources, and intuition would suggest that this extra information would increase our ability to infer the source behavior.

Let $\mathbf{x}$ represent the n-dimensional vector of mixtures, and denote an individual mixture as x_i, where x_i is understood to be the time-series data received by detector i. Likewise denote the n-dimensional vector of sources as $\mathbf{s}$, and an individual source as s_i, where s_i is understood to be the time-series representing the signal emitted by source i. Then the probability of the mixing matrix and the sources, given the data and any prior information, I, is given by Bayes Theorem:

$$P(\mathbf{A},\mathbf{s}\,|\,\mathbf{x},I) \propto P(\mathbf{x}\,|\,\mathbf{A},\mathbf{s},I)\cdot P(\mathbf{A}\,|\,I)\cdot P(\mathbf{s}\,|\,I). \tag{4}$$

Since we do not know much about the mixing matrix, A, we can indicate our ignorance by assigning a flat prior that is constant for all appropriate matrices,

$$P(\mathbf{A}\,|\,I) = \text{constant}\,, \tag{5}$$

resulting in

$$P(\mathbf{A},\mathbf{s}\,|\,\mathbf{x},I) \propto P(\mathbf{x}\,|\,\mathbf{A},\mathbf{s},I)\cdot P(\mathbf{s}\,|\,I). \tag{6}$$

Since there is a relationship between the mixing matrix, **A**, the sources, **s**, and the mixtures, **x**, we do not need to find both **A** and **s**. We can simplify the problem by treating **s** as a nuisance parameter and marginalizing over **s** by integrating the posterior probability with respect to **s**:

$$P(\mathbf{A}\,|\,\mathbf{x},I) \propto \int d\mathbf{s}\; P(\mathbf{x}\,|\,\mathbf{A},\mathbf{s},I)\cdot P(\mathbf{s}\,|\,I). \tag{7}$$

Now we make the assumption that the sources are statistically independent so that $P(\mathbf{s}|I)$ can be represented as the product of the probabilities of the independent sources, $p_j(s_j)$. In addition, assuming the noiseless case, we can write the likelihood of **x** as a product of □$(x_i - A_{ik}\,s_k)$ for each of the mixtures, where the Einstein summation convention is used to denote the matrix multiplication:

$$P(\mathbf{A}\,|\,\mathbf{x},I) \propto \int d\mathbf{s} \prod_i \delta(x_i - A_{ik}\,s_k)\cdot \prod_l p_l(s_l) \tag{8}$$

$$\propto \frac{1}{\det \mathbf{A}} \prod_l p_l(A^{-1}_{lk}\,x_k) \tag{9}$$

Looking at the logarithm of the probability we get

$$\log P(\mathbf{A}\,|\,\mathbf{x},I) = -\log\det\mathbf{A} + \sum_l \log p_l(A^{-1}_{lk}\,x_k) + C \tag{10}$$

where the constant, C, is the logarithm of the proportionality constant implicit in eqn. (9). The results are simplified if we write **W** □ $\mathbf{A}^{-1}$:

$$\log P(\mathbf{A}\,|\,\mathbf{x},I) = \log\det\mathbf{W} + \sum_l \log p_l(W_{lk}\,x_k) + C \tag{11}$$

To obtain the best estimate of the mixing matrix **A** we want to find the maximum of the log of the posterior probability, $P(\mathbf{A}|\mathbf{x},I)$. We proceed by taking the derivative of the log of the posterior probability with respect to the matrix elements of **W**. We define as in eqn. (3), $u_i = W_{ij}\, x_j$.

$$\frac{\partial}{\partial W_{ij}} \log P(\mathbf{A} \mid \mathbf{x}, I) = \frac{\partial}{\partial W_{ij}} \left[\log \det \mathbf{W} + \sum_l \log p_l(W_{lk}\ x_k) + C \right] \tag{12}$$

$$= A_{ji} + \frac{\partial}{\partial W_{ij}} \left[\sum_l \log p_l(W_{lk}\ x_k) \right] \tag{13}$$

$$= A_{ji} + \frac{\partial u_i}{\partial W_{ij}} \frac{\partial}{\partial u_i} \left[\sum_l \log p_l(u_l) \right] \tag{14}$$

$$= A_{ji} + x_j \frac{\partial \log p_i(u_i)}{\partial u_i} \tag{15}$$

$$= A_{ji} + x_j \left(\frac{p_i'(u_i)}{p_i(u_i)} \right)_i \tag{16}$$

This can be expressed in matrix form

$$\frac{\partial}{\partial \mathbf{W}} \log P(\mathbf{A} \mid \mathbf{x}, I) = \mathbf{W}^{-T} + \left(\frac{p_i'(u_i)}{p_i(u_i)} \right)_i \mathbf{x}^T, \tag{17}$$

where the ratio of the probability density of u_i to the probability distribution of u_i is a column vector.

Due to the high dimensionality of the space, (n □ n), we search for the maximum of the log of the posterior probability using a stochastic gradient method. We begin by making an initial guess as to the value of **W**, and ascend the gradient while updating our estimate by adding a term proportional to the derivative in eqn. (17):

$$\mathbf{W}_{i+1} = \mathbf{W}_i + l\,\Delta\mathbf{W} \tag{18}$$

where $\Delta\mathbf{W}$ is $\frac{\partial}{\partial \mathbf{W}} \log P(\mathbf{A} \mid \mathbf{x}, I)$ as in eqn. (17), and l is the proportionality constant or learning rate. However, the equation above is not covariant since $^{\mathrm{I}}_{\mathrm{N}}\mathbf{W}$ is not actually a matrix, but is a gradient of a scalar field, or a one-form. The above equation can be made covariant by post-multiplying the gradient by $\mathbf{W}^{\mathrm{T}}\mathbf{W}$ (Amari et al. 1996; MacKay 1996) resulting in:

$$\Delta \mathbf{W} = \mathbf{W} + \left(\frac{p_i'(u_i)}{p_i(u_i)} \right)_i \mathbf{u}^T \mathbf{W} . \tag{19}$$

The data is used to provide an unbiased estimate of the probability distribution and density above, and the algorithm is iterated until it converges on a solution. This result is identical to the Bell & Sejnowski neural network implementation of the ICA algorithm.

The advantage of the Bayesian treatment over other approaches is that the formalism makes explicit the assumptions that go into the mathematical description of the problem. In our example we assume nothing about the form of the mixing matrix, which is made explicit by the fact that the prior $P(\mathbf{A}|I)$ is uniform. We assumed that the data is noise-free which was made explicit by the use of a delta function to express the likelihood of the data given the model. Finally, the sources were assumed to be independent which is represented by the fact that we wrote the joint source prior as a product of the individual source priors, $p_i(u_i)$, whose forms can then be explicitly described.

As we will show, the choice of the source prior can be a delicate matter. In ICA the most commonly used priors are the typical threshold functions used in neural models such as the sigmoid, hyperbolic tangent, and arctangent. If one can estimate the mean and variance of the sources, we may be tempted to use a Gaussian MaxEnt prior. However, use of the Gaussian prior can be shown to only diagonalize the covariance matrix of the data, which results in an algorithm akin to principal component analysis. Another notable approach (Pearlmutter and Parra 1996) uses a more elaborate prior that amounts to a sum of sigmoid functions.

In the case that there is noise present in the data, one can replace the delta function representation of the likelihood in eqn. (8) with a Gaussian distribution:

$$P(\mathbf{x} \mid \mathbf{A}, \mathbf{s}, I) = \prod_i \frac{1}{\sqrt{2\pi}\sigma} Exp\left[\frac{(x_i - A_{ik}s_k)^2}{2\sigma^2} \right] \tag{20}$$

where σ is the variance of the noise. In general this technique is useful, but depending on the form of the source priors the integrals may not be analytically solvable. In the case of EEG or MEG data, event-related potentials have been averaged over many stimuli and the variance of the noise in the channels can be estimated from the pre-stimulus portion of the recordings. This variance could be used in conjunction with eqn. (20) above to account for the noise in EEG or MEG data.

4 Demonstration of the Bell-Sejnowski Algorithm using Audio Signals

In this section we apply the algorithm to a set of artificially mixed audio signals. The set is comprised of sounds from the television show "Star Trek" and is composed of four speech sounds from four different speakers and two artificial sounds, a hail (whistle), and a photon torpedo blast (a noisy frequency glide). No artificial noise has been added to the signals. In keeping with Bell and Sejnowski's original implementation, we choose a sigmoid function as the source prior distribution (Figure 1)

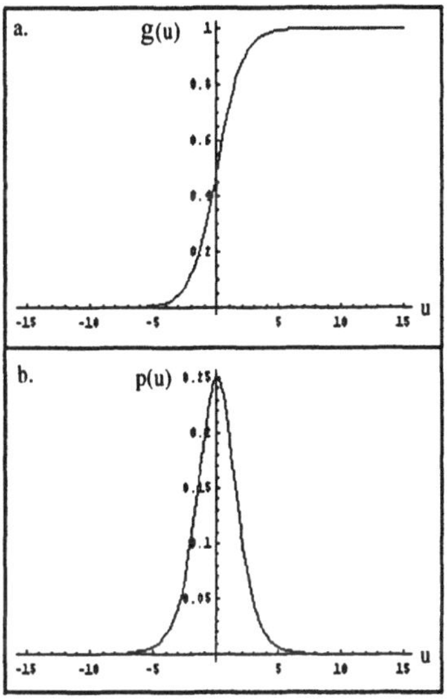

Figure 1. (a) The sigmoidal probability distribution, (b) and its probability density.

$$g(u)=\frac{1}{1+e^{-u}}, \tag{21}$$

so that the source prior density is

$$p_i(u_i)=\frac{dg(u_i)}{du_i}. \tag{22}$$

By choosing this source prior we are assuming that the probability density functions (p.d.f.s) of the sources are more sharply peaked than a Gaussian density, i.e. the source p.d.f.s have a high kurtosis.

Figure 2 shows the original signals on the left, the artificially mixed signals in the middle and the unmixed signals on the right. Note that the speech sounds are completely unmixed, although permuted, whereas the other two sounds (the "Hail" and the "Photon Torpedo") are still mixed with one another (Unmix 1 and Unmix 3). This pathological solution is called a diagonal solution, and it can be shown to consist of the sum and difference of re-scaled versions of the two source waveforms, as represented by the matrix multiplication:

$$\begin{pmatrix} u_1 \\ u_2 \end{pmatrix}=\begin{pmatrix} 1 & 1 \\ -1 & 1 \end{pmatrix}\begin{pmatrix} a & 0 \\ 0 & b \end{pmatrix}\begin{pmatrix} s_1 \\ s_2 \end{pmatrix}. \tag{23}$$

The original waveforms, up to a scale factor, can be restored by taking the sum and the difference of the two diagonal solutions. Interestingly, the incorrect diagonal solutions have p.d.f.s (Figure 3) which better match the source prior in equation (22). The diagonal solution is well described by Bell & Sejnowski (1995), and it is easily visualized by considering two uniformly distributed sources. By taking the sum and difference of two uniformly distributed sources one obtains mixtures that have triangular p.d.f.s that more closely match a peaked density like the sigmoidal source prior. This is an excellent example of how the choice in the source prior can affect the results of the separation.

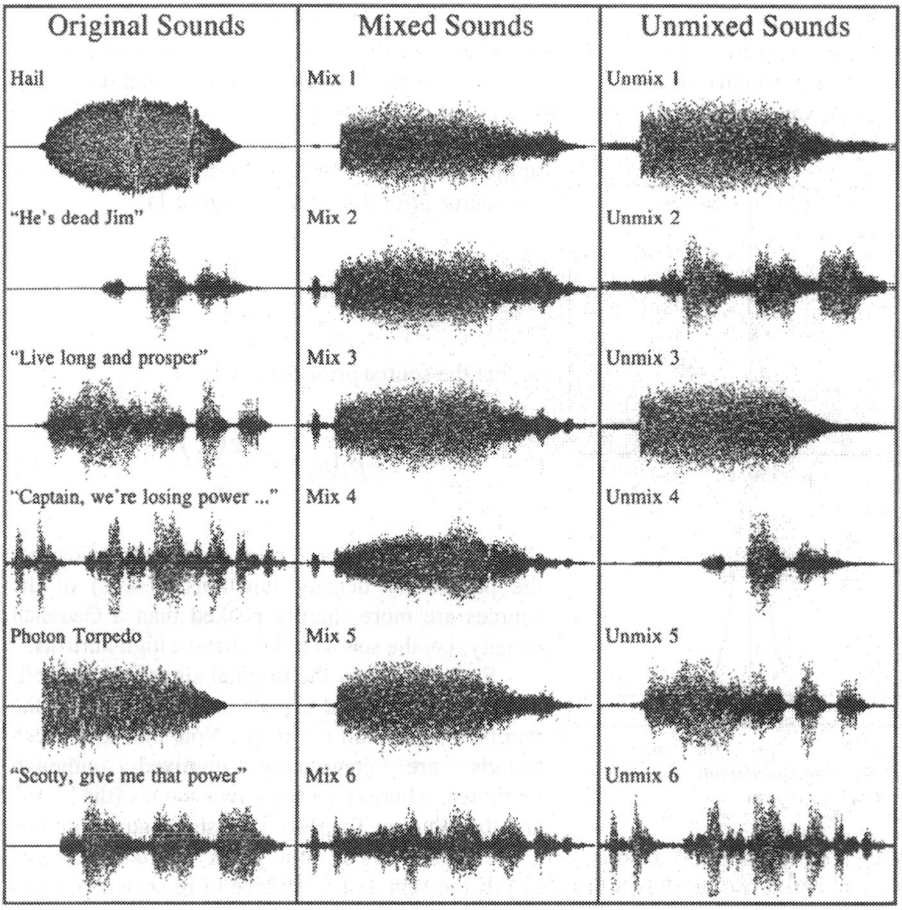

Figure 2. (Left) Original source waveforms. All are speech sounds except the Hail and the Photon Torpedo, which are a whistle and a noisy frequency glide. (Center) Six distinct mixtures of the original sources used to test the ICA algorithm. (Right) The six waveforms unmixed by the ICA algorithm. Note that the speech sounds are completely unmixed, however, the Hail and the Photon Torpedo are not (Unmix 1 and Unmix 3). These waveforms, Unmix 1 and Unmix 3, are interesting because they represent the sum and difference of the original sources.

5 Application of the Technique to EEG and MEG Signals

The electric and magnetic signals detected by EEG and MEG are generated by highly correlated activity in the brain involving hundreds of thousands of synchronously active neurons. The signals are lower in frequency than are audio signals, with a bandwidth ranging from about 1-100 Hz. Thus the EEG/MEG signals exhibit less variation than audio signals and are much less rich in appearance.

One common technique is the acquisition of event related potentials (ERPs) in which one records electrical events time-locked to a sensory stimulus. By repeatedly presenting the stimulus and averaging the electrical responses, one obtains waveforms consisting of mixtures of signals correlated to the stimulus. These signals represent the average activity from the various neural generators that are evoked by the stimulus. Application of these source separation techniques to EEG and MEG signals would be advantageous. By isolating the responses from different neural generators one could not only accurately locate the anatomical origin of the signals, but also study the dynamical behavior of the generators. In addition, one could also separate the noise from the signal without applying Fourier techniques that can distort the data.

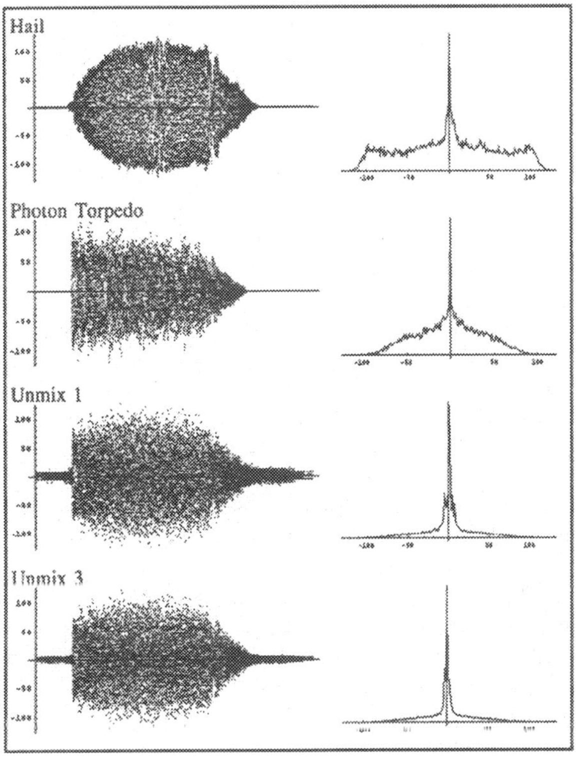

Figure 3. (Left) The Hail, Photon Torpedo, and their diagonal solutions, which represent the sum and difference of the original waveforms. (Right) Histograms of the amplitudes of the waveforms, which represent their probability density functions. Note that the diagonal solutions have p.d.f.s that better match the source prior density. In addition, the large amount of baseline noise also assists in matching the p.d.f.s of the diagonal solutions with the density assumed by the source prior.

We now demonstrate the application of the algorithm on an auditory event related response recorded using MEG. The response was evoked using a 0.5 ms square-wave condensation click presented to the right ear. The radial component of the magnetic field was recorded over the left hemisphere of the subject's head using a 37-channel biomagnetometer. Responses were averaged 512 times to extract the signals that are time-locked to the stimulus. The data was sampled at a sampling rate of 1041.7 Hz, and was conditioned between 0.1 and 400 Hz by using digital high and low pass filters. It can be shown that filtering does not affect the mixing matrix; this is because filtering is a linear operation that acts on each channel identically. This particular data set was chosen because the stimulus delivery was time-locked to a 60 Hz cycle, which prevented the 60 Hz line noise from being cancelled out during averaging. This 60 Hz signal provides us with an extra source, which can be used to evaluate the success of the technique. By applying singular value decomposition to the data we estimated that there were about 7 sources responsible for the data. Greater and fewer numbers of sources were tested without an increase of success.

The left side of Figure 4 shows 7 of the 37 original recorded waveforms. These waveforms were selected because they spanned the set of typical waveforms observed across the detector. Physiologically significant peaks in the responses can be seen clearly in the waveform recorded from channel 35. The two small peaks between 0 and 0.1 s are the magnetic equivalent of the electrical middle latency response. The large negative peak just after 0.1 s is known as the M100, and two similar peaks can be seen at 0.2 s and 0.3 s. Channels 5, 11, and 34 show a preponderance of 60 Hz noise, which is known to come from a source independent of the recorded brain activity.

The right side of Figure 4 shows the waveforms that result from applying the ICA algorithm to the data. At first glance the waveforms seem to be relatively independent of one another, however, one can see several peaks that are replicated throughout many of the "unmixed" waveforms. One would also expect that the "unmixed" waveforms should have less noise than the original waveforms and this is obviously not the case. Examination of the power spectra of the "unmixed" waveforms reveals that there is still a large amount of 60 Hz noise in "unmixed" waveforms 2, 5, 6, and 7 demonstrating that the 60 Hz noise has not been extracted from the data.

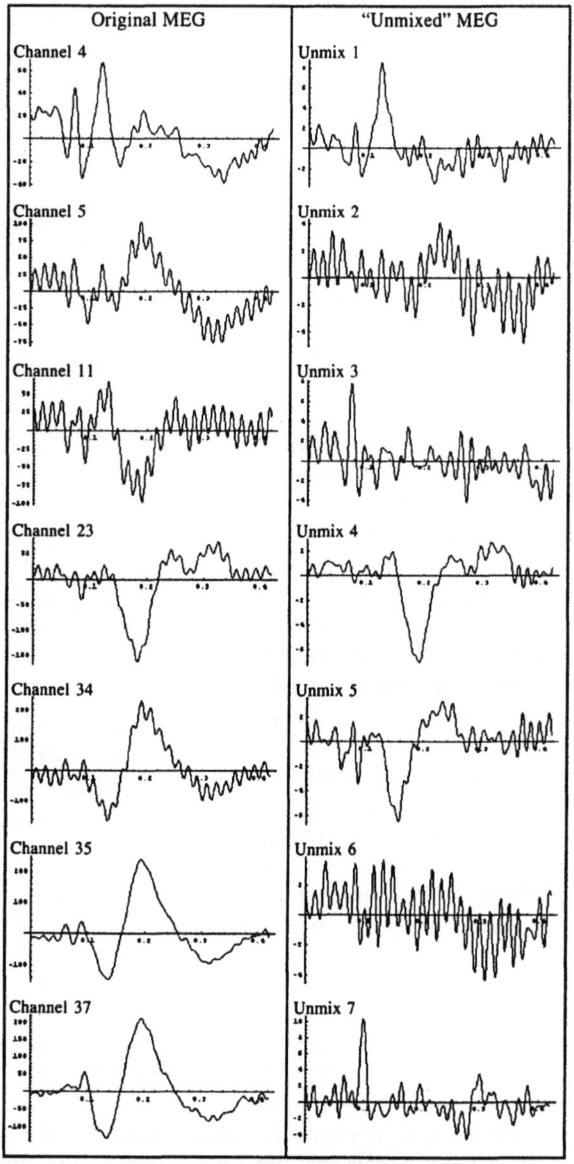

Figure 4. (Left) Original MEG signals (auditory evoked fields). Note the 60 Hz noise in channels 5, 11, and 34. (Right) "Unmixed" MEG signals. At first glance the waveforms may appear to be independent, but several peaks are still replicated throughout the waveforms. In addition, there is still a prevalence of 60 Hz noise in the "Unmixed" waveforms 2, 5, 6, and 7.

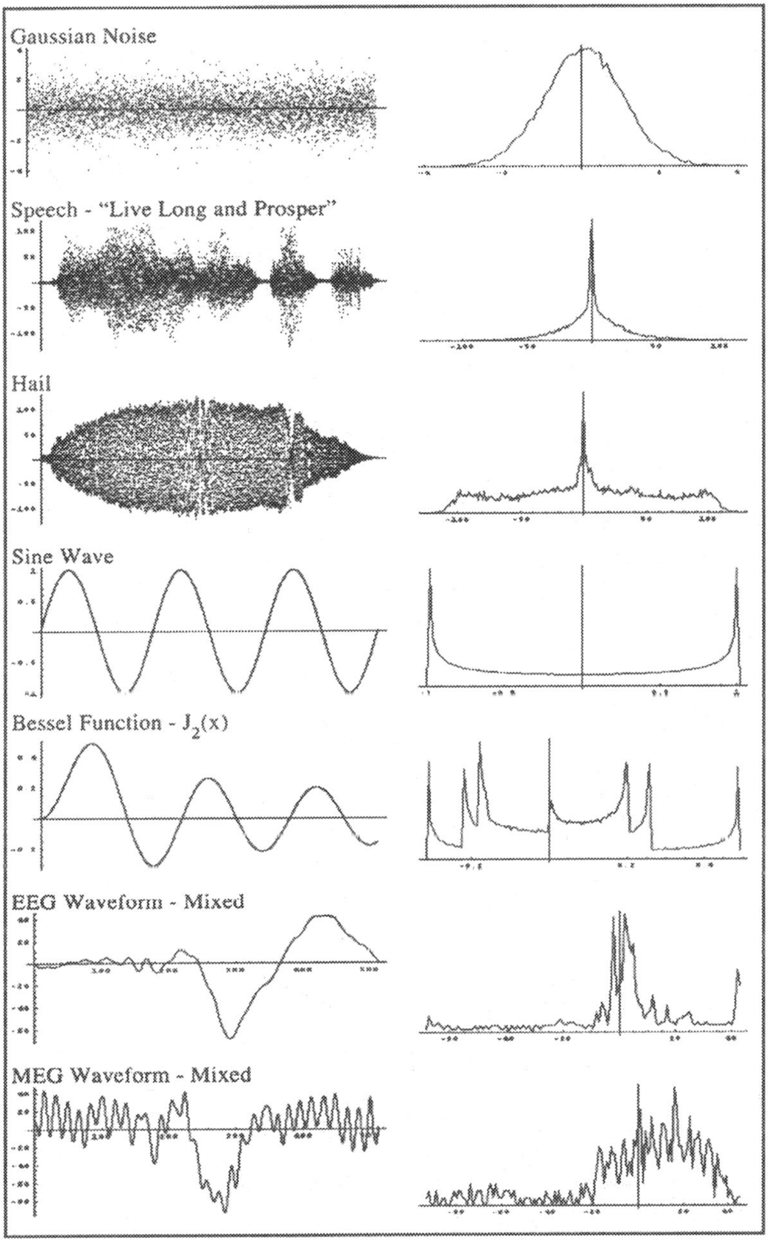

Figure 5. (Left) Various signals represented by time-series data. (Right) Histograms of the amplitudes of the corresponding time-series. In the limit these histograms approach the probability density functions of the signals' amplitudes. (The scale on the vertical axis is absent for clarity. It is understood that the areas are normalized to unity.)

6 Probability Density Functions of Various Signals

A better appreciation for the blind source separation problem can be obtained by examining the p.d.f.s of some of the signals in the examples above. Figure 5 shows waveforms represented by time-series data and the histograms of their amplitudes, which approximate the p.d.f.s of the waveforms. The first waveform shown is Gaussian noise. Contrast this Gaussian p.d.f. with the p.d.f. of the spoken phrase, "Live long and prosper", which has a high kurtosis, visible from the sharp central peak and heavy tails. The broad, flat tails of the p.d.f. of the "Hail" waveform was the cause of the pathological solution described above. From these three examples and the discussion on diagonal solutions one can see how the sigmoid prior is well adapted for separating speech sounds, yet fails to separate sources with broad p.d.f.s. The following waveforms are more similar to the EEG/MEG signals of interest. The sine wave is a good example as it can be shown analytically that its p.d.f. is a scaled secant function with peaks at ± 1. These peaks come from the fact that the derivative of the waveform is zero at amplitudes of ± 1. Another illustrative example is the Bessel function, $J_2(x)$. This waveform is qualitatively more similar to what we would expect from a neural source. Again it demonstrates that the p.d.f. will have peaks at amplitudes where the derivative of the waveform is small. These multi-modal distributions have a low kurtosis and it can be shown that the ICA algorithm as such cannot properly separate them.

The last two waveforms are of EEG and MEG data from the experiment described above. It is important to remember that these waveforms are mixtures of sources and thus the p.d.f.s are derived from a convolution of the p.d.f.s of the original sources. With this in mind the overall Gaussian-like shape of the p.d.f.s is not surprising since we expect the Central Limit Theorem to be at work. In addition, the strong spikes in the p.d.f. are due to the spikes in the original source p.d.f.s. This is especially obvious in the MEG waveform where there is sinusoidal 60 Hz noise. The convolution of the secant function, representing the p.d.f. of the noise source, with the MEG source p.d.f.s results in a p.d.f. with a series of strong spikes. Figure 6 shows the p.d.f.s of the first three "unmixed" MEG waveforms. Note that even waveforms 1 and 3, which seem to show rather distinct source-like behavior, have p.d.f.s that are to be expected from mixtures.

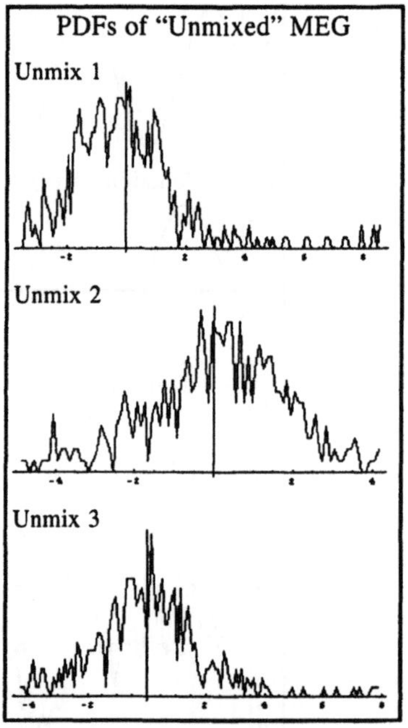

Figure 6. Histograms of the amplitudes of the first three "Unmixed" MEG waveforms. Note the spiky Gaussian-like shape that fits the assumed source prior density.

The fact that one expects mixtures to have Gaussian-like p.d.f.s explains why the simple technique of PCA fails to separate EEG signals. In addition, one can see that in the context of ICA, where the source prior densities have a high kurtosis, one also expects mixtures of EEG/MEG sources to better fit the source priors. This effect is similar to the pathological diagonal solutions that result when sources have broad p.d.f.s.

7 Discussion

It is often true in Bayesian analysis that the assignment of the prior probabilities is not a critical issue, and this is demonstrated by the impressive ability of ICA to separate speech sounds. Indeed, it is conjectured that the typical source priors utilized in ICA, such as the sigmoid, hyperbolic tangent, and arctangent functions, are sufficient to separate any sources with large kurtosis. A partial proof of this conjecture is given by MacKay (1996) for the case of two sources having the same high kurtosis distributions. Pathological solutions, however, are known to exist for cases where the source p.d.f.s are broad compared to the sharply peaked source prior p.d.f.. We have demonstrated that the separation of EEG and MEG signals is not possible using sharply peaked source priors, and these results suggest that multiply peaked source densities may need to be utilized to obtain proper separation.

It is important to focus on several of the assumptions made in the derivation and application of this technique. Some of these assumptions facilitate the mathematical description of the problem, while others reflect our ignorance about the nature of the signals and the sources.

Three assumptions deal with the mathematical description of the problem. First, we assume the same numbers of sources as detectors. It is expected that with more detectors we have more information about the sources and separation would be facilitated. In this case, however, we are no longer dealing with square mixing matrices, and we would need to develop a solution using pseudoinverse matrices. Second, there is a difficulty in estimating the number of sources. One of the most popular techniques is to use singular-value decomposition, but it is known that the technique is not robust to noise. The advantage of the Bayesian approach here again is that the ratio of probabilities obtained from models using different source numbers can be computed allowing us to choose the optimal model. Third, we have demonstrated the necessity for finding generic source priors that are effective in separating signals with multimodal p.d.f.s. The high-kurtosis source priors that are typically used in ICA result in producing mixtures rather than sources. Generic source priors that can accommodate multimodal p.d.f.s will most likely require many adjustable parameters. Pearlmutter and Parra (1996) suggest using sums of sigmoid functions. However, it is often the case that algorithms become unstable as the number of parameters to be fit becomes large.

We briefly address another assumption that deals with a more difficult issue, that of statistical independence of the sources. Physiologically, strong dynamical coupling is likely to exist between many sources, and this may pose a problem for the separation algorithm. Although these algorithms can often separate sources exhibiting some dependence, this is probably a matter of degree. A more difficult case would be one where a spatially extended active region exhibits spatially-varying dynamical behavior. This is a continuous generalization of the case of discrete coupled sources, and brings into question the validity of the notion of neural generators.

Acknowledgements

I would like to thank the following people for their assistance, advice, and support: John Broadhurst, Steve Cobb, Joslyne Foley, Irv Hochberg, Lacey Kurelowich, Barry Schwartz, Patti Quint, and Herb Vaughan. This work was supported in part by Albert Einstein College of Medicine, Biomagnetic Technologies Inc., City University of New York Graduate Center, Scripps Research Institute, and NIH NIDCD 5 T32 DC00039-05.

References

Bell, A.J. and Sejnowski, T.J. (1995) An information-maximization approach to blind separation and blind deconvolution, *Neural Computation*, **7**, 1129-1159.

Cardoso, J.-F. (1995) A tetradic decomposition of 4th-order tensors: application to the source separation problem, in M. Moonen and B. de Moor (eds.), *Algorithms, Architectures and Applications, Volume III of SVD and Signal Processing*, Elsevier, Amsterdam, pp. 375-382.

Cardoso, J.-F. (1997) Infomax and maximum likelihood for source separation, *IEEE Letters on Signal Processing*, **4**(4), 112-114.

Comon, P. (1994) Independent component analysis, a new concept?, *Signal Processing*, **36**, 287-314.

MacKay, D.J.C. (1996) Maximum likelihood and covariant algorithms for independent component analysis, Draft Paper, http://wol.ra.phy.cam.ac.uk/mackay/.

Pearlmutter, B.A. and Parra, L.C. (1996) A context-sensitive generalization of ICA, 1996 International Conference on Neural Information Processing, Hong Kong.

Penrose, R. (1955) A generalized inverse for matrices, *Proceedings of the Cambridge Philosophical Society*, part 3, pp. 406-413.

Yang, H.H. and Amari S. (1997) Adaptive on-line learning algorithms for blind separation – maximum entropy and minimum mutual information, To appear in *Neural Computation*.

THE HISTORY OF PROBABILITY THEORY

ANTHONY J.M. GARRETT
63 High Street, Grantchester
Cambridge CB3 9NF, United Kingdom

Abstract. An outline is given, essentially from the Bayesian point of view, of the history of probability. Because probability theory is mathematical today, histories have tended to suppose it began with the first deeply mathematical exchange, between Pascal and Fermat in the 17th century. In fact Pascal and Fermat simply led the translation of pre-existing ideas into mathematics. Special attention is paid to where these ideas came from in the prehistory of quantitative probability theory. There are acknowledged sources in philosophy but the strongest connection is with law, through the notion of the extent to which guilt is implied by evidence.

Key words: probability, induction, inductive logic, history

1. The Aim of a History

In line with their purpose, expositions of probability are ordered for logical clarity. This order does not coincide with the history of the subject; in probability theory, ontogeny and phylogeny are profoundly dissimilar. Here, an outline history of the subject is given. The history of ideas is not less contingent than the history of events, and things might well have been otherwise – even if we lack the imagination to see how. Therefore a history picks out a story. This story serves to remind us that we are in a tradition, for ideas that today seem obvious were once pioneering advances – a truth extending to the most basic of notions and even beyond, into things we take so much for granted that our conscious minds have not yet learned of their presence in the subconscious. The evolution we trace in the meaning of the word 'probable' serves to remind us that other words will have changed in meaning, too, and that historians must learn to understand the mentality of the ages they study: must *indwell* their subject.

Of course, the past needs continual re-appraisal in the light of the present, varying from a pruning of ageing fine detail to complete reinterpretation in the light of some huge event. The study of history has been well described as a dialogue between past and present. The outline given here proceeds, naturally, from the author's point of view (outlined in the next section). In acknowledging my point of view I am simply making explicit what many authors leave implicit; but I do not accept the contemporary claim that the writing of history is solely an exercise in power ('by the winners'). For, if it is asserted that there is no truth, how can such an assertion be true? In particular, there are good reasons to choose one view of probability over another, and so to emphasise certain historical observations. But how to justify *those* reasons? There is a hierarchy of levels ('meta-languages'), each capable of weighing the previous, extending at least to the cultural level; this hierarchy is capable of termination only by acceptance of something that cannot be proved: by faith. All have such a faith, of some sort – though not all acknowledge it.

2. Probability Theory

I take probability theory to be a universal calculus for handling uncertainty, involving a logical measure of the extent to which the truth of one binary proposition (A, say) is implied by the reported truth of another (B), according to the ontological relations between the things to which A and B refer; call this measure ***implicability***. Logic is the programme of making formal and explicit the processes of human reasoning; and in human reasoning we can discern this idea of implicability. For example, how strongly does the proposition that the results of tossing a coin are as stated imply that the next toss will show heads? How strongly does the proposition, that the experimental data take the values observed, imply that a parameter of interest takes a particular value? A little introspection, moreover, indicates that implicability can be placed on a scale: the more bad apples are pulled out of a sack, the more strongly this implies intuitively that the next apple will be bad. Therefore implicability satisfies an ordering, and is transitive, and therefore may be represented by a real number; denote this number $p(A|B)$. This notion of implicability is, as the examples indicate, precisely what is needed in problems where all agree that 'probability theory' is relevant. On these grounds I shall henceforth call implicability 'probability'; but it is the idea that counts, not the name, and anyone disagreeing with this definition of probability is free to hold another (and make a global change of probability to implicability in this paper from now on). Since implicability is always what is actually needed, however, other definitions will not be useful.

Another view is that probability theory deals with how strongly it is appropriate to believe that one proposition is true, supposing another is true. This might be called the 'believability' view; as with implicability it can stand alone in order to defuse accusations that probability *must* be identified with believability. Believability is the so-called Bayesian view of the subject, and operationally it is the same as implicability; interpretationally, it carries excess psychological baggage that makes it controversial.

3. Where to Look

We have seen that the study of probability is the making explicit of principles for reasoning about one thing from another, when there is absence of certainty. While humans often find themselves needing to act in the absence of certainty, pressure to make explicit the reasoning for their actions comes through having to explain or justify those actions to others. Moreover the deepest and most subtle situations in which people act are those involving other human beings; and the pressure to justify is greatest in adversarial situations. Hence law, which is how an organised society formalises adversarial situations, is the principal spur to development of notions of probability. More specifically, it is the reduction of a legal case to a proposition which one party holds is true and the other holds is false, and the motivated searching for evidence (including witness statements), followed by evaluation of that evidence to find the probabilities that the proposition is true or false – a classic inverse problem. Since the same evidence is often available to both parties, the sum rule is already on the horizon. The development of legal systems is therefore the first place to look for the roots of probability, together with guides for legal argumentation.

As law provided the stimulus to probability theory, ethics provided the stimulus to decision theory, through the question of what the individual is supposed to do, and why, in well-defined and reasonably common moral quandaries. Of course, law is secondarily relevant to decision theory (discretion in sentencing; innocent or guilty), and ethics to probability theory (since decision theory under uncertainty presupposes probability theory).

The urge to quantify the underlying principles came from problems, involving uncertainty, in which sums of money are involved: from risk management, in particular the setting of insurance premiums and annuities. Risk management took on increasing importance as commerce grew; it is the need to solve problems of cultural concern, rather than philosophising *per se*, which drives genuine innovation. Gaming, which is often taken to be a spur to quantification, is more accurately seen as a source of model problems.

4. Probability in the Strands of Western Culture

The Western tradition in which quantitative probability theory originated is a confluence of two streams: the 'classical' tradition which coalesced in Greece, and which the Roman empire took as its own culture on conquering the eastern Mediterranean; and the Hebrew tradition which mingled with classical ideas in Rome as the Christian church there grew to dominance. The legal connection is immediate in Rabinovitch's study (1973) of probabilistic ideas in Jewish writings; examples almost always involve the making of just decisions in which parties have differing interests. In ancient Greece, there was a great flowering of civilisation in the 5th century BC. Aristotle (384–322BC), writing more than a generation after the decline marked by the Peloponnesian war, wrote an encyclopaedia of works which summarised and enormously extended the knowledge of the day. This remains a staggering achievement. Aristotle's treatment of logic was contained in several works, which centuries later was collected into the *Organon*. Of these the work he calls ***Analytics***, in particular the part we call the ***Prior Analytics***, contains his comprehensive treatment of syllogistic logic. Syllogistic logic concerns membership of categories, in which a third statement (for example, "Socrates is mortal") is implied from two others which have something in common ("Socrates is a man" and "all men are mortal" – a pointed example). Aristotle recognised that not all reasoning is of this sort, but it was left to the later Stoic school to phrase logic in terms of propositions. (A useful guide to classical Greek probability is by Sambursky, 1956.) Aristotle certainly understood that non-tautological reasoning went on in the absence of certainty, and in his *Rhetoric* – a practical guidebook for argumentation – there is recognition that the aim is to get people to alter what might be called their probabilities of something; and to do it by systematic argument rather than tricks with words as was then common, particularly in courts of law. Aristotle goes into these systematic arguments in depth; even in this philosopher *par excellence*, deployment of probabilistic ideas is treated best in his work of immediate relevance to the life of his civilisation.

The word *eikos* (in our alphabet) is used to denote what we mean by (fairly) probable; later on, *eulogian*, originally 'reasonable', was to mean much the same. Another Greek word, *pithanon*, means 'persuasive'.

5. Fall and Rise of the West

The classical era came to an end when, for reasons still debated, the Western half of the Roman empire lost its unity and identity and collapsed. The spread of illiteracy is documented in the signing of documents with the symbol of the cross (and only influential people had to sign documents); cultural achievement fell dramatically. This poorly documented time is the age in which the conversion of Europe to Christianity was carried through. When thought about the world was able to flower afresh in the 11th century, it was in a Christian setting. From this point classical authors became available anew, having been translated through Arabic, and universities were founded to study them. Arabic scholars had discussed the fair division of inheritance and wagering; although Islam banned actual games of chance.

The new ideology, working to bear fruit through the scholars of a reunified European civilisation, was this. First, that everything had a cause because everything was part of the divine plan; and that, because the author of that plan had made a full revelation of himself and of where and in what form ultimate truth lay, causes were in principle knowable by man who is in the image of that author. The Christian manifesto set forth by Augustine of Hippo, as the empire was collapsing, lays great stress on this. (A discussion is given by Cochrane, 1944.) This belief is a necessary precursor to the systematic character of Western science, since if you do not believe that you are able to find out the truth you will never look for it, nor recognise it should you stumble upon it. (Advocacy of a secular version of this idea, as mechanical determinism, was to bring unjust derision on Laplace; determinism can never be proved but must always be assumed.) Another new ingredient was the doctrine of creation, establishing sovereignty of creator over created – a distinction which makes possible a description of the created using only descriptive tools that are themselves created: such is science. These ideas together drive out pantheistic explanations involving reference to capricious or limited deities such as classicism's *fortuna*, and justify the seeking of explanations for 'signs' – claims of empirical correlations between things seen and things not yet seen. Another

Christian idea is the notion of contingency, that the omnipotent creator was free to create as he chose and might have created differently, so that to see how things were you had to interrogate them. Here begins the systematic dialogue between theory and designed experiment that is unique to post-classical Western science; although it took many centuries more before the capacity for visualisation that is involved in designing an experiment was developed sufficiently. That this capacity was absent in classicism is evident from the Greek failure to see that, for example, two similar bodies attached by a loose piece of string, comprising a composite body of twice the mass of either, would fall at the same rate as either one alone; the Greeks believed that rate of fall depended directly on mass. The air of classicism is more different than we often suppose.

From here the line points ineluctably to the Bayesian viewpoint that probability is needed because of human ignorance of something that might be known, and away from the frequentist viewpoint that nature is intrinsically unknowable ('random'). However it took centuries more before quantification could come about; and frequentism had not been extinguished before the decline of Western Christian faith began, and before frequentism rose to dominance in the 19th century. But this is to anticipate.

6. Pre-quantitative Probability

In mediaeval documents – written in Latin, which was more developed than vernacular tongues and was comprehended Europe-wide – it is therefore to the areas of law, ethics, insurance and gaming that we should look in order to see how the foregoing ideas transmuted classical thought towards the present view. Mediaeval law developed from commentaries on a rediscovered copy of the Emperor Justinian's *Digest* of Roman law first published in 533. Like law in other early civilisations, the *Digest* is authoritative on what to do in clear-cut situations – sentences for particular crimes, or how to divide property, for example – but is largely silent on the evaluation of evidence, which was left to the judge. Mediaeval commentary on the *Digest*'s legal prescriptions has played a great role in developing the morality of Western Europe. It is mediaeval development of evidence evaluation (probability assignment) and evidence combination (relating probabilities), so as to make explicit what a judge does in his head or a jury in private debate, which is the direct precursor of modern probability. The connection has been traced by Franklin (1991).

There is likewise an enormous amount of writing on mediaeval moral theology; in particular, the education of the conscience that took place in Confession – a legal term – carries something of legal interrogation. (Confession was also the first time an entire civilisation had been ordered to *think*.) Both probability and decision theory are involved here. The connection of casuistry to probability is indicated by the term *probabilism*, meaning adherence to a course which has been approved by some authority (even though other courses might be approved by other authorities). Casuistry in the immediate pre-quantitative era is discussed by Bellhouse (1988). There was also 'probabiliorism', which is closer to what we recognise as appropriate. On the emergence of scientific thought, a notable scholar is Alistair Crombie; chapter 17 of his monumental *Styles of Scientific Thinking in the European Tradition* (1994) concerns the prehistory, and the rest of Part VI the early history, of quantitative probability theory. Other relevant references, showing how diverse are the strands which came together in probability theory, are Byrne (1968), Sheynin (1974), Garber and Zabell (1979), Kassler (1986), and Kantola (1994).

Though classicism recognised the need to reason under uncertainty, the actual word *probabilis* (or *verisimilis*) originally meant an opinion that was supposedly provable, or was approvable by authority. This usage persisted in the Middle Ages, which were deeply hierarchical: there was a tendency to argue simply by counting authorities on either side, weighting them perhaps by their eminence but not by the logic of what they said – because that would be to judge them, and by their eminence *they* were the ones fit to judge. It was the gradual overturning of this viewpoint, together with the making explicit of a notion of degree, which signified advance. John of Salisbury (*Metalogicon*, 1159: bk. 2 ch. 14) understood that the probability of truth of a proposition was something that could increase or decrease – implying a scale – according to the evidence. The commentators on Justinian's *Digest* had begun the building of that scale by asking what the testimony of a single witness to murder was worth if two were required for conviction, as the Bible demanded (Numbers 35:30; Deuteronomy 17:6). By the late 14th century the scale had been filled, and subdivided as far as is legally meaningful, in the works of Baldus de Ubaldis (1327–1400) (J. Franklin, unpublished). By the Middle Ages the word 'probability' had picked up two of its present-day colloquial connotations: the old classical idea of inferiority to deductive certainty ('merely probable') and the abbreviation of 'highly probable'.

Both probability as relative frequency, stemming originally from gaming, and probability as implicability, stemming from law (continuing its stimulus to the exposing of probabilistic and decision principles), are visible at once in this (translated) quote from Thomas Aquinas (*Summa Theologiae*, written 1265–74, II-II q. 70 art. 2): "a probable certainty is sufficient, which attains the truth in most cases, even though in a few cases it does not. It is probable that what a number says has more truth than what one says." The Latin includes *ut in pluribus*, which might generally be rendered as 'frequently'. Thomas it was who, in his famous distinction between things that can be known by faith and things that can be known by reason, played a large part in re-opening the gap between faith and reason that Augustine had sealed, so making possible the Enlightenment view that they are antagonistic. The Augustinian position is summed up in the phrase *crede ut intelligas* – believe in order to know (Augustine, *d.* 430, sermon 43). For you cannot avoid starting from somewhere, and that starting position is by definition held by faith. (If you argue for it then it is not really your starting position – the stance from which you argue is, and also the tools by which you argue.) Reason is then a way by which you progress from your starting point in making sense of the world as it comes to you. It is Enlightenment championing of reason against faith and hence belief (the more common phrasing of the implicability idea) that is another underlying issue in the frequentist–Bayesian dispute. The dispute clearly taps into deep undercurrents – also involved is the relation between thought and the subject of that thought – and it is plausibly Enlightenment hubris which accounts for the antagonism of frequentists to the Bayesian position.

The next part of the story runs from the Renaissance to the emergence of a quantitative theory that was understood to be applicable across different types of problem. An accessible guide to who did what when from here on is by Hacking (1975; neglect his claim that the idea itself did not exist prior to its quantification). An important book on games of chance was written in the mid-16th century by Cardano (though published only in 1663). Renaissance Italy was the source of many innovations in mathematics, and also in capitalism; insurance is the field of relevance of the special-case connection of implicability to observed frequency. For a history of the financial strand, of how questions like insurance and investment – 'risk management' – has driven probability theory, see Bernstein (1996).

Gaming also poses the probabilistic question of how to divide the stakes in an uncompleted game, according to the state of the game at that point; probability was sometimes *interpreted* in this way (as it was sometimes made meaningful using legal analogy). Combinatorics, too, was studied, originating out of alchemy and divination – the latter having some Biblical precedent in the drawing of lots. Indications of a numerical way of expressing uncertainty in English culture – such as Shakespeare's quoting of odds – are traced by Bellhouse and Franklin (1997). The urge to quantify was famously made explicit by Galileo, the great pioneer of experimental method in physical science, in his statement (1623) that "...this grand book – I mean the universe... cannot be understood unless one first learns to comprehend the language... in which it is written. It is written in the language of mathematics..." (meaning geometry, whose optimal expression in algebra is the main stream in the history of mathematics). Around 1660 there was an explosion of quantitative applications. This was nourished, and in turn fed, by the generality in the abstract character of mathematical expression – as with mechanics in the same era.

7. Quantification

Pascal (1654) and Fermat (1654) had a famous correspondence over a gaming problem which, for a short while, put them clearly ahead in quantifying the idea. Huygens (1657) then wrote a tract devoted to probability in gaming, for which task his native tongue was now better developed than Latin. The influential text *La Logique, ou l'Art de Penser* (1662), known in English as the Port-Royal Logic since it came from that centre of learning (Pascal's), began the advocacy of comparative testing of hypotheses against each other and made quantitative the notion of probability in matters of reasoned belief (though translating it into gaming terminology for comprehension). This work, written by Antoine Arnauld and Pierre Nicole, also advocated the joint use of utility and probability in decision making. Leibniz (*d.* 1716), also writing philosophically in this era, explicitly took the view espoused in the present book, that probability is logic (for details see Hacking, 1975). Significantly, most of these pioneers either had legal training (Fermat, Huygens, Leibniz) or came from a family that dealt in law (Cardano, Pascal, Arnauld). Also, the exploitation of population statistics to work out annuities, and other matters of interest (estimation of the infectiousness of a disease from its fatality statistics per week and per area, for example), came of age in the second half of the 17th century. Life annuities – which become complicated where two or more people are joint recipients – were studied by Hudde and by Jan de Witt (1671) in Holland, both men of public affairs with mathematical competence, and population statistics by Graunt (1662) in England.

After a hiatus came the first great textbook of quantitative probability, summarising the past and pointing to the future, Jacques Bernoulli's *Ars Conjectandi*. It was published posthumously in 1713. Its most important result is a proof of the special-case relation between implicability and relative frequency called the weak law of large numbers. It was published in a burst which included Montmort's book of 1708 and de Moivre's book of 1718 and others, applying the new ideas in the quantitative areas which had spawned them, particularly gaming where probabilities on an appropriate space were trivially equal, so that only the sum and product rules were involved and combinatorics was able to develop. De Moivre was the most powerful probability theorist of his time, and this book and his later one on annuities (1725) became 18th century standards.

8. Enlightenment Probability; Bayes and Laplace

From here on, fruitful advance lay in continuing these scholars' development and application of the quantitative theory, rather than discussion of what it meant. The latter came rather to a dead end with the philosopher David Hume (1739–40, 1748), who rejected inductive logic outright on the *a priori* grounds that the unseen was unknown, and could not be reasoned about from the related seen or from anything else: Hume denied the basic intuition that seeing ever more bad apples pulled from a sack helps us say something about the next apple. The precise relation between induction, implication, belief and probability was not then fully clear (the philosophical tradition debates it to this day), but Hume was an Enlightenment figure; the deeper strands of the Enlightenment were hostile to the notion of faith, and it seems that the relation between faith and belief was close enough for this hostility to be the underlying cause of Hume's scepticism. From Descartes on philosophy had demanded certainty or nothing, and was therefore hostile to the notion of quantifying uncertainty.

The Enlightenment also gave rise to the philosophical view called utilitarianism: that the procedures of assigning numerical probability and desirability, and deploying them in decision theory, are in principle applicable to every decision faced by man. Qualitative decision theory may have begun in an attempt to tackle moral problems, but the stronger claim that *all* matters of morality are capable of being reduced to numbers remains controversial.

Patey (1984) shows that explicitly probabilistic reasoning was seen by the time of the Enlightenment as a normal strand of culture. More technically the Reverend Thomas Bayes of Kent published posthumously in 1763 a paper which by difficult reasoning inverts the binomial distribution – that is, a special case of the theorem which now bears his name. Bayes' name is also attached to the 'believability' view of probability; though it is not clear that his work warrants it.

In the middle of the 18th century, Euler successfully tackled many problems in probability (collected, 1923). The greatest advance, though, came with the work of Pierre Simon, ennobled as the Marquis de Laplace, who did for probability theory as he did for mechanics: he used it to tackle real problems, and in doing so extended it enormously. Before quantification it was problems in law which drove the theory; now it was problems in science. Beginning in 1772 Laplace wrote many papers involving probability and a great textbook, *Théorie Analytique des Probabilités* (1812), the first to fully integrate formalism and application, quantitative analysis and explanatory wordage. The third edition (1820) includes a long introductory essay first published independently. Laplace understood inverse reasoning as probabilistic and used it to tackle diverse inverse problems. He also understood the principle of indifference for distributing probabilities when the conditioning information is indifferent among the possibilities.

9. The Rise of Frequentism and Sampling Theory

The history of modern probability theory published by Isaac Todhunter (1865), which like many begins with the Fermat–Pascal quantification, ends at this point – perhaps because extension to further applications seemed to be all of substance that remained. (The maximum entropy principle was not yet in place, but there was no great awareness of something missing because problems seen as probabilistic did not need it; the application which did finally generate it was not understood as related.) From the middle of the 19th century, the subject was again given over to those of a philosophical bent, only now with mathematical facility. The tone became anti-Bayesian – which in practice meant anti-Laplacean – and took the frequentist view that probability was proportion. Prominent texts were by George Boole (1854), a notable worker in the calculus of propositions also, and John Venn (1866); it is a pity that Laplace had not been translated into English to argue for himself. The importance of getting the underlying ideas right is indicated by the technical errors to which these authors were led (see Jaynes 1976, p.240). An authoritative history, concentrating on the 19th century, is by Stigler (1986).

It was in that century that more sophisticated inference was sought by collecting and exploiting data on how certain things were distributed among samples of a population, extending insurance and annuity calculations and the like into quantitative social studies. Unfortunately the refusal to countenance either Bayesian inversion, or the notion of prior probabilities based on anything more complicated than a discrete symmetry (when the prior is obvious), led to a search for *ad hoc* techniques which invert by stealth; for example, estimation of the infectiousness of a disease from statistics like Graunt's. The collection of methods that results is known as *sampling theory* and is seldom equivalent to Bayes' theorem; moreover, if the prior information is sufficiently cogent for the prior probability density to vary considerably across the peak in the likelihood, its neglect leads to even worse answers. The notion of an ensemble of imaginary worlds, reflecting what you believe could happen, became accepted in an attempt to widen the applicability of the frequentist view back to what Bayesians were able to tackle routinely. (The idea dates much further back than von Mises, with whom it is often associated.)

The most standard 19th century method, least squares, can be seen as a generalisation of the earlier intuitive idea of ironing out errors by taking the average of a series of noisy observations (which is the least squares solution); prediction of the motion of heavenly bodies from observations was the problem which historically drove such developments. The techniques of sampling theory generally work well in the problems they were invented to tackle, meaning that they give answers close to the Bayesian. This is a tribute to the insight of their inventors, men like R.A. Fisher, E.S. Pearson and Jerzy Neyman. But a diversity of methods gives, unsatisfactorily, a diversity of answers to the same question based on the same data. Ultimately these methods are either equivalent to use of Bayes' theorem or they are wrong. Also, their failure to incorporate relevant prior information is a weakness; that Bayesians have not yet learned how to do this in every problem is ground for research rather than blanket condemnation. These weaknesses in the frequentist position led to disputes among its sub-schools which, from the present point of view, are reminiscent of those among mediaeval scholastic theologians.

10. The Maximum Entropy Principle

Meanwhile another advance was coming from a different direction. The physicist Ludwig Boltzmann (1868) wrote down, using a combinatorial argument, the canonical distribution of energy in statistical mechanics. This is the distribution of maximum (information) entropy. The meaning of what Boltzmann did as inference was not to emerge for many decades – it hardly could under frequentism – and Boltzmann himself subsequently pointed statistical mechanics down a different path. But his idea was taken up and extended in the work of Willard Gibbs (1874–8), still without making plain the inferential nature of the calculations (although Gibbs' last and perforce unpolished book (1902) hints at it). Statistical mechanics is still in conceptual confusion today, and it is necessary to take with caution what others say Gibbs said. In particular, he never advocated an ergodic viewpoint; he never mentioned it, because he had all the tools he needed to solve problems without it.

The derivation and justification of the maximum entropy principle as a way of distributing probability when symmetry between the possibilities, in the conditioning information, is broken by direct knowledge about the distribution (for example, the value of the mean) came half a century later, with Claude Shannon's famous foundation of information theory (1948). Even here the justification is phrased in a frequentist way, and Bayesian re-interpretation is due to the physicist E.T. Jaynes (see the collection, Jaynes 1983).

(Jaynes has also re-exported the resulting understanding back into statistical physics to give a predictive formalism for statistical mechanics (developed in full by Grandy, 1987–8). Resolved here is the notorious paradox of microscopic reversibility yet thermodynamic irreversibility.)

11. The New Bayesianism; to the Present

Study of the specialised techniques of sampling theory widened the gap between theory and application and, with the abstract axiomatic approach of Kolmogorov, diminished understanding in the importance of propositions as the arguments of probabilities. These changes led also to the notion of the professional statistician. A good deal of technical development has taken place that does not rest on the meaning of probability, but it is not a coincidence that the Bayesian counter-revolution had at its forefront mostly working scientists who wanted to apply probability. The economist J.M. Keynes (1921) wrote a textbook which expressed probability as a degree of partial implication – the present view. A distinguished Bayesian text, dissenting from the prevailing frequentism, came from the geophysicist Harold Jeffreys in 1939. In 1946 came the landmark paper by the physicist R.T. Cox, deriving the 'laws of probability', which relative frequency satisfies trivially, as a calculus of partial implication: the calculus for implicability follows from the Boolean calculus of the propositions which are arguments of implicability. The first textbook based on the propositional view, and incorporating the maximum entropy principle, was written by an engineer, Myron Tribus (1969). There was an increasing acceptance of the Bayesian notion of probability among statisticians from the 1960s onwards, although with little recognition of the crucial role played by propositions as the arguments of probabilities, or of the maximum entropy principle. This acceptance seems to have stemmed from the demonstration by Wald (1950) that Bayesian results were equivalent to his treatment of probabilistic inference as a facet of decision theory (which is not the relation between them understood here). Also related to decision-making was von Neumann's ground-breaking text *Theory of Games and Economic Behaviour* (1944), indicating that gaming and finance remain powerful forces driving theoretical innovation. And, since quantification can clarify logical principles and so act beneficially back on qualitative reasoning, it is pleasing to report a Bayesian legal textbook (Robertson and Vignaux, 1995). DNA sampling will make numeracy more important in law.

The computer revolution opened up new fields of applied science to probabilistic analysis. Inverse problems as complex as image reconstruction could now be tackled; a picture speaks a thousand words in demonstrating the superiority of Bayesian algorithms. Theory, too, continues: Jaynes (1983) has described methods for extending Laplace's principle of indifference, especially to continuous spaces. The result, in combination with the principle of maximum entropy, is to justify some distributions used as priors in Bayesian inversion and generate others. Yet legal argumentation reminds us that we have only just begun the quantification of uncertain reasoning. Also, new misunderstandings have entered just where Bayesianism could help most, in Artificial Intelligence where *ad hoc* rules for reasoning such as fuzzy logic are regularly proposed which are inequivalent to the present ones. It was a neural network with its parameters tuned in Bayesian fashion that won a recent real-world prediction competition against dozens of alternative routines (MacKay 1996), and it will most likely be the application of such routines to quantitative problems of pressing human interest, together with the drive to quantify ever more complex matters of principle and human interest, that will generate the next advances.

References

Many of the historical references cited are available in volumes of collected works of their authors. Many are also available on microfilm in research libraries which do not hold printed copies.

Aristotle. *d.* 322BC. *The Art of Rhetoric.* The Greek, side by side with a modern English translation, by J.H. Freese, can be found in the corresponding volume of the Loeb Classical Library, Heinemann, London, UK, 1926. Aristotle's *Rhetoric* was finalised in the last thirteen years of his life in Athens, and reads like lecture notes.

Aristotle. *d.* 322BC. *Prior Analytics.* The Greek, side by side with a modern English translation, by H. Tredennick, can be found in the corresponding volume of the Loeb Classical Library, Heinemann, London, UK, 1938.

Arnauld, A. & Nicole, P. 1662. *La Logique, ou l'Art de Penser.* Charles Savreux, Paris, France. The 'Port-Royal Logic'. English translation of the fifth edition by T.S. Baynes, second edition published by Sutherland & Knox, Edinburgh, UK, 1851.

Augustine of Hippo. *d.* 430. Complete works: *Patrologiae Cursus Completus* (Latin series), volumes 32–46, editor J.-P. Migne, Paris, France, 1842–45. English translation by M. Dods, T. & T. Clark, Edinburgh, UK, 1871–76.

Baldus de Ubaldis. (*d.* 1400) 1586. *Opera* (nine volumes), apud Iuntas, Venice.

Bayes, T. 1763. An essay towards solving a problem in the doctrine of chances. *Philosophical Transactions of the Royal Society* **53**, 370–418. Reprinted in: Studies in the History of Statistics and Probability, editors E.S. Pearson & M.G. Kendall. Griffin, London, UK, 1970.

Bellhouse, D.R. 1988. Probability in the sixteenth and seventeenth centuries: an analysis of Puritan casuistry. *International Statistical Review* **56**, 63–74.

Bellhouse, D.R. & Franklin, J. 1997. The language of chance. *International Statistical Review* **65**, 73–85.

Bernoulli, J. 1713. *Ars Conjectandi.* Thurnisiorum, Basle, Switzerland.

Bernstein, P.L. 1996. *Against the Gods: the Remarkable Story of Risk.* Wiley, New York, USA.

Boltzmann, L. 1868. Studien über das Gleichgewicht der lebendige Kraft zwischen bewegten materiellen Punkten. *Sitzungsberichte der kaiserliche Akademie der Wissenschaften in Wien, Mathematisch–Naturwissenschaftliche Klasse IIa* **58**, 517–560.

Boole, G. 1854. *An Investigation of the Laws of Thought.* Macmillan, London, UK.

Byrne, E.F. 1968. *Probability and Opinion: a Study in the Medieval Presuppositions of Post-Medieval Theories of Probability.* Martinus Nijhoff, The Hague, Netherlands.

Cardano, G. 1663. *De Ludo Aleae*. In: *Opera Omnia*, cura Caroli Sponii, Lyons, France, volume 1. Cardano was a popular lecturer in his time but publication is a century later. An English translation is included in Ø. Ore, *Cardano, the Gambling Scholar*, Princeton University Press, Princeton, New Jersey, USA, 1953.

Cochrane, C.N. 1944. *Christianity and Classical Culture: A Study of Thought and Action from Augustus to Augustine*. Oxford University Press, Oxford, UK.

Cox, R.T. 1946. Probability, frequency and reasonable expectation. *American Journal of Physics* **14**, 1–13.

Crombie, A.C. 1994. *Styles of Thought in the European Tradition* (three volumes). Duckworth, London, UK.

Euler, L. (*d.* 1783) 1923. *Opera Omnia*, series prima: opera mathematica, editors F. Rudio, A. Krazer & P. Stäckel, volume 7: Theory of Combinatorics and Probability (volume editor: L. du Pasquier), published for Societatis Scientiarum Naturalium Helveticae by B. Teubner, Leipzig, Germany (volume 1: 1911).

Fermat, P. de. 1654. Correspondence with Pascal. In: *Oeuvres de Fermat*, editors P. Tannery & C. Henry, Gauthier-Villars, Paris, France, volume 2 (1894), pp.288–331. (The letters of Pascal are included.) English translation in *Games, Gods and Gambling* by F.N. David, Griffin, London, UK, 1962, pp.229–253. Also set out in Pascal (1654).

Franklin, J. 1991. The ancient legal sources of seventeenth century probability. In: *The Uses of Antiquity*, editor S. Gaukroger, Kluwer, Dordrecht, Netherlands, pp.123–144 & 251.

Galilei, Galileo. 1623. *Il Saggiatore* [The Assayer.] In: *Le Opere di Galileo Galilei*, edizione nazionale, editor A. Favaro, Barbèra, Florence, Italy, volume 6 (1896). The quote is on p.232, lines 11–15. English translation in: *The Controversy on the Comets of 1618*, editors and translators S. Drake & C.D. O'Malley, University of Pennsylvania Press, Philadelphia, Pennsylvania, USA, 1960; the quote is on pp.183–4.

Garber, D. & Zabell, S. 1979. On the emergence of probability. *Archive for History of Exact Sciences* **21**, 33–53.

Gibbs, J.W. 1874–8. On the equilibrium of heterogeneous substances. *Transactions of the Connecticut Academy of Arts and Sciences* **3**, 108–248 & 343–524. Reprinted in: The Collected Works of J. Willard Gibbs, Longmans Green, New York, USA, 1928, volume 1.

Gibbs, J.W. 1902. *Elementary Principles in Statistical Mechanics*. Yale University Press, New Haven, Connecticut, USA.

Grandy, W.T., Jr. 1987–8. *Foundations of Statistical Mechanics, Volume I: Equilibrium Theory* (1987), *Volume II: Nonequilibrium Phenomena* (1988). Reidel, Dordrecht, Netherlands.

Graunt, J. 1662. *Natural and Political Observations Mentioned in a Following Index, and Made upon the Bills of Mortality*. Printed by T. Roycroft for J. Martin, J. Allestry & T. Dicas, London, England. (Books were often then available from a single outlet only.) Reprinted in *Natural and Political Observations Made upon the Bills of Mortality by John Graunt*, editor W.F.

Willcox, Johns Hopkins Press, Baltimore, Maryland, USA, 1939. The fifth and definitive edition (published by J. Martyn, London, 1676) is reprinted in *The Economic Writings of Sir William Petty*, volume II, editor C.H. Hull, Cambridge University Press, Cambridge, UK; reprinted by Kelley, New York, USA, 1963–4. Petty was closely involved with Graunt.

Hacking, I. 1975. *The Emergence of Probability*. Cambridge University Press, Cambridge, UK.

Hume, D. 1739–40. *A Treatise of Human Nature*. Originally published in three volumes: volumes 1 & 2 published by John Noon, London, England, 1739; volume 3 by Thomas Longman, London, 1740. The anti-induction material occupies volume I, part III (*Of knowledge and probability*). The work is seldom out of print: a recent single-volume edition is published by Penguin, London, UK, 1985.

Hume, D. 1748. *An Enquiry Concerning the Human Understanding*. A. Millar, London, England. The original title was *Philosophical Essays Concerning Human Understanding*. The anti-induction material occupies sections (essays) IV, V. The work is seldom out of print: a recent printing (of the third edition) is by Oxford University Press, Oxford, UK, 1975.

Huygens, C. 1657. *De Ratiociniis in Ludo Aleae*. In: *Exercitionum Mathematicorum*, volume 5, editor F. van Schooten, Elsevir, Amsterdam, Netherlands, pp.521–534. The work is phrased essentially as a set of problems and answers. English translation by J. Arbuthnot, *Of the Laws of Chance, or a Method of Calculation of the Hazards of Game*. B. Motte (publisher) & R. Taylor (seller), London, England, 1692.

Jaynes, E.T. 1976. Confidence intervals *vs* Bayesian intervals. In: *Foundations of Probability Theory, Statistical Inference and Statistical Theories of Science*, editors W.L. Harper & C.A. Hooker, Reidel, Dordrecht, Netherlands, pp.175–257. Largely reprinted as chapter 9 of Jaynes (1983).

Jaynes, E.T. 1983. *E.T. Jaynes: Papers on Probability, Statistics and Statistical Physics*. Synthese Library **158**. Editor R.D. Rosenkrantz, Reidel, Dordrecht, Netherlands.

Jeffreys, H. 1939. *Theory of Probability*. Oxford University Press, Oxford, UK. Third edition published 1961.

John of Salisbury. 1159. *Metalogicon*. In: *Patrologiae Cursus Completus* (Latin series), volume 199, editor J.-P. Migne, Paris, France, 1855, pp.823–946. English translation by D.D. McGarry, University of California Press, Berkeley, California, USA, 1955.

Justinian, 533. *Digest* (or *Pandects*). Constantinople, in the eastern half of the divided Roman empire. An authoritative Latin version is given side by side with an English translation by A. Watson, University of Pennsylvania Press, Philadelphia, Pennsylvania, USA, 1985. The *Digest* is the largest part of Justinian's comprehensive *Corpus of Civil Law*.

Kantola, I. 1994. *Probability and Moral Uncertainty in Late Medieval and early Modern Times*. Luther-Agricola-Society, Helsinki, Finland.

Kassler, J. 1986. The emergence of probability reconsidered. *Archives Internationales d'Histoire des Sciences* **36**, 17–44.

Keynes, J.M. 1921. *A Treatise on Probability*. Macmillan, London, UK.

Laplace, P.S., Marquis de. 1812. *Théorie Analytique des Probabilités*. Courcier, Paris, France. The third edition (Courcier, Paris, 1820) has, as introduction, the *Essai Philosophique sur les Probabilités* (Laplace 1814). No English translation of the *Théorie* has been published.

Laplace, P.S. 1814. *Essai Philosophique sur les Probabilités*. Courcier, Paris, France. English translation of the sixth edition (Bachelier, Paris, 1840) by F.W. Truscott & F.L. Emory, Wiley, New York, 1902; revised translation in second edition, Wiley, New York, 1917. Reprinted by Dover publications, New York, USA, 1951.

Leibniz, G.W. *d.* 1716. *Oeuvres* (seven volumes), editor L.A. Foucher de Careil, republished by G. Olms, Hildesheim, Germany, 1969. A definitive series of Leibniz' works has been in progress since 1923 under the title *Sämtliche Schriften und Briefen*, and G. Olms of Hildesheim have republished many older collections of Leibniz' work. There is no collection in English but the more important works have been translated individually.

MacKay, D.J.C. 1996. Bayesian non-linear modeling for the prediction competition. In: *Maximum Entropy and Bayesian Methods, Santa Barbara, California, USA, 1993*, editor G.R. Heidbreder, Kluwer, Dordrecht, Netherlands, pp.221-234.

de Moivre, A. 1718. *The Doctrine of Chances, or a Method of Calculating the Probability of Events in Play*. W. Pearson, London, England. Second edition published by Woodfall, London, England, 1738; third edition by Millar, London, UK, 1756; successive editions were greatly expanded. The second edition was reprinted by Cass, London, UK, 1967, and the third edition by Chelsea publishing, New York, USA, 1967.

de Moivre, A. 1725. *Annuities on Lives*. Printed by W. P[earson], sold by F. Fayram, London, England.

Montmort, P.R. de, 1708. *Essay d'Analyse sur les Jeux de Hazard*. J. Quillau, Paris, France. Second edition, much expanded, published by Quillau, Paris, 1713.

von Neumann, J. & Morgenstern, O. 1944. *Theory of Games and Economic Behaviour*. Princeton University Press, Princeton, New Jersey, USA.

Pascal, B. 1654. Correspondence with Fermat. In: *Blaise Pascal, Oeuvres Complètes*, volume 2, editor J. Mesnard, published by de Brouwer, Bruges, Belgium and Paris, France, 1970, pp.1132–1158 (volume 1: 1964). (The letters of Fermat are included.) English translation in *Games, Gods and Gambling* by F.N. David, Griffin, London, UK, 1962, pp.229–253. Also set out in Fermat (1654).

Patey, D.L. 1984. *Probability and Literary Form: Philosophic Theory and Literary Practice in the Augustan Age*. Cambridge University Press, Cambridge, UK.

Rabinovitch, N.L. 1973. *Probability and Statistical Inference in Ancient and Medieval Jewish Literature*. University of Toronto Press, Toronto, Canada.

Robertson, B. & Vignaux, G.A. 1995. *Interpreting Evidence: Evaluating Forensic Science in the Courtroom*. Wiley, Chichester, UK.

Sambursky, S. 1956. On the possible and the probable in ancient Greece. *Osiris* **12**, 35–48.

Shannon, C.E. 1948. A mathematical theory of communication. *Bell System Technical Journal* **27**, 379–423 & 623–659. Reprinted in: *The Mathematical Theory of Communication*, editors C.E. Shannon & W.W. Weaver, University of Illinois Press, Urbana, Illinois, USA, 1949.

Sheynin, O.B. 1974. On the prehistory of the theory of probability. *Archive for the History of Exact Sciences* **12**, 97–141.

Stigler, S.M. 1986. *The History of Statistics: the Measurement of Uncertainty before 1900.* Belknap Press of Harvard University Press, Cambridge, Massachusetts, USA.

Thomas [of] Aquinas. 1265–74. *Summa Theologiae*. Multi-volume Latin with modern English translation, Eyre & Spottiswode, London, UK; II-II q. 70 is in volume 38, published 1975.

Todhunter, I. 1865. *A History of the Mathematical Theory of Probability from the Time of Pascal to that of Laplace.* Macmillan, London, UK.

Tribus, M. 1969. *Rational Descriptions, Decisions and Designs.* Pergamon Press, New York, USA.

Venn, J. 1866. *The Logic of Chance.* Macmillan, London, UK.

Wald, A. 1950. *Statistical Decision Functions.* Wiley, New York, USA.

de Witt, J. 1671. *Waerdye van Lyf-renten naer Proportie van Los-renten* [Value of Life Annuities in Proportion to Redeemable Rents.] J. Scheltus, s'GravenHage, Netherlands. Reprinted in *Feest-gave van het Wiskundig Genootschap te Amsterdam*, published by J. Enschedé, Haarlem, Netherlands, 1879. Translated in: R.G. Darnwell, *A Sketch of the Life and Times of John de Witt, Grandpensionary of Holland, to which is added his Treatise on Life Annuities*, New York, USA, 1856.

WE MUST CHOOSE THE SIMPLEST PHYSICAL THEORY: LEVIN-LI-VITÁNYI THEOREM AND ITS POTENTIAL PHYSICAL APPLICATIONS

D. FOX, M. SCHMIDT, M. KOSHELEV
V. KREINOVICH, L. LONGPRÉ
Department of Computer Science
University of Texas at El Paso
El Paso, TX 79968, USA[‡]

AND

J. KUHN
National Solar Observatory/Sacramento Park,
Sunspot, NM 88349-0062, USA[§]

Abstract. If several physical theories are consistent with the same experimental data, which theory should we choose? Physicists often choose the *simplest* theory; this principle (explicitly formulated by Occam) is one of the basic principles of physical reasoning. However, until recently, this principle was mainly a *heuristic* because it uses the *informal* notion of simplicity.

With the explicit notion of simplicity coming from the Algorithmic Information theory, it is possible not only to *formalize* this principle in a way that is consistent with its traditional usage in physics, but also to *prove* this principle, or, to be more precise, *deduce* it from the fundamentals of mathematical statistics as the choice corresponding to the least informative prior measure. Potential physical applications of this formalization (due to Li and Vitányi) are presented.

In particular, we show that, on the *qualitative* level, most fundamental ideas of physics can be re-formulated as natural steps towards choosing a theory that is the simplest in the above *precise* sense (although on the intuitive level, it may seem that, e.g., *classical* physics is easier than quantum physics): in particular, we show that such ideas as Big Bang cosmology, atomism, uncertainty principle, Special Relativity, quark confinement, quantization, symmetry, supersymmetry, etc. can all be justified by this (Bayesian justified) preference for formalized simplicity.

Key words: Kolmogorov complexity, Algorithmic Information theory, Occam razor, Bayesian statistics, Fundamental physics

[‡]Emails: {dfox,mschmidt,mkosh,vladik}@cs.utep.edu.
[§]E-mail: jkuhn@sunspot.noao.edu

1. The problem

The problem: what theory should we choose? If several physical theories are consistent with the same experimental data, which theory should we choose?

How is a theory chosen now? Physicists often choose the *simplest* theory.

This principle (explicitly formulated by Occam) is one of the most basic principles of physical reasoning.

The idea of choosing the simplest theory was actively advocated by Einstein. The most famous promoter of this idea was A. Einstein. For example, in his lecture "On the method of theoretical physics", he said that "It is the grand object of all theory to make ... irreducible elements as simple and as few in number as possible" ([3], p. 272).

This principle used to be heuristic. The principle of choosing the simplest physical theory was, until recently, only *heuristic*, because until recently, there was no well-accepted definition of simplicity.

Simplicity can now be defined formally. Now, with the advent of Algorithmic Information Theory (see, e.g., [16]), there are formal (and well-accepted) definitions of simplicity. The first such definition was proposed, in mid-60s, practically simultaneously, by A. Kolmogorov, R. Solomonoff, and G. Chaitin; this notion is called *Kolmogorov complexity*. Crudely speaking, Kolmogorov complexity of a text is the shortest length of a program (in some universal language) that generates this text. There are also several modifications of this definition. For different problems, different formalizations turn out to be more adequate.

This formalization leads to the following two (natural and related) questions:

- First, which of the definitions of simplicity is more adequate for choosing a physical theory?
- Second, now that the idea of choosing the simplest theory becomes formal, the question is: is this principle mathematical or physical? In other words:
  - does this formal principle *follow* from the traditional principles of mathematical *statistics* (that are used to process physical data), or
  - is this principle *independent* from the general principles of mathematical statistics, and somehow reflecting the specific structure of *our* physical world (as opposed to other mathematically possible "worlds")?

What we are planning to do. In Section 2, we describe the result by Li and Vitányi (which is, in its turn, based on Levin's theorem) that the basic principles of mathematical statistics do indeed lead to a formalization of the principle of choosing the simplest theory, and we explicitly describe the formalization of complexity that this theory corresponds to. Thus, we get answers to both questions.

The main ideas of our exposition are from Li and Vitányi [16]. However, since our main goal is formalizing this principle with respect to *physical theories*, our exposition will be specifically oriented towards the physics readers and therefore, our formulation and our exposition will be somewhat different from [16]. In particular, we will give physical motivations for all definitions and requirements that are used in the proof.

In Section 3, we show that the resulting formalization is consistent with the traditional use of this principle in theoretical physics. Hopes and speculations are described in the final Section 4.

2. Justification of the principle of choosing the simplest physical theory

Basic statistical approach: brief reminder. According to the fundamentals of mathematical statistics, in order to choose a theory based on the experimental data, we must do the following:

- find the prior probabilities $P_0(H_i)$ of different hypotheses H_i;
- use the experimental data E to compute the posterior probabilities $P(H_i)$ of different hypotheses; to compute these probabilities, we can use the Bayes rule
$$P(H_i) = \frac{P(E|H_i) \cdot P_0(H_i)}{\sum P(E|H_j) \cdot P_0(H_j)},$$
where $P(E|H)$ is the probability of observing the data E in case the hypothesis H is true;
- and, finally, select a hypothesis whose probability is the largest $P(H_i) \to \max$.

The posterior probability $P(H_i)$ of a hypothesis H_i is also called its *likelihood*, and correspondingly, the above basic statistical approach is also called the *maximum likelihood* (ML) approach.

From the computational viewpoint, it is often convenient to use *negative logarithms* $-\log(P(H_i))$ of the probabilities: e.g., for the most widely used Gaussian distribution, the probabilities $\exp(-c \cdot x^2)$ require calling an exponential function and are, therefore, much more difficult to compute than their negative logarithms $-c \cdot x^2$. Since negative logarithm is a decreasing function, in terms of negative logarithms, the criterion $P(H_i) \to \max$ takes the form $-\log(P(H_i)) \to \min$.

At first glance, this approach may sound somewhat *subjective* because the resulting choice of a hypothesis depends on the original (rather subjective) choice of prior probabilities. However, it is known that in the long run (as we get more and more experimental data), this dependence disappears. To be more precise, let us assume that the actual theory is H_a. Then:

- If the prior probability of H_a is 0 (i.e., $P_0(H_a) = 0$), then no matter how many experimental data we get, if we apply the above Bayes formula, we always end up with $P(H_a) = 0$.
- However, if we make sure that the prior probability of the correct hypothesis is positive ($P_0(H_0) > 0$), then, as the number of experimental data increases, the probability $P(H_a)$ of this hypothesis will increase, while the probabilities of other hypotheses will decrease and eventually, we will get $P(H_a) \gg P(H_i)$ for all $H_i \neq H_a$ and therefore, eventually, the hypothesis H_a will be chosen.

Since we do not know which hypothesis is correct, we must require that a prior probability of each hypothesis is positive.

This mathematically sounding result is actually very intuitively clear: if we do not know which of the hypotheses is true, we must not throw away any of them, otherwise, we may throw away the only theory that is consistent with the observations.

How to apply this general approach to fundamental physics (first approximation). One of the major goals of physics is to predict the results of different experiments. Whatever *mathematical* language a theory or hypothesis is formulated in, eventually, we are interested in what *observation* results this theory predicts. These results can also be described in different terms, but since most measurements and practically all data processing is done by using computers, we can use the fact that inside the computers, everything is represented by 0's and 1's. Each 0 or 1 is called a *binary unit*, or, for short, a *bit*. Therefore, for simplicity of analysis (and without losing any generality), we can assume that all possible experimental results form a (potentially infinite) binary sequence $X = x_1x_2 \ldots x_n \ldots$

Our goal is to predict either the results of *all* the experiments, or at least the results of *some* of them. So, our goal is to predict either the entire (potentially infinite) sequence X, or a finite portion of this sequence. Therefore, to apply the standard statistical methodology, we need a prior measure on the set of all such sequences.

To describe this measure, it is sufficient to describe, for each finite sequence $x = x_1 \ldots x_n$, the probability $p(x)$ that the (possible infinite) sequence X of measurement results will start with x. These probabilities must satisfy the two natural requirements:

- Since every sequence starting with x must continue either with 0, or with 1, we get the *additivity* condition for probabilities: $p(x) = p(x0) + p(x1)$.
- Also, the total probability must be equal to 1, i.e., $p(0) + p(1) = 1$.

The second requirement can be formulated as a particular case of the additivity requirement, if we allow an empty string $x = \Lambda$ and assume that

$$p(\Lambda) = 1 \qquad (1).$$

Comment. In the following text, we will see that we need to make a (minor) modification to these formulas; that is why we called this description the *first approximation.*

The prior probabilities must be computable (in some reasonable sense). The above-described statistical methodology requires us to *compute* the posterior probabilities based on the prior ones. Bayes' formula is explicit, but in order to use it, we must assume that the prior probabilities $P_0(H)$ are, in some reasonable sense, computable. What does this "computable" mean?

Crudely speaking, a prior probability of an event means the probability of this event that is computed based on some *a priori* ideas, i.e., probabilities computed based on some prior "theory".

If an event is simple enough, then, in this prior "theory", we can simply compute its probability *directly*. To be more precise, we can compute this probability with an arbitrary *accuracy*, i.e., there exists an algorithm that, given a positive integer k, computes a rational number r_k that is 10^{-k}-close to the desired probability

p (i.e., for which $|r_k - p| \leq 10^{-k}$). Real numbers p for which such an algorithmic approximation is possible are called *computable.* If we have an algorithm that, given x and k, returns a 10^{-k}-approximation to $p(x)$, then $p(x)$ is called a *computable function.*

Often, however, the desired event can have many different reasons; it is *possible* to compute the probabilities related to *each reason* but it is *difficult* to compute the *total probability.* This effect is well known to experimental physicists who estimate the reliability of their experimental results (i.e., the probability that the observed results are caused not by the analyzed phenomena, but rather by noise). There are well-known examples when experimental claims, that were initially thought to be correct, later on turned out to be false because of some additional factors: e.g., Weber's experimental discovery of gravitational waves turned out to be an observation of a different phenomenon [17].

If we take this feature of experimental physics into consideration, we will arrive at the conclusion that for each finite sequence x, the desired probability $p(x)$ is not necessarily a computable real number, but that it is the *limit* $p(x) = \lim p_N(x)$, where $p_N(x)$ is the probability that only takes first N factors into consideration. The sequence $\{p_N(x)\}$ is:

- *non-decreasing* in the sense that $p_1(x) \leq p_2(x) \leq \ldots \leq p_N(x) \leq p_{N+1}(x) \leq \ldots$, and
- *computable* in the sense that there is an algorithm that, for every x, N, and k, returns a 10^{-k}-approximation to $p_N(x)$.

Functions that can be represented as limits of non-decreasing computable sequences are called *enumerable* (crudely speaking, this name comes from the fact that after we *enumerate* all factors, we get the desired probabilities). Thus, we can re-formulate the requirement on the function $p(x)$ by saying that p should be an *enumerable* function.

How to distinguish a reasonable prior probability measure from other possible probability measures. When choosing prior probabilities $p_0(x)$, the main requirement is not to assign probability 0 to an event E that could end up with a non-zero probability if we use a different prior probability measure $p(x)$. In particular, if, as the event E, we take the statement that X must start with a given finite sequence x, this requirement means that we must avoid the situations when $p_0(x)/p(x) = 0$, i.e., we must require that $p_0(x)/p(x) > 0$ for all words x.

Events E can be more complicated than that; for example, we may have events that are defined as *limits* of the above simple events. To cover all possible more complicated events, it is reasonable to require that not only all the ratios $p_0(x)/p(x)$ be positive, but that also the *limits* $p_0(x^{(N)})/p(x^{(N)})$ of such ratios, corresponding to different sequences $x^{(N)}$, should also be always positive.

In mathematical terms, this requirement can be formulated as follows: we have a set S of all possible values of the ratio $p_0(x)/p(x)$, and we require that if a sequence of real numbers from this set S has a limit, then this limit should be positive. For an arbitrary set S of real numbers, the smallest possible limit of sequences from S is known to be equal to the *infimum* $\inf(S)$ of this set S (i.e., to its *greatest lower bound* (g.l.b.)). Thus, the requirement that *all* the limits are positive is equivalent to requiring that *the smallest* of these limits is positive, i.e., that $\inf(p_0(x)/p(x)) > 0$. This, in turn, means that there exists a positive number $c > 0$ for which, for all x, $p_0(x) \geq c \cdot p(x)$.

Thus, we can say that an enumerable probability measure $p_0(x)$ is an ***ideal prior measure*** if for every other enumerable probability measure $p(x)$, there exists a constant $c > 0$ for which $p_0(x) \leq c \cdot p(x)$ for all x.

Modification is needed. The above idea of an "ideal" prior probability measure was proposed, in the early 60s, by R. Solomonoff, one of the authors of Algorithmic Information Theory. However, Solomonoff himself showed that, in effect, the seemingly reasonable formalization of this idea (the one we have just described) does not work, in the sense that the above-defined "ideal" prior measure cannot exist (for detailed and precise history, see [16]).

In view of this impossibility, we must ***modify*** our definitions. Such a modification has actually been proposed. The modification, as we'll see in a second, is very natural from the computer science viewpoint (and from the common sense viewpoint as well).

Final definitions. The prior probability $p(x)$ can be obtained, e.g., if we poll several experts and take, as $p(x)$, the fraction $N(x)/N$ between the total number $N(x)$ of experts who believe that x will occur in the measurement, and the total number N of experts. If every expert is ***a priori*** definite in his beliefs, i.e., if he believes, for every word x, either that x will occur, or that x will not occur, then the resulting values $p(x)$ indeed form a probability measure: e.g., among all the experts who believe in $x = x_1 \ldots x_n$, some believe in $x0 = x_1 \ldots x_n 0$ and some in $x1 = x_1 \ldots x_n 1$ (but everyone does believe in one of these two), and therefore, $N(x) = N(x0) + N(x1)$ and $p(x) = p(x0) + p(x1)$.

In reality, however, experts may be ***undecided*** about some measurement results. So, among $N(x)$ experts who initially believe in x, some will believe in $x0$, some in $x1$, while some will be undecided about what will follow x. In databases and in other areas of computer science, the case when we do not know a certain value is called a ***wild card*** and is denoted by $*$. Thus, we can express the situation in which an expert has no opinion about the continuation of x by $x*$, and conclude that $N(x) = N(x0) + N(x1) + N(x*)$ and therefore, $p(x) = p(x0) + p(x1) + p(x*)$.

We are interested only in the probabilities $p(x)$ of ***definite*** sequences (i.e., sequences that contain only 0's and 1's but not the wild card $*$). For these values, the above complicated ***additivity*** requirement turns into an ***inequality***:

$$p(x) \geq p(x0) + p(x1). \tag{2}$$

It is known that other methods of soliciting the degree of belief (e.g., methods that take into consideration not only who voted for and who against, but also to what extent each expert believes in x) also lead to numerical measures that satisfy the inequality (2).

Thus, instead of probability measures, we must consider functions p that maps finite binary strings into real numbers and that satisfy the properties (1) and (2). In Algorithmic Information theory, such functions are called ***semi-measures***.

Main results. Using the same arguments as above, we conclude that it is reasonable to require:

- that the function $p(x)$ is *enumerable*, and
- that the *ideal* prior semi-measure $p_0(x)$ must be such that for any other enumerable semi-measure $p(x)$, there exists a constant $c > 0$ for which for all x, $p_0(x) \geq c \cdot p(x)$.

For thus modified definition (with probability measure replaced by semi-measure), it is already possible to prove that the ideal prior semi-measure exists (and that it is, in some reasonable sense, unique; this theorem was first proven by L. Levin; see [16] for a detailed history). The resulting ideal prior semi-measure is usually denoted by $\mathbf{M}(x)$.

Thus, according to the statistical methodology, we must choose, for each n, the sequence $x = x_1 \ldots x_n$ for which $\mathbf{M}(x) \to \max$, or, equivalently, for which $KM(x) \to \min$, where we denoted $KM(x) = -\log_2(\mathbf{M}(x))$.

It turns out that this negative binary logarithm $KM(x) = -\log_2(\mathbf{M}(x))$ is closely related to one of the versions of *Kolmogorov complexity*: namely, to the *monotone complexity* $Km(x)$ that is defined (crudely speaking) as the length of the shortest program that computes a sequence x or a sequence that starts with x on a universal *monotone* Turing machine (i.e., on a computer with a one-way read-only input tape, a one-way write-only output tape and some work tapes).

The close relationship between $Km(x)$ and $KM(x)$ is described, e.g., by the following two properties:

- $KM(x) \leq Km(x) \leq KM(x) + Km(KM(x)) + O(1)$;
- for almost all infinite sequences $X = x_1 \ldots x_n \ldots$, the difference between $Km(x_1 \ldots x_n)$ and $KM(x_1 \ldots x_n)$ grows slower than any unbounded computable function.

Due to this close relationship, the negative logarithm $KM(x)$ of a universal semi-measure is also considered one of the versions of Kolmogorov complexity.

Conclusion. *From the basic statistical methodology, we can deduce that, if there are several physical theories consistent with the known experimental data, we must always choose the simplest theory x, i.e., a theory for which $KM(x) \to \min$, where $KM(x)$ is the above formalization of complexity.*

Historical comment. This derivation was first proposed by M. Li and P. Vit'anyi; see [14–16,26,27].

Why is this particular formalization of Kolmogorov complexity the most adequate for choosing a theory? Physical explanation. Let us give a physical explanation of why not the *original* Kolmogorov complexity, but its *monotone* version turned out to be more adequate for choosing physical theories. Indeed, the main difference between $Km(x)$ and the original Kolmogorov complexity $C(x)$ is that:

- in the definition of original Kolmogorov complexity, we only consider programs that produce exactly x, while
- in the definition of $Km(x)$, we also consider programs whose output contains x followed by something else.

This difference has a simple physical interpretation: In physics, often, when we want to predict a certain value of the field at some specific moment of time, a natural way is:

- to find a *general* solution of the corresponding system of partial differential equation, and then
- to extract the desired values from the general solution (this "general" solution is, frequently, only applicable to a certain area, e.g., a vacuum electromagnetic field solution only holds outside the bodies).

The resulting algorithm consists of two parts:

- solving the system of equations, and
- extracting the desired values from the general solution.

From the viewpoint of the original Kolmogorov complexity, we have to count the complexity of *both* parts when we estimate the complexity of the solution. The complexity of the first part is indeed indicative of the complexity of the physical world, while the complexity of the second part (extraction) has nothing to do with the physical world and is caused solely by the fact that we are not currently interested in *all* the solution, only in the *part* of it. Therefore, if we are interested in characterizing the physical world itself (and not in our own goals), we should neglect the complexity of the second (extraction) part. This is exactly what $Km(x)$ is doing.

This idea has been successfully used in many applications. In the above derivation, we did not use many specifics of physics as opposed to data processing in general. It will, therefore, not be a surprise to a reader to know that this idea (of choosing the hypothesis for which some version of Kolmogorov complexity takes the smallest possible value) has been originally proposed and successfully used in data processing. Namely, it was first described, under the name of the *minimum description length principle*, by J. J. Rissanen in [20]; for a more recent modifications and updates, see, e.g., [16,21,22].

This idea has been applied to various areas such as handwriting recognition [9], surface reconstruction in computer vision [19], economic forecasts [11], cognitive psychology [2], biology, etc. [16]. These successes make us believe that this formalized principle will be useful in fundamental physics as well.

3. Qualitative physical examples

3.1. MAIN IDEA BEHIND THE EXAMPLES

Our intention. In this section, we will show that, on the *qualitative* level, most fundamental ideas of physics can be re-formulated as natural steps towards choosing a theory that is the simplest in the above precise sense.

Is this intention consistent with our intuition? At first glance, this claim may seem counter-intuitive, because the advance of physics usually means going to more and more complicated theories. For example:

– a simple pre-physical theory that everything in this world is determined by the good and evil spirits and is, therefore, completely unpredictable, a theory that uses no mathematics at all, seems to be very simple,
– while Newtonian mechanics, a theory that uses complicated differential equations, seems to be much more complicated.

This seeming contradiction with the intuition only comes from the fact that the word "simple" is very ambiguous. If we use the word "simple" in the precise way indicated in the previous section, then Newtonian mechanics is no longer simpler that the belief in a completely unpredictable world. Indeed, according to this definition, a complexity of a theory is, crudely speaking, the smallest length of the program that is able to predict (according to this theory) the results of all possible observations and measurements.

– By the intuitive understanding of a completely unpredictable world, the only way for a program to *predict* all the measured values is to actually store them. (If we would be able to store *some* of these values and predict the others, the theory would *not* be completely unpredictable.) Thus, within this theory, we must store *all* the values to predict them. The complexity of this theory is (in the formal sense described in the previous section) the largest possible.
– On the other hand, in Newtonian mechanics, e.g., for particles, we only need to know the positions and velocities of all the particles at the initial moment of time t_0; then we will be able, by integrating the equations of motion, to predict the values at all other moments of time. Thus:
    - instead of storing the values of the particles' coordinates and velocities at *all* possible moments of time,
    - it is sufficient to store these values only for a *single* moment of time.

Thus, when we go from a lawless world to a world described by Newtonian mechanics, we get a drastic *decrease* in the (formally understood) complexity.

Comment. The idea that progress in physics actually leads to theories that are *simpler* in some reasonable sense was emphasized and advocated by Einstein; e.g., he wrote that as "our experience grows larger and larger ... the simpler the logical structure ⟨of the physical theory⟩ becomes – that is to say, the smaller the number of logically independent conceptual elements which are found necessary to support the structure" ([3], p. 273).

How to minimize the complexity: a general idea. One of the major goals of a physical theory is to describe what is happening everywhere in the Universe. In other words, we must describe the values of all possible physical quantities in different points of space at different moments of time.

In the above-mentioned hypothetical completely unpredictable and lawless world, to describe all these values, we would actually need to describe the value of each quantity at each point of space-time. (Since all these values are unlimited and unrelated, if we miss one of these values, we will not be able to reconstruct it from the other stored ones, and therefore, if we want to be able to reconstruct the state of the Universe, we must actually store *all* the values.) In other words, the

Kolmogorov complexity of this description is equal to the total length of all the values stored. How can we decrease the complexity of this description?

In order to answer this question, let us estimate how many bits we need to store the complete information about the lawless world. By using b bits, we can store $V = 2^b$ possible values. Therefore, if a certain physical quantity has V possible values, we must use $\approx \log_2(V)$ bits to store its value. Therefore, if we have T moments of time, S points in space, if we have the total number Q of measurable quantities, and if we have V possible values of each quantity, we need to store $T \cdot S \cdot Q$ different values, each of which requires $\log_2(V)$ bits. Therefore, totally, we need $B = T \cdot S \cdot Q \cdot \log_2(V)$ bits to store all this information. (If one of these numbers is infinite, then, of course, we must store infinitely many bits.)

From the purely *mathematical* viewpoint, to decrease the complexity B of this description, we must, therefore, do one (or several) of the following things:

- decrease the number T of moments of time;
- decrease the number S of possible points in space;
- decrease the number Q of possible quantities;
- decrease the number V of possible values of each quantity, and
- introduce *dependency* between:
  - values of different quantities,
  - values at different moments of time, and/or
  - at different spatial points.

3.2. HOW TO MINIMIZE COMPLEXITY: EXAMPLES FROM FUNDAMENTAL PHYSICS

Let us show that, on the qualitative level, the above possibilities are exactly what major fundamental physical ideas have been achieving.

Restricting T: Big Bang. Restricting the "number of different moments of time" means, in effect, restricting the lifetime of the Universe. Thus, we naturally arrive at the idea of a Universe that has a beginning.

Restricting S: atomism. Similarly, restricting the "number of different spatial points" means that, according to our theory, most of the spatial points inside the Universe have no objects in them, and the matter is concentrated in a few places while others are filled with vacuum. Thus, we get the idea of *atomism.*

Restricting Q: uncertainty principle. Since the total number of *possible* physical quantities is given, the only way to restrict the total number of *measured* quantities is to impose restriction according to which measuring *one* of the properties restricts or even prohibits the measurement of the other quantities.

Restricting V: Special Relativity Theory, quark confinement, quantization. Similarly to the above two methods of decreasing the total number S of spatial points, there are two ways to decrease the total number of possible values V of a physical quantity:

- First, we can *bound* the possible values of this quantity.

- The best known example of such bounding is the bounding of possible velocities which is the basis of ***special relativity***.
- Another example, also well known but much less fundamental, is the restriction on the relative location of two quarks known as *quark confinement* (see, e.g., [28]).

– Second, we can assume that not ***all*** values within the bound are possible, but only ***some*** of these values. This is the original fundamental idea of ***quantization***: in quantum mechanics, many quantities can also take values from a certain discrete set (spectrum).

Dependency between different values: general idea. Finally, we can assume some *dependency* between the values of different quantities, and between the values of the same quantity at different moments of time, and at different spatial points.

Dependency between the values of different quantities: symmetry, supersymmetry, Dirac's large numbers. Dependency between the values of different components of a multi-component physical field (e.g., a vector, a spinor, or a tensor field) corresponds, crudely speaking, to the fundamental notion of ***symmetry*** of a physical theory (see, e.g., [7,8,10,12,18,25]).

The relationship between the Kolmogorov-complexity based principle and symmetries is not surprising to us, because in [13], we used ***Kolmogorov complexity*** (and related notion of randomness) to ***explain*** why ***symmetries*** are a universal language of physics.

Specifically, dependency between components of fields of different types corresponds to the idea of ***supersymmetry***.

If we do not have a complete relation between different fields, we can at least get ***some*** relation, e.g., in terms of equality of some numerical characteristics corresponding to different fields. Such equalities have been discovered since Dirac, in 1937, first observed that the ratio between the electromagnetic and gravitational forces between, say, two protons ($\approx 10^{40}$), is almost equal to the ratio of the Universe's size and the time during which light passes through a proton. This equality is not very easy to explain within specific theories of modern physics (see detailed discussion in [17]), but it is quite in line with our general principle of choosing the simplest theory.

4. Speculations and hopes

In this paper, we have shown two things:

– First, we have described, for a physics reader, how the principle of choosing the simplest physical theory, the idea that is usually formulated in ***informal, qualitative*** terms, can be reformulated in ***precise, quantitative*** terms; this formalization has been successfully used in various areas of data processing.

– Second, we have shown that this formalization explains, on the ***qualitative*** level, many fundamental principles of physics.

The success of formalizing the principle itself makes us hope that it may be possible to formalize these ***explanations*** as well, and thus, come out with a new universal physical theory in which the choice of the simplest hypothesis (or, equivalently, algorithmic information theory) is the sole axiom.

This possibility is in good accordance with the vision of Einstein who said ([3], p. 274) that "Our experience hitherto justifies us in believing that nature is the realization of the simplest conceivable mathematical ideas."

This hope of ours, that *physics* may be eventually reduced to *algorithmic* information theory, is also in clear agreement with the physically motivated ideas of J. A. Wheeler of logic as pre-geometry and pre-physics (see, e.g., [17] and references therein).

Acknowledgment

This work was supported in part by NASA under cooperative agreement NCCW-0089. It was also partially supported by NSF under grants No. DUE-9750858 and EEC-9322370, and by the Future Aerospace Science and Technology Program (FAST) Center for Structural Integrity of Aerospace Systems, effort sponsored by the Air Force Office of Scientific Research, Air Force Materiel Command, USAF, under grant number F49620-95-1-0518.

The authors are thankful to all participants of the MaxEnt'97 Workshop, especially to Peter Cheeseman, Anton Garrett, Steve Gull, and John Skilling, for important discussions.

References

1. L. Brink and M. Henneaux, *Principles of string theory*, Plenum Press, N.Y., 1988.
2. N, Chater, "Reconciling simplicity and likelihood principles in perceptual organization", *Psychological Reviews*, 1996, Vol. 103, pp. 566–581.
3. A. Einstein, "On the method of theoretical physics", The Herbert Spencer Lecture delivered at Oxford on June 10, 1933. Reprinted in: A. Einstein, *Ideas and opinions*, Crown Publishers, N.Y., 1954, pp. 270–276.
4. A. Finkelstein, O. Kosheleva, and V. Kreinovich, "Astrogeometry, error estimation, and other applications of set-valued analysis", *ACM SIGNUM Newsletter*, 1996, Vol. 31, No. 4, pp. 3–25.
5. A. Finkelstein, O. Kosheleva, and V. Kreinovich, "Astrogeometry: towards mathematical foundations", *International Journal of Theoretical Physics*, 1997, Vol. 36, No. 4, pp. 1009–1020.
6. A. Finkelstein, O. Kosheleva, and V. Kreinovich, "Astrogeometry: geometry explains shapes of celestial bodies", *Geombinatorics*, 1997, Vol. VI, No. 4, pp. 125–139.
7. A. M. Finkelstein and V. Kreinovich, "Derivation of Einstein's, Brans-Dicke and other equations from group considerations," *On Relativity Theory. Proceedings of the Sir Arthur Eddington Centenary Symposium, Nagpur India 1984*, Vol. 2, Y. Choque-Bruhat and T. M. Karade (eds), World Scientific, Singapore, 1985, pp. 138–146.
8. A. M. Finkelstein, V. Kreinovich, and R. R. Zapatrin, "Fundamental physical equations uniquely determined by their symmetry groups," *Lecture Notes in Mathematics*, Springer-Verlag, Berlin-Heidelberg-N.Y., Vol. 1214, 1986, pp. 159–170.
9. Q. Gao and M. Li, "An application of minimum description length principle to online recognition of handprinted alphanumerals", In: *Proc. 11th Int's Join Conferences on Artificial Intelligence IJCAI*, Morgan Kaufmann, San Mateo, CA, 1989, pp. 843–848.
10. *Group theory in physics: proceedings of the international symposium held in honor of Prof. Marcos Moshinsky, Cocoyoc, Morelos, Mexico, 1991*, American Institute of Physics, N.Y., 1992.

11. H. A. Keuzenkamp and M. McAleer, "Simplicity, scientific inference, and econometric modelling", *The Economic Journal*, 1995, Vol. 105, pp. 1–21.
12. V. Kreinovich. "Derivation of the Schroedinger equations from scale invariance," *Theoretical and Mathematical Physics*, 1976, Vol. 8, No. 3, pp. 282–285.
13. V. Kreinovich and L. Longpré, "Unreasonable effectiveness of symmetry in physics", *International Journal of Theoretical Physics*, 1996, Vol. 35, No. 7, pp. 1549–1555.
14. M. Li and P. M. B. Vitányi, "Inductive reasoning and Kolmogorov complexity", *J. Comput. System. Sci.*, 1992, Vol. 44, No. 2, pp. 343–384.
15. M. Li and P. M. B. Vitányi, "Computational machine learning in theory and practice", In: J. van Leeuwen (ed.), *Computer Science Today, Recent Trends and Developments*, Springer Lecture Notes in Computer Science, Springer-Verlag, Berlin-Heidelberg-N.Y., 1995, Vol. 1000, pp. 518–535.
16. M. Li and P. M. B. Vitányi, *An Introduction to Kolmogorov Complexity and its Applications*, Springer-Verlag, N.Y., 1997.
17. Ch. Misner, K. S. Thorne, and J. A. Wheeler, *Gravitation*, W. H. Freeman and Co., San Francisco, 1973.
18. P. J. Olver, *Equivalence, invariants, and symmetry*, Cambridge University Press, Cambridge, N.Y., 1995.
19. E. P. D. Pednault, "Some experiments in applying inductive inference principles to surface reconstruction", In: *Proc. 11th International Joint Conferences on Artificial Intelligence IJCAI*, Morgan Kaufmann, San Mateo, CA, 1989, pp. 1603–1609.
20. J. J. Rissanen, "Modeling by the shortest data description", *Automatica*, 1978, Vol. 14, pp. 465–471.
21. J. J. Rissanen, *Stochastic complexity and statistical inquiry*, World Scientific, Singapore, 1989.
22. J. J. Rissanen, "Fisher information and stochastic complexity", *IEEE Transactions on Information Theory*, 1996, Vol. IT–42, No. 1, pp. 40–47.
23. I. Rosenthal-Schneider, "Presuppositions and anticipations", In: P. A. Schlipp (ed.), *Albert Einstein: philosopher-scientist*, Tudor Publ., N.Y., 1951.
24. W. Siegel, *Introduction to string field theory*, World Scientific, Singapore, 1988.
25. *Symmetries in physics: proceedings of the international symposium held in honor of Prof. Marcos Moshinsky, Cocoyoc, Morelos, Mexico, 1991*, Springer-Verlag, Berlin, N.Y., 1992.
26. P. M. B. Vitányi and M. Li, "Ideal MDL and its relation to Bayesianism", In: D. Dowe, K. Korb, and J. Oliver (eds.), *Proc. ISIS: Information, Statistics, and Induction in Science Conference*, World Scientific, Singapore, 1996, pp. 282–291.
27. P. M. B. Vitányi and M. Li, *Minimum description length induction, Bayesianism, and Kolmogorov complexity*, Manuscript, CWI, Amsterdam, 1996.
28. F. J. Yndurain, *Quantum chromodynamics: an introduction to the theory of quarks and gluons*, Springer-Verlag, N.Y., 1983.

MAXIMUM ENTROPY AND ACAUSAL PROCESSES: ASTROPHYSICAL APPLICATIONS AND CHALLENGES

M.V. KOSHELEV

Computer Science Department
University of Texas at El Paso
El Paso, TX 79968, USA†

Abstract. In Newtonian physics, if we know the state of the world at some moment of time, then we can precisely predict the state of the world at all future times. In this sense, Newtonian physics is ***deterministic***. In modern physics (starting with quantum mechanics), theories are usually ***non-deterministic*** in the sense that even if we know exactly the initial state of the world, we cannot uniquely predict the future state of the world. In quantum mechanics (and in most modern quantum-based physical theories), the best we can get is ***probabilities*** of different future states. In this sense, in the majority of modern physical theories, uncertainty is of statistical nature.

Lately, a new area of *acausal* (causality violating) processes has entered mainstream physics. This area has important astrophysical applications. It can be shown that if an acausal process is possible, then some events will take place, whose probability is normally extremely low to prevent this acausal influence from happening.

According to statistical physics, the probability of a macro-state monotonically depends on its entropy, and therefore, the state most probable to occur is the one with the ***maximal entropy***. In this paper, we use MaxEnt to show how the possibility of acausal processes explains the isotropy of the universe and the overabundance of particles over anti-particles. We will also describe the open MaxEnt problems arising from acausal applications and show that the possibility of acausal processes leads to a field of physics where probabilities are non-frequential.

Key words: Maximum entropy, acausal processes, astrophysics, isotropization, particle-antiparticle asymmetry, future computers

†Email: mkosh@cs.utep.edu

1. Acausal Processes

1.1. WHAT ARE ACAUSAL PROCESSES

Traditional physics is *causal* in the sense that future events are determined by the past state of the Universe. This dependence can be deterministic (as in classical, pre-quantum physics), or stochastic (as in quantum physics).

There has been for some time an idea of the possibility of *acausal* processes, in which the influence can go in the opposite direction: future can influence the past. Such processes are called *acausal*.

1.2. ACAUSAL PROCESSES LEAD TO PARADOXES

The idea of acausal processes was, for a long time, mainly part of science fiction, because this idea is paradoxical. The most well known *father* paradox is most convincingly described in terms of the actual *time travel* (travel to the past):

The time traveler paradox occurs when a time traveler goes to the past and shoots his own father to death before he himself was conceived. Then:

- On one hand, the time traveler *is* still *alive*, because he was alive before the killing, and he did not harm himself in any way.
- On the other hand, since his father has died, he could not have conceived the time traveler, and hence, the time traveler cannot be born. So, he is at the same time *not alive*.

A similar paradox occurs if we cannot actually *travel* to the past, only *influence* it; also, it occurs even if we have no human beings at all, simply physical processes. The reason why we (and other authors) present this paradox in its time-traveler form rather that in the form of differential equations of physics is that when we have a problem with differential equations, there can be many reasons for that (wrong equations, wrong method of solution, etc.), while the *time-traveler* paradox reveals the paradoxical character of acausal processes themselves.

1.3. A BRIEF HISTORY OF ACAUSAL PROCESSES IN PHYSICS

The idea that the speed of all particles cannot exceed the speed of light (and that, therefore, it is impossible to influence the past) was one of the main ideas of Einstein's Special Relativity Theory.

In quantum mechanics, due to its probabilistic character, many deterministic restrictions of pre-quantum physics become somewhat "blurred" in the sense that they are no longer prohibiting some events completely, but simply telling that these formerly prohibited events have small probability. For example, in classical physics, a particle cannot penetrate the potential barrier if the energy of this barrier exceeds the initial energy of the particle; in quantum physics, however, it is quite possible (although not highly probable) that a particle "tunnels" through this barrier and end up on the other side of it. This is not simply a theoretical conclusion, this "tunnel effect" is the basis of "tunnel diodes" that are extensively used in nowadays electronics.

Uncovered possibility that quantum mechanics can make pre-quantum restrictions "soft" lead to a possibility that causality may also be violated in quantum processes. Such violations were first discovered by Einstein, Podolsky, and Rosen in their famous paradox (physicists call it EPR paradox for the first letters of the authors' names; for details, see, e.g., [36]). Einstein, who was not a great fan of quantum mechanics, proposed this paradox as a way of disproving this theory. (It is worth noticing at this point that all experiments so far seem to confirm quantum mechanics.)

EPR paradox does not lead to a real time travel: it simply shows that in the resulting quantum formalism, the future state influences the past one; however, all attempts to extract a real time travel from it turned out to be futile because, crudely speaking, the resulting influence on the past is so small that, when we try to measure it, it "drowns" in the inevitable quantum uncertainty of measurements.

This fact does not mean that causality is true in quantum physics. In the last decade, several more sophisticated schemes have been proposed that, in principle, can lead to the actual time travel [37–39].

In addition to ***physical*** arguments in favor of possible causality violations, there exist ***biological*** motivations for such processes: Rosen [29] suggests that the living beings can use physical processes that influence the current events depending on the future ones (he calls such acausal processes ***anticipatory***; see also [28,30–32]).

Until 1988, these processes have been mainly considered as one of many possibilities, not the most probable possibility, and not part of the mainstream physics. In 1988, the situation radically changed.

1.4. ACAUSAL PROCESSES BECOME MAINSTREAM PHYSICS

Physicists' attitude to acausal processes changed when Kip S. Thorne, the world's leading astrophysicist, published several papers in the leading physical journal ***Physical Reviews*** in which he showed that within the existing quantum physics and cosmology, acausal processes are highly probable; these publications lead to several other serious research results [1,10,21,22,24–26,35]. As a result of this research, three basic types of acausal processes have been discovered; these processes are summarized in Thorne's recent monograph [36] (for more popular expositions, see, e.g., [5,6,17,18,23,27,40]).

These acausal processes are based on the mainstream physics approach to gravity and, in particular, to quantum gravity. However, acausal processes also appear in alternative gravitational theories (see, e.g., [7,15,16]). Moreover, it looks like some current physical observations cannot be explained without taking acausal processes into consideration [11].

1.5. ACAUSAL PROCESSES DUE TO QUANTUM EFFECTS

In order to create an acausal process due to quantum effects, there must be a very large object (e.g., a balloon) that can be instantly shrunk and expanded to any size, and which is capable of containing an object inside it. As all objects, this balloon exerts gravity. If it is expanded, it exerts the same amount of gravity over a bigger area, and thus exerts less gravity on the object inside it than if it is shrunk.

Normally, the difference between the gravity exerted on the object when the balloon is expanded and the gravity exerted on the object when the balloon is shrunk is very small. However, if one were to make the balloon a *quantum* balloon, i.e., make the size of the balloon determined by, e.g., the energy level of a particle, then the balloon would simultaneously assume all sizes if the particle was not observed. This would create an enormous gravitational field which could create a very large distortion in time which would transport the object inside the balloon to the past.

1.6. ACAUSAL PROCESSES CAUSED BY WORMHOLES

Wormholes are shortcuts in space between two distant points, whose entrances usually lie in black holes. Under normal circumstances, wormholes tend to collapse as soon as they form because of the gravitational conditions in black holes. Under these conditions, only something traveling faster than the speed of light could pass through them. However, a possible method exists of keeping the wormhole open. Gravity is exerted by both mass and pressure. Unlike mass, pressure can be both positive and negative. Negative pressure would cause an object to exert negative gravity, i.e., to "antigravitate." The existence of such negative gravity is predicted, e.g., by the inflationary universe theory (see, e.g., [12], pp. 172–173).

Normally, the mass of the object would still exert enough positive gravity to cancel out the negative gravity. However, it is possible that an exotic material exists for which its negative gravity exceeds the gravity created by the mass of the object. How can this exotic material lead to actual time travel? A wormhole's ends are connected when the same amount of time has passed for both ends, even if one is in the past relative to the other. Thus, if one end of a wormhole is kept stationary while another is moved away from the first end and then back to its previous position at near the speed of light, the moved end would be ahead in time of the stationary end. Then, if a spaceship traveled at the speed of light from the stationary end to the moved end and enter the wormhole at the moved end, it would travel to its original position to a time before it started its journey.

1.7. ACAUSAL PROCESSES CAUSED BY COSMIC STRINGS

Cosmic strings are extremely long and thin strings of pure energy which survived in their original state rather than cooling off with the rest of the universe after the Big Bang. They are very dense, approximately a thousand trillion tons of mass per inch, and thus create a drastic space-time distortion. To use cosmic strings to travel back in time, it is necessary to have two cosmic strings of infinite length moving towards each other at close to the speed of light. To travel back in time, a spaceship would have to move around one string in the direction opposite to that string's direction of travel, then turn around and move around the other string, opposite to its direction of travel.

Acausal processes caused by cosmic strings naturally appear even in alternative gravitational theories in which no acausal processes are caused by quantum effects or by wormholes [7].

1.8. HOW ARE PARADOXES OF ACAUSAL PROCESSES RESOLVED NOW?

For clarity of exposition, we will describe the current solution of the paradoxes of acausality on the same time-traveler example as we described the paradox itself. Of course, similar to the fact that the paradox occurs even when there is no time traveler at all, the described solution is also applicable to the case of purely physical acausal processes.

This solution is described in [13,14,19,20].

Since the time traveler is alive at the time when he starts the shooting, this means that he was conceived after all, and therefore, that his attempt to kill his father has failed. Why could it have failed? Well, the gun may have malfunctioned, or he might have missed, or a brick (or a meteorite) might have fallen on the time traveler's head at the very moment when he was ready to shoot, or a policeman of the past has stopped him, etc.

Some of these possibilities are quite realistic, some (like a meteorite) have an extremely low probability. The time traveler can prepare for some of these possibilities: he can check his gun before going to the past, use automatic weapons, wear a hard hat against falling bricks, a fake police uniform to prevent an interference of the past's police, etc. In principle, whatever possibility we describe, the time traveler can take care of it. However, he cannot take care of them all: for example, in principle, the gun can malfunction simply due to some unexpected (but probable) random Brownian motion of its molecules.

If the time traveler takes care of all possibilities with reasonable (sufficiently high) probability, this still leaves other possibilities, with extremely low probability, that normally do not occur, but that would have to occur because otherwise, we would have a paradox.

Summarizing: ***if an acausal process is possible, then some events will take place, whose probability is normally extremely low to prevent this acausal influence from happening.*** This conclusion is true not only for a time traveler, but for an arbitrary acausal process.

1.9. MAXIMUM ENTROPY

In physics, the entropy S of a given macro-state is defined as

$$S = -\log_2 N,$$

where N is the number of micro-states each of which can lead to this macro-state. This formula can be reformulated in terms of the frequencies p_i defining this macro-state as

$$S = -\sum_{i=1}^{n} p_i \log p_i.$$

According to statistical physics, the probability of a macro-state occurring is an increasing function of its entropy S. Therefore, in any given situation, the macro-state most likely to occur is the one with the ***maximal entropy***. Thus, the method of maximal entropy can be used to determine the macro-state which is most likely to occur, even when acausal processes are taken into consideration.

2. Possible Applications of Acausal Processes to Mainstream Astrophysics

2.1. A GENERAL IDEA OF THESE APPLICATIONS

As we have mentioned, acausal processes lead to highly unprobable events. According to statistical physics, if Nature has a choice, it would rather prefer situations where these highly unprobable events do not occur. Therefore, if there is a random (statistical physics-type) process that can either lead to an acausal process or not, then, the ***actual*** probability of this process resulting in acausality is very low, ***much lower*** than it would have been if we did not take the possibility of acausal processes into consideration (practically, thermodynamically impossible).

In this paper, we only show this idea on two main possible applications, whose description enables us to avoid technical details; other applications are also possible (see, e.g., [14]).

2.2. THE ISOTROPIZATION OF THE UNIVERSE

One of the main problems of modern cosmology (see, e.g., [36]) is that the Universe is too isotropic. On large scale, in all directions in which we look, we see the same statistical distribution of matter. The initial state of the Universe was, according to the modern physical viewpoint, ***random***, and therefore, far from being isotropic. Hence, the observable isotropization is due to some physical processes. Many physical processes shuffle matter around and thus, contribute to the isotropization, but calculations show that during the lifetime of our Universe, these processes are not sufficient to explain the current isotropy; to be more precise, for ***random*** initial conditions, the probability of the initial conditions that lead to the observed isotropy is very low.

This phenomenon is explained in modern physics by the inflationary universe theory ([12], pp. 180–187). However, this explanation has two problems:

1. The inflationary universe succeeds in explaining the isotropization of the universe in ***almost all*** random initial conditions, but not in ***all*** initial conditions ([12], pp. 277).
2. Superstring theory, a serious candidate for a correct fundamental physical theory, may not allow for inflation at all, and thus that the inflationary universe theory may be false ([12], pp. 239).

The explanation of this phenomenon in acausal physics is as follows: Anisotropy means that different distant areas of the Universe will have radically different matter densities. For acausal processes, there is no speed restriction; therefore, since there is an excess of matter in one area and abundance in another area, acausal processes will re-shuffle the matter from the dense area to the area where matter is scarce. Such a process is, as we have mentioned, thermodynamically improbable and therefore, it is much more possible that the random initial conditions are chosen in such a way that prevents these acausal re-shufflings, i.e., that the initial conditions lead to the observable isotropic Universe.

What this explanation does is show that the probability of an initial state of the Universe leading to isotropization, the probability that is small if we do not take acausal processes into consideration, becomes much larger if we consider the possibility of acausal processes.

2.3. SEPARATION OF PARTICLES AND ANTI-PARTICLES IN THE EARLY UNIVERSE

Equations of physics do not show any preference to particles (electron e^-, proton p, neutron n, etc.) or to their anti-particles (positron e^+, anti-proton $\tilde{p}$, anti-neutron $\tilde{n}$, etc.). So, if the initial condition is truly random, we would expect to see approximately as many particle as anti-particles. However, in the observed Universe, there are mainly particles, anti-particles are rare. Why?

The probability that a random initial condition leads to such a dis-balance is extremely small. There exist several hypothetical explanations for this dis-balance (see, e.g., [12], pp. 106–113, [2,3,9,42,33,41]). However, they are not completely satisfying from the viewpoint of modern physics ([12], pp. 238).

Acausal physics explains this visible asymmetry as follows. The exact formulation of the particle-anti-particle symmetry is made by the so-called *CPT theorem*, according to which physical equations remain invariant if we apply three transformations:

C: replacing particles with anti-particles and vice versa;
P: *parity* transformation: replacing coordinates $\vec{r}$ by $-\vec{r}$;
T: *time reversal*: $t \to -t$.

As a result, when we turn to anti-particles, in order to make the equations work, we need to change the orientation of time: when in normal particles, past events determine the future one, in the anti-particles, vice versa, future events influence the past ones.

Comment. This idea that, e.g., positron can be considered as an analogue of an electron that is going *back* in time was actively promoted by R. Feynman and was part of the motivating ideas behind *Feynman diagrams*, one of the major tools of modern theoretical physics (see, e.g., [8]).

According to our general idea, if we take into consideration that the presence of anti-particles leads to acausal processes, the probability of *not* having anti-particles in a random initial state increases and thus, becomes quite reasonable.

3. Possible Applications of Acausal Processes to Computer Science

According to the above analysis, the very possibility of an acausal process leads to the implementation of highly improbable events. Let us assume that we have organized such an acausal process in such a way that if we switch it on, it will lead to an implementation of a highly improbable event with a probability $p_0 \ll 1$.

Let us show how this device can be used to solve a typical hard-to-compute problem of ***propositional satisfiability***: given a Boolean (propositional) formula $F(x_1, \ldots, x_n)$ with n Boolean variables $x_1, \ldots, x_n$, find the values (if any) for which the resulting formula is true. This problem can be easily solved by trying all 2^n possible combinations of n "true" and "false" values. Unfortunately, this ***exhaustive search*** becomes non-feasible even for $n \approx 300$, when the resulting computation time exceed the lifetime of the Universe. It is known that this problem is computationally hard (the precise term is ***NP-hard***) in the sense that if we can solve it in reasonable time (i.e., time bounded by a polynomial of n), then we would be able to solve ***all*** problems from a large class (called NP) in reasonable time, and this most computer scientists consider impossible.

To find the values x_i using ***acausal processes***, we can set up n quantum random number generators that generate n random bits. Then, we check whether the results $x_1, \ldots, x_n$ of these bits satisfy a given formula. If they do, these values are the desired answer; if they do not, we switch the above-mentioned acausal process on. Nature has two choices:

- It can generate the desired solution in the random generators. If the formula has only one satisfying combination of variables (out of 2^n), the probability of this event is 2^{-n}.
- It can also generate a vector that ***does not*** satisfy the given formula. In this case, the switched-on acausal process makes nature implement the highly improbable event, with probability p_0.

Therefore if $p_0 \ll 2^{-n}$, it is much more probable that Nature will prefer the first alternative.

4. Acausal Processes Lead to Non-Frequential Uncertainty

So far, we have talked in terms of probabilities. At first glance, it may seem that the solution to the paradox (described above) is inside the paradigm of frequential statistics. Alas, we will see in a moment that the proposed solution is ***not*** exactly frequential.

Indeed, according to this solution, the "father"-type paradox is resolved by saying that in actuality, to avoid inconsistency, nature would choose some highly improbable events to happen. If this paradox-causing situation happens repeatedly, the actual ***frequencies*** of these "highly improbable" events will be much higher (closer to 1) than their "probabilities" (which are closer to 0).

Hence, in the presence of acausal processes, the "probabilities" computed by physical equations are ***not*** real ***objective*** probabilities (= frequencies) as in frequential interpretations of statistical and quantum physics. In other words, in the presence of acausal processes, ***uncertainty is non-frequential.***

The acausal interpretation of probability has actually been proposed as a general foundation of quantum mechanics (see, e.g., [4]). We hope that this and other formalisms for describing non-frequential uncertainty that have been developed in the past will help in formalizing physical non-frequential uncertainty as well.

Acknowledgments

This work was partially supported by NASA Grant No. NCC 5-97.

The author is greatly thankful to Kip S. Thorne for his inspiring talks, papers, and books, and for his encouragement and to all the participants of the 1997 MaxEnt Workshop, especially to Steve Gull, Anton Garrett, Carlos Rodriguez, and John Skilling, for their ideas and suggestions.

References

1. B. Allen and J. Simon, "Time travel on a string", *Nature*, 1992, Vol. 357, May 7, pp. 19–21.
2. R. A. Alpher and R. Herman , "On Nucleon-Antinucleon Symmetry in Cosmology," *Science*, 1958, Vol. 128, pp. 904.
3. J. H. Christenson, J. W. Cronin, V. L. Fitch, and R. Turlay, "Evidence for the 2π Decay of the K_2^0 Meson," *Physical Review Letters*, 1964, Vol. 13, pp. 138–140.
4. J. G. Cramer, "The Transactional Interpretation of Quantum Mechanics," *Reviews of Modern Physics*, 1986, Vol. 58, pp. 647–687.
5. M. Davidson, "Is time travel possible?" *USA Today (Magazine)*, Jan. 1990, pp. 64–67.
6. P. Davies, "Wormholes and time machines," *Sky & Telescope*, Jan. 1992, pp. 20–23.
7. C. J. L. Doran, A. N. Lasenby, and S. F. Gull, "The Physics of Rotating Cylindrical Strings," *Phys. Rev. D.*, 1996, Vol. 54, pp. 6021–6031.
8. R. P. Feynman, R. B. Leighton, and M. L. Sands, *The Feynman Lectures On Physics*, Addison-Wesley, Redwood City, CA, 1989.
9. H. Georgi and S. L. Glashow, "Unity of All Elementary-Particle Forces," *Physical Review Letters*, 1974, Vol. 32, pp. 438-441.
10. J. R. Gott, "Closed timelike curves produced by pairs of moving cosmic strings: exact solution", *Physical Review Letters*, 1991, Vol. 66, pp. 1126–1129.
11. S. F. Gull, "Charged Particles at Potential Steps," In: A. Weingartshofer, D. Hestenes (eds.), *The Electron*, Kluwer Academic, Dordrecht, 1991.
12. A. H. Guth, *The Inflationary Universe*, Addison-Wesley, Reading, MA, 1997.
13. O. M. Kosheleva and V. Kreinovich, "What can physics give to constructive mathematics," *Mathematical Logic and Mathematical Linguistics*, Kalinin, 1981, pp. 117–128 (in Russian).
14. V. Kreinovich and R. Mignani, "Noncausal quantum processes and astrophysics", *Bolletino della Sociatá Italiana di Fisica*, 1977, Vol. 112, August 29, p. 88.
15. A. N. Lasenby, C. J. L. Doran, Y. Dabrowski and A. D. Challinor, "Rotating Astrophysical Systems and a Gauge Theory Approach to Gravity," In: N. Sanchez, A. Zichini (eds.), *Advances in Astrofundamental Physics*, Erice, 1996.
16. A. Lasenby, C. Doran, and S. Gull, "Gravity, Gauge Theories and Geometric Algebra," *Phil. Trans. R. Soc. Lond. A.*, 1997 (to appear).
17. M. D. Lemonick, "How to Go Back in Time", *Time*, 1991, May 13, p. 74.
18. J. W. Macvey, *Time Travel: a guide to journey in the forth dimension*, Scarborough House, Chelsea, MI, 1990.
19. S. Yu. Maslov, *Theory of deductive systems and its applications.*, MIT Press, Cambridge, MA, 1987.
20. H. Moravec, *Time travel and computing*, Carnegie-Mellon University, Computer Science Department, Preprint, 1991.
21. M. S. Morris and K. S. Thorne, "Wormholes in spacetime and their use for interstellar travel: a tool for teaching general relativity", *American Journal of Physics*, 1988, Vol. 56, May, pp. 395–412.
22. M. S. Morris, K. S. Thorne, and U. Yurtzever, "Wormholes, time machines, and the weak energy condition", *Physical Review Letters*, 1988, Vol. 61, pp. 1446–1149.

23. P. Nahin, *Time machines: time travel in physics, metaphysics, and science fiction*, American Institute of Physics, N.Y., 1993.
24. I. D. Novikov, "Analysis of the operation of a time machine", *Soviet Physics JETP*, 1989, Vol. 68, pp. 439–443.
25. I. D. Novikov, "Time machines and self-consistent evolutions in problems with self-interaction", *Physical Review D*, 1992, Vol. 45, pp. 1989–1994.
26. A. Ori, "Rapidly moving cosmic strings and chronology protection", *Physical Reviews D*, 1991, Vol. 44, pp. 2214–2215.
27. B. Parker, "Tunnels through time", *Astronomy*, June 1992, pp. 28–36.
28. R. Rosen, "Church's thesis and its relation to the concept of realizability in biology and physics", *Bull. Math. Biophysics*, 1962, Vol. 24, pp. 375–393.
29. R. Rosen, *Anticipatory systems*, Pergamon Press, N.Y., 1985.
30. R. Rosen, In: R. Herken (ed.) *The universal Turing machine: a half-century survey*, Kammerer & Unverzagt, Hamburg, 1988, pp. 523–537.
31. R. Rosen, *Life itself: a comprehensive inquiry into the nature, origin, and fabrication of life*, Columbia University Press, N.Y., 1991.
32. R. Rosen, "What can we know?", In: J. L. Casti, A. Karlqvist (eds.), *Beyond belief: randomness, prediction, and explanation in science*, CRC Press, Boca Raton, FL, 1991, pp. 1–13.
33. A. D. Sakharov, "Violation of CP Invariance, C Asymmetry, and Baryon Asymmetry of the Universe," translated in *Soviet Physics, JETP Letters*, 1967, Vol. 5, pp. 32–35.
34. L. Smolin, *The Life of the Cosmos*, Oxford University Press, New York, 1997.
35. K. S. Thorne, "Do the laws of physics permit closed timelike curves?", *Annals of the New York Academy of Sciences*, 1991, Vol. 631, pp. 182–193.
36. K. S. Thorne, *From black holes to time warps: Einstein's outrageous legacy*, W. W. Norton & Company, New York, 1994.
37. F. J. Tipler, "Rotating cylinders and the possibility of global causality violation", *Physical Review D*, 1974, Vol. 9, pp. 2203–2206.
38. F. J. Tipler, "Causality violation in asymptotically flat space-times", *Physical Review Letters*, 1976, Vol. 37, pp. 879–882.
39. F. J. Tipler, "Singularities and causality violation", *Annals of Physics*, 1977, Vol. 108, pp. 1–36.
40. J. Travis, "Could a pair of cosmic strings open a route into the past?", *Science*, 1992, April 10, pp. 179–180.
41. M. Yoshimura, "Unified Gauge Theories and the Baryon Number of the Universe," *Physical Review Letters*, 1978, Vol. 41, pp. 281–284.
42. S. Weinberg, "The Quantum Theory of Massless Particles," In: S. Deser, K. Ford (eds.), *Lectures on Particles and Field Theory: 1964 Brandeis Summer Institute in Theoretical Physics, Volume 2,* Prentice-Hall, Englewood Cliffs, NJ, 1965.

COMPUTATIONAL EXPLORATION OF THE ENTROPIC PRIOR OVER SPACES OF LOW DIMENSIONALITY

HOLLY E. FITZGERALD AND EVERETT G. LARSON
Department of Physics and Astronomy
Brigham Young University
Provo, Utah 84602, USA

Abstract. Bayesian methods provide an objective analysis for problems with incomplete information. Yet, the use of Bayesian methods requires the assigning of an a priori probability, or prior. The prior should be assigned to contain the least information, while at the same time being consistent with the statistical parameters of the problem. To do this correctly, a complete integration of Information Theory with Bayesian Estimation methods is necessary. When finding probability distributions over the probability assignments, traditional methods are not entirely self-consistent. In papers presented at the 1996 MaxEnt Workshop, Larson, Evenson, and Dukes demonstrated that commonly used methods minimize only part of the total information. Minimizing the total information produces an entropic prior. Implementing this complete method to find the best probability distribution over probability assignments for three- to five-sided dice showed that the entropic prior Bayesian method gives results which differ significantly from these standard approaches. This is especially apparent at low dimensions.

1. Introduction

Bayesian methods provide a valid and objective method for assigning probabilities[1]. The essence of Bayesian statistics is contained in the Bayesian definition that a probability distribution function is a mathematical representation of a state of knowledge. A Bayesian analysis requires the use of an a priori probability, or a prior, for a given event before any data is taken. Since this a priori probability distribution function represents the state of knowledge of the observer, the assignment of the prior is one of the most difficult aspects of a Bayesian analysis. Jaynes states that "in the absence of any principle for setting up prior distributions. . . problems cannot be treated mathematically at all [1]." One of the standard ap-

[1]The words *valid* and *objective* carry different definitions in this context than they might otherwise. *Valid* means self–consistent and *objective* means that the probability assignments contain the minimum possible information while remaining consistent with the statistical parameters of the problem.

proaches is to use a "uniform" prior, one that assumes that all equal volumes of the probability space are equally likely.

Developed historically via statistical mechanics and information theory, the Principle of Maximum Entropy states that probabilities should be assigned to contain the least information, while being consistent with the statistical parameters of the problem. The method of Maximum Entropy[2] chooses the one probability assignment in the probability space which minimizes the information. This research seeks to find not the one probability assignment but the probability distribution over the probability assignments in the probability space which minimizes the information consistent with the Principle of Maximum Entropy. Information theory must be used to help assign the prior for such a problem. This research explores the implications of fully using information theory to minimize the information in selecting a prior for a Bayesian analysis.

To analyze the implications of a complete integration of information theory with Bayesian statistics, the best probability distribution over probability assignments was examined for problems in which the number of outcomes is known, probabilities are assumed to be normalizable, and an expectation value is known, but for which no other information is available. To clarify this process, the geometrical structure of probability spaces is utilized [2]. Since the probabilities are limited to values between zero and one, the parameter space for an n-dimensional die is the n-dimensional unit hypercube. The normalization condition, that the probabilities add to one, forms a hyperplane. The intersection of the hypercube with the hyperplane produces a simplex, or hypersurface. The measure associated with the simplex is described by the original pi's. But this simplex now is described by only (n-1) independent coordinates, which form a non-orthogonal coordinate system for the simplex. The differential volume associated with the simplex is a uniform constant, the Jacobian, times the product of independent differentials. Because all calculations involve a normalization procedure, this constant does not appear in the final results and has been disregarded in the calculations. The linearity of the pi's ensures that the measure of points (i.e., the density of states) is uniform throughout the simplex. It is also the linearity of the pi's that indicates that it is p and not some other function of p that ensures that the geometric distance is the appropriate mathematical distance[3].

When no other information about the values of the probabilities is known, then the parameter space contains a uniform density of states, every measure or equal volume of the space being then equally available to the system [2]. To find the probability distribution function which assumes the least information, a prior probability distribution function must be chosen such that the information is minimized. A standard Bayesian approach to do this would assign a "uniform" prior, a prior for which every point in the system is equally available and likely.

[2]The method of Maximum Entropy should not be confused with the Principle of Maximum Entropy. The Principle of Maximum Entropy states that probabilities should be assigned to minimize the information. The method of Maximum Entropy seeks to satisfy this principle by using the peak of the entropy curve as an estimator.

[3]This is an important point which can not be discussed thoroughly in this paper but will be discussed in a future paper.

This approach does minimizes the information contained in the prior itself, as a mathematical function. However, it ignores the information theory entropy, which accounts for the information missing from each of the individual probability assignments, associated with each of the points in the probability space. To correctly assign the prior, the total information must be minimized. Larson, Evenson, and Dukes have demonstrated that when done properly, every equal volume of the parameter space is not equally likely [2,3]. Each point in the space is weighted by a factor of e^S where S is the entropy, $\sum_i -p_i \log[p_i]$ [3]. Since standard approaches fail to include this e^S weighting, they assign a probability distribution function which implicitly assumes additional information that is unnecessary and unjustified. Larson, Evenson, and Dukes demonstrate that to correctly integrate information theory with Bayesian methods, an e^S prior must be used in assigning probability distributions for problems of this nature [2,3].

The work of Larson, Evenson, and Dukes seeks to expand upon the basic principles of Maximum Entropy to find the a priori probability distribution over the probability assignments that is necessary for a full Bayesian analysis [3]. The expectation value over the prior for the entropy is given by

$$< S >= \int_v S(p_1...p_n)P(p_1...p_n)dV_p. \tag{1}$$

Here $< S >$ represents the expectation value for the information missing in the probability assignments of the parameter space. There is also an information contribution from the prior itself, as a mathematical function. The missing information in the prior is:

$$H = -\int_v P(p_1...p_n)\log(\frac{P(p_1...p_n)}{P_0})dV_p, \tag{2}$$

where P_0 is a function which ensures that H is invariant to a general change of parameter. The total missing information for a probability distribution over probability assignments must include the information from both the prior and the probability assignments [3]. Thus, the total missing information is I, where $I = S + H$. When this quantity is maximized, the total information is minimized. Larson, Evenson, Dukes, show that the prior which minimizes the total information is

$$P(p_1...p_n) = \frac{e^{S(p_1...p_n)}}{\int_v e^{S(p_1...p_n)}}dV_p. \tag{3}$$

This prior is an entropic prior and is invariant over any choice of parameterization [2,3]. In addition, it satisfies the Principle of Maximum Entropy by minimizing the total information and allows for a Bayesian analysis.

2. Description of Research

The objective of this research was to explore the computational implications of this result. To do this, several simple die problems have been examined. In each case, the number of sides on the die (hence, the number of outcomes) is known but the loading of the die is not, so that the die could be either loaded or asymmetrical. A normalization condition, $\sum_i p_i = 1$, is imposed on the probabilities. An additional constraint, that of an expectation value for the face of the die, $\sum_m m p_m =< m >$, is also imposed[4]. This research explores how using a complete analysis to assign the best probability distribution over probability assignments for three- to five-sided dice compares to other methods currently used.

Each problem considered was approached using three different methods, each of which leads to a different result. The first method is a Bayesian estimation using the Larson, Evenson, Dukes entropic prior. Next, a Bayesian estimation using a uniform prior is considered. Finally, a Maximum Entropy method of taking the peak of the entropy curve as an estimator, is used. Since Bayesian estimations result in probability distributions and the method of Maximum Entropy results in expectation values for the probabilities, expectation values for the probabilities of the Bayesian probability distributions were found, so that adequate comparisons could be made between the three methods.

In this research, the expectation values of the probabilities of the probability distributions over the probability assignments ($< pi >$) were calculated for three- to five-sided dice when a constraint in the form of an expectation value for the face of the die ($\sum_m m p_m =< m >$) is given. This was done for each of the three estimation methods: Maximum Entropy, Bayesian estimation with uniform Prior, and Bayesian estimation with entropic prior.

3. Method of Analysis

In comparing results from each of the methods with each other, in the case where an expectation value has been given, two methods were used. First, the percent difference in the expectation values for the probabilities was calculated between each method and the entropic prior Bayesian method[5] Or in other words, the quantity:

$$\frac{< p > - < p >_{ent}}{< p >_{ent}} \times 100\%. \tag{4}$$

Second, mean square distances were found that described how well the expectation values represent a given probability distribution. The value

$$< |\mathbf{p} - \mathbf{p}^o|^2 >= \int P(p_1 \ldots p_n) \sum_m (p_m - p^o_m)^2 dV_p^m \tag{5}$$

represents the mean square distance of p^o from the prior probability distribution. The symbol V_p^m is the hypersurface in $(p_1, .., p_n)$ space satisfying both the

[4]It is not necessary to limit oneself to an expectation value for the face of the die. Other constraints could be imposed.

[5]It is important to note that all of the percent differences cited in the text are percent differences between an estimator for a given method and the estimator given by the entropic prior.

normalization condition and the $< m >$ constraint. The quantities $\{p_1^o, p_2^o, ..., p_n^o\}$ are the expectation values for the probabilities. The mean square distance of the expectation values to the distribution gives a measure of how well the expectation value of the probability represents the distribution.

This quantity can be evaluated using any set of numbers as the expectation values and would indicate how well that particular set of numbers represents the probability distribution function. If, for example, the $< p_i >' s$ for the uniform prior were used with the entropic prior, this quantity would represent how well the expectation values for the uniform prior represent the entropic prior distribution. The difference between the quantities $< |\mathbf{p} - \mathbf{p}_{\mathbf{uni}}^0|^2 > P_{ent}$ and $< |\mathbf{p} - \mathbf{p}_{\mathbf{ent}}^o|^2 > \mathbf{P}_{\mathbf{ent}}$ represents a measure of the "badness" or how badly the results from the uniform prior represent those of the entropic prior.

4. The Three-Sided Die

A three-sided die has a three-dimensional unit cube parameter space associated with it. The normalization condition is a plane and the intersection of this plane with the unit cube forms a triangular 2-dimensional parameter space. The prior in this parameter space is continuous. Each corner of the triangle represents a probability of $p = 1$ for one of the sides of the die and $p = 0$ for the other two. Each point on the triangle is associated with a triplet (p_1, p_2, p_3). The distance from tho i^{th} corner to the opposing side is proportional to the probability (p_i). Lines parallel to the side of the triangle represent lines of constant probability (i.e., lines along which the value of p_i does not change). Vertical lines in this representation are lines of constant $< m >= p_1 + 2p_2 + 3p_3$. This line can be expressed in terms of $< m >$ and p_2 and used to solve for $< p_i >$.

There is a significant difference between the results of the three different methods. For example, at $< m >= 2$, the entropic prior Bayesian estimation gives $< p_2 >= .451114$. The uniform prior Bayesian estimation for this same $< m >$ is $< p_2 >= .5$, a difference of around 11%. The Maximum Entropy estimation is 26% lower than that of the entropic prior Bayesian and has a value of $< p_2 >= .33$. Large percent differences for both methods were also found for the expectation values of the probabilities for sides one and three.

The results of the three-sided die were also used to determine the so-called measure of badness as described earlier. The Maximum Entropy method's measure of badness was higher than that of the uniform prior Bayesian for all values of $< m >$. At $< m >= 2$, for example, the Maximum Entropy method gives a measure 5.91 times higher than the uniform prior method gives.

5. The Four-Sided Die

A four-sided die has associated with it a four-dimensional outcome space. Normalization reduces the parameter space to a three-dimensional tetrahedron. Each corner of the tetrahedron represents $p = 1$ for one of the sides of the die and $p = 0$ for all of the others. Triangular sections perpendicular to a line drawn from a given corner to the opposing face have pi constant. When the additional constraint of an expectation value for the face, $< m >= \sum_i p_i$, is imposed, the parameter space is further reduced to two-dimensional triangles imbedded in the tetrahedron. On the face of this triangle $< m >$ is constant. The expectation value of the probability is found by integrating p_i over the face of this triangle.

As with the three-sided die, there is a significant percent error associated with the results of each method as compared with the entropic prior Bayesian method. However, the percent error is in general smaller for the four-sided die than it was for the three-sided die. In examining side 1, for example, the largest percent difference for the Maximum Entropy method is 16.7% and for the uniform prior Bayesian method is 3.6%. So, while a significant difference occurs in both cases the Maximum Entropy results are certainly much worse than the uniform prior Bayesian results. Similarly, large differences also occur for the probabilities for the other sides of the die, with those of the uniform prior Bayesian being smaller than those of the Maximum Entropy estimation. The measure of badness is also much larger for the Maximum Entropy method than for the uniform prior Bayesian method. For example at $< m >= 2.125$, the Maximum Entropy method's measure of badness is 7.35 times larger than that of the entropic prior Bayesian.

6. The Five-Sided Die

For a five-sided die, the outcome space is five dimensional. The normalization condition produces a four-dimensional parameter space. As before, the inclusion of an expectation value, $< m >$, reduces the space again by one, to a three-dimensional space. To find the expectation values of the probabilities, it is necessary to integrate the probabilities over this volume, weighted by the prior.

The percent differences associated with each of the methods for the five-sided die are much smaller than they were for the other two die. At $< m >= 3$, for example, $< p_3 >$ is .25, .239187, and .2 for the uniform prior Bayesian, entropic prior Bayesian, and Maximum Entropy estimations respectively. So here there is only a 4.5% difference between the Uniform and Entropic Prior Bayesian estimations and a 16.4% difference between the Maximum Entropy estimation and the entropic prior Bayesian estimation. While still a significant difference, the differences for both methods have decreased from the results for the three-sided die. A complete analysis of the percent differences for all values of $< m >$ shows that this is generally true.

The measure of badness shows, however, that the Maximum Entropy method is increasingly worse at higher dimensions (in comparison with the uniform prior Bayesian). At $< m >= 3$, for example, the uniform prior Bayesian method gives a measure of .000160197. At this same value of $< m >$, the measure of badness for the Maximum Entropy method is .00335688. So here the Maximum Entropy method is 20.95 times worse than the Uniform Prior method is.

7. Conclusions

Research done by Larson, Evenson, and Dukes shows that to correctly find the a priori probability distribution over the probability assignments, a complete integration of Bayesian Estimation methods with Information Theory is necessary [2,3]. They show that to minimize the information contained in the problem consistent with the statistical constraints present, the information missing from both the probability assignments and the prior, as a mathematical function, must be maximized. The result of this maximization is the entropic prior [3]. From the results obtained in this research, a picture can be drawn of just how important it is to use this prior and under what circumstances another approach may suffice. From the two other approaches examined, the Bayesian estimation with a uniform prior and a Maximum Entropy method, three conclusions can be drawn. First, a significant difference occurs between the entropic prior Bayesian estimation and the other methods. Second, in all cases a uniform prior Bayesian approach gives results closer to the entropic prior Bayesian results than does the Maximum Entropy approach. Finally, as the dimensionality of the die increases, both methods give results closer the entropic prior Bayesian results. By solving these die problems, several basic questions have been answered. These die problems demonstrate what a complete representation of the problems gives, they show how a complete representation differs from standard methods, and they also show how this difference changes with higher dimensions.

References

[1] E.T. Jaynes, IEEE Transactions on Systems Science and Cybernetics, 227 (1968).
[2] E.G. Larson, W.E. Evenson, and P.R. Dukes, "An Invariant Form for the Prior Probability II," Proceedings of the MaxEnt Workshop, 1996.
[3] W.E. Evenson, E.G. Larson, and P.R. Dukes, "An Invariant Form for the Prior Probability I," Proceedings of the MaxEnt Workshop, 1996.

ENVIRONMENTALLY-ORIENTED PROCESSING OF MULTI-SPECTRAL SATELLITE IMAGES: NEW CHALLENGES FOR BAYESIAN METHODS

S.A. STARKS AND V. KREINOVICH
NASA Pan American Center
for Earth and Environmental Studies
University of Texas at El Paso
El Paso, TX 79968, USA†

Abstract. Remotely sensed images from new generation satellites present an opportunity for scientists to investigate problems in environmental and earth science which have been previously intractable. The magnitude of data that will arise from these hyperspectral instruments create the need for innovative techniques to accomplish data reduction. This paper presents an algorithm which shows promise as a tool for reducing the dimensionality of data resulting from remote sensing. The optimality criteria for the algorithm is the Bayes Risk in the reduced dimension space.

Key words: satellite imaging, multi-spectral satellite data, environmental applications, Bayes risk

1. Introduction

Recently, the University of Texas at El Paso, through a grant from the National Aeronautics and Space Administration (NASA), created the Pan American Center for Earth and Environmental Studies (PACES). As part of its goals, PACES seeks to establish a repository of remotely sensed and ground-based data to enable scientific studies of the geologic, ecologic and environmental processes taking place in the area of the Southwestern United States and Northern Mexico [1].

Current remote sensing imagery result from the processing of data obtained from sensors mounted on satellites with the capability of taking measurements in a limited number of spectral bands. A well known and widely used source of remotely sensed imagery is the Landsat series of satellite launched by NASA [2]. The maximum number of bands associated with the Landsat series of platforms is seven. As a result, for each point on the earth which we wish to study, we are limited to seven measurements arising from different regions of the electromagnetic

†Emails: {sstarks,vladik}@utep.edu

spectrum. This limitation on the number of measurements often prevents the study of interesting phenomena because the information which we can obtain via remote sensing is not sufficient to discriminate different earth features. To improve the ability to gather information about our planet, NASA is in the process launching new generation satellites under its Mission to Planet Earth strategic enterprise [3]. Many of these systems will be based upon hyperspectral sensing which will enable the collection of data in the order of several hundred spectral bands [4].

The drastic increase in the amount of data that will be collected make it extremely important to look for new data compression and feature extraction techniques. Currently, a mix of techniques are employed in this area ranging from the *heuristic* to the highly mathematical. Some are based upon elements drawn from soft computing techniques such as neural nets. In this work, we present an algorithm which is based upon the minimization of the Bayes risk of misclassification which we intend to investigate as a means for reducing the dimensionality of hyperspectral imagery.

2. Mathematical Background

In the study of statistical pattern recognition, much attention has been focused on the field of dimensionality reduction. Numerous algorithms have been developed which are optimal with respect to a variety of criteria. Some criteria include measures of pattern class separability such as the Bhattacharyya distance and the average divergence. Other techniques have their bases in the field of discriminant analysis. Still other methods are based upon the Karhunen-Loeve expansion and factor analysis. Since the ultimate test of any practical pattern recognition system is its ability to perform accurate classification in the reduced-dimension or feature space, it seems natural that a desirable figure of merit for the dimensionality reduction algorithm is the Bayes risk of misclassification in the feature space. The mathematic ground work for this work was laid in a paper by de Figueiredo [5].

In this report, an algorithm for determining an optimal linear transformation $\widetilde{A}$ from a real n-dimensional Euclidean measure space E^n, in which the raw data to be classified into M ($M \geq 2$) Gaussian pattern classes appear, to a feature space E^m of pre-described dimension $m \leq n$, in which classification is to be performed. We denote the latter space by $E^m(\widetilde{A})$ to emphasize that the probability measures in E^m depend upon the transformation $\widetilde{A}$.

The case in which the dimensionality of the feature space m is unity warrants special treatment and is described in [6]. In the present paper, we describe an approach for the case in which $m \geq 2$.

Let $H^1, \ldots, H^M$ denote the M pattern classes under consideration. We assume that the class conditional probability density functions $f_X(x \mid H^j)$, $j = 1, \ldots, M$ on E^n to be of the form

$$f_X(x \mid H^j) = (2\pi)^{-n/2} |\widetilde{R}^j|^{-1/2} \exp\left[-\frac{1}{2}(x - \bar{x}^j)^T (\widetilde{R}^j)^{-1} (x - \bar{x}^j)\right] \qquad (1)$$

for $j = 1, \ldots, M$. Here, $x = col(x_1, \ldots, x_n)$, and the $(n \times 1)$ vector $\bar{x}^j$ and the $(n \times n)$ symmetric positive definite matrix $\widetilde{R}^j$ are the mean and covariance of H^j

in E^n. The superscript T denotes the transpose operator and $|\widetilde{R}^j|$ denotes the determinant of $\widetilde{R}^j$.

Upon application of the linear dimensionality reduction transformation $A : E^n \to E^m$, the class conditional probability density functions given in (1) are converted to the corresponding ones in E^m:

$$f_Y(y \mid H^j) = (2\pi)^{-n/2}|R^j|^{-1/2} \exp\left[-\frac{1}{2}(y-\bar{y}^j)^T(R^j)^{-1}(y-\bar{y}^j)\right], \tag{2}$$

where

$$y = Ax \in E^m(A), \tag{3}$$

$$\bar{y}^j = A\bar{x}^j, \tag{4}$$

and

$$R^j = A\widetilde{R}^j A^T. \tag{5}$$

We assume that the classification strategy in the feature space $E^m(A)$ is Bayes.

In what follows, we present an algorithm which, for a given dimensionality $m \geq 2$ of the feature space $E^m(A)$, obtains the optimal transformation $\widehat{A}$ by minimizing the Bayes risk of misclassification $R_m(A)$ in $E^m(A)$ over the set of all linear transformations $A : E^n \to E^m$.

3. Description of the Algorithm

The approach used in the formulation of the algorithm under discussion is based upon minimizing the Bayes risk in the feature space under a constraint placed upon the linear transformation matrix A. We may write the Bayes risk in as

$$R_m(A) = \sum_{i=1}^{M}\sum_{\substack{j=1\\ j\neq i}}^{M} P_j C_{ij} \int_{\Omega_i(A)} f_Y(y \mid H^j, A)\, dy, \tag{6}$$

where the probability density functions f_Y are expressed as in (2), $\Omega_i(A)$ is the Bayesian decision region in $E^m(A)$ for class H^i, and the non-negative numbers C_{ij} constitute the elements of the Bayes cost matrix.

If

$$C_{ij} = 1 - \delta_{ij}, \tag{7}$$

where δ_{ij} is the Kronecker delta, (6) represents the probability of misclassification in $E^m(A)$.

Without loss of generality, we require that the norm of the transformation matrix A remain constant for the duration of the optimization procedure. With this constraint, we construct from (6), under (7), an expression for the Lagrangian

$$R_m(A) = \sum_{i=1}^{M}\sum_{\substack{j=1\\ j\neq i}}^{M} P_j \int_{\Omega_i(A)} f_Y(y \mid H^j, A)\, dy + \lambda\left(\frac{1}{2}\operatorname{trace}(AA^T) - \alpha\right), \tag{8}$$

where λ is the Lagrange multiplier for the problem, and α is a positive constant usually set equal to unity.

The algorithm determines the optimal dimensionality reduction transformation $\widehat{A}$, by calculating the extremum of (8) with respect to A. This is done by the iterative procedure presented in the following text. At each step of the approach an estimate of the gradient of (8) with respect to the elements of the transformation matrix A is needed. We may express this gradient as

$$\nabla_A Q_m(A,\lambda) =$$

$$\sum_{i=1}^{M}\sum_{\substack{j=1\\ j\neq i}}^{M} P_j\left\{\left[((R^j)^{-1}D^{ij} - E^{ij}I\right]\cdot(R^j)^{-1}A\widetilde{R}^j + (R^j)^{-1}F^{ij}(\bar{x}^j)^T\right\} + \lambda A \quad (9),$$

where

$$D^{ij} = \int\limits_{\Omega_i(A)} (y-\bar{y}^j)(y-\bar{y}^j)^T f_Y(y\,|\,H^j,A)\,dy, \tag{10}$$

$$E^{ij} = \int\limits_{\Omega_i(A)} f_Y(y\,|\,H^j,A)\,dy, \tag{11}$$

$$F^{ij} = \int\limits_{\Omega_i(A)} (y-\bar{y}^j) f_Y(y\,|\,H^j,A)\,dy, \tag{12}$$

and I is the $(m\times m)$ identity matrix.

The derivation of the above expression (9) is lengthy and is presented elsewhere (see, e.g., [5]).

For a given dimensionality $m\geq 2$ for the feature space, our approach replaces the integrals expressed in (10)–(12) by their estimates, which are provided below:

$$D^{ij} = \frac{1}{N_j}\sum_{y_k^j\in\Omega_i(A)\cap S^j} (y-\bar{y}^j)(y-\bar{y}^j)^T, \tag{13}$$

$$E^{ij} = \frac{1}{N_j}\sum_{y_k^j\in\Omega_i(A)\cap S^j} 1, \tag{14}$$

$$F^{ij} = \frac{1}{N_j}\sum_{y_k^j\in\Omega_i(A)\cap S^j} (y-\bar{y}^j), \tag{15}$$

where

$$S^j = \{y_k^j \in E^m(A)\,|\,k=1,\ldots,N^j\} \tag{16}$$

is a set of N^j labeled samples drawn from the class H^j.

4. Implementation of the Algorithm

The algorithm for determining the optimal transformation matrix $\widehat{A}$ is based upon an iterative procedure which makes use of the expressions for the gradient of the criterion expressed in (9). The algorithm makes use of a set of labeled n-dimensional samples drawn from a training set to calculate the value of the criterion function as well as that of its gradient for a particular choice of the transformation matrix. First the labeled samples are transformed into the m-dimensional feature space using an initial guess for the transformation matrix A_0. The samples are then classified in the reduced-dimension space $E^m(A_0)$, according to the Bayes decision scheme. The value for the probability of error is then estimated using the formula

$$\widetilde{P}_E(A_0) = \sum_{i=1}^{M} \sum_{\substack{j=1 \\ j \neq i}}^{M} P_j \frac{\beta_{ij}}{\delta_j} \tag{17}$$

where β_{ij} is equal to the number of samples from the training set representative of class H^j which are classified in $E^m(A_0)$ as being from class H^j, and δ_j represents the total number of samples taken from the set representative of class H^j.

The value for the criterion function is then calculated by replacing the integral in (8) by $\widetilde{P}_E(A_0)$. The value for the gradient of (8) is also calculated using the samples from the training set and the formulas (9), (13), (14), and (15).

An unconstrained multivariate optimization procedure such as that developed by Fletcher and Powell is then invoked to minimize the criterion function by adjusting the elements of the matrix A_0 based upon the value of the gradient. This process of repeated until convergence is obtained.

The basic algorithm for computing the optimal transformation $\hat{A}$ and its corresponding Lagrange multiplier $\hat{\lambda}$ which minimize (8) is presented as follows:

Basic Optimal Linear Feature Extraction Algorithm:

(a) Enter.
(b) Obtain class statistics through training.
(c) Obtain initial A_0 and λ_0; set $k \leftarrow 0$.
(d) Optimize $Q_1(A, \lambda_k$ by varying A to form A_{k+1}.
(e) Convergence test.

- *If convergence is obtained, we return $\widehat{A} \leftarrow A_{k+1}$ and $\widehat{\lambda} \leftarrow \lambda_k$.*
- *If convergence is not obtained, we go to block (f).*

(f) Update λ_k to form λ_{k+1}.
(g) Set $k \leftarrow k+1$, and go to block (d).

It is important to note that at each iterative step, the value of the Lagrange multiplier is updated by the formula

$$\lambda_{k+1} = (\langle A_{k+1}, A_{k+1} \rangle)^{-1} \left\{ \frac{1}{2}(\|A_{k+1}\|^2 - 1) - \langle A_{k+1}, \nabla_A f(A_{k+1}) \rangle \right\}, \tag{18}$$

where $\langle .,. \rangle$ denotes the inner product in E^n, $\|.\|$ represents the Frobenius norm in E^n, and

$$f(A) = Q_m(A, \lambda) - \lambda \left(\frac{1}{2} \text{trace}(AA^T) - \alpha \right). \tag{19}$$

Also, it should be noted that in block (d) of the above algorithm, a single iteration of the Fletcher-Powell minimization routine is performed by leaving the $(m \times n)$ components of the transformation matrix A_k free to vary and holding λ_k constant. A convergence test may be applied to check whether the optimal A_k and λ_k have been found. The convergence test bases its result on the change in A_k from one step to the next and on the norm of the gradient.

5. Summary and Future Work

An iterative algorithm has been presented for the development of a linear dimensionality reduction transformation for the case that the pattern classes are multi-variate Gaussian and the decision strategy is Bayes. Current work focuses upon other parametric forms of transformation where the class-conditional probability density functions are non-Gaussian. Application of the current approach to hyperspectral remotely sensed imagery is under development.

Acknowledgment

This work was supported in part by NASA under cooperative agreement NCCW-0089. It was also partially supported by NSF under grants No. DUE-9750858 and EEC-9322370, and by the Future Aerospace Science and Technology Program (FAST) Center for Structural Integrity of Aerospace Systems, effort sponsored by the Air Force Office of Scientific Research, Air Force Materiel Command, USAF, under grant number F49620-95-1-0518.

References

1. S.A. Starks, G.R. Keller, and D.E. Cooke, "Earth and Environmental Remote Sensing at PACES," *Geocarto, Int'l.*, 1997, Vol. 12, No. 3 (Sept. 1997).
2. T.M. Lillesand and R.W. Kiefer, *Remote Sensing and Image Interpretation*, J. Wiley, New York, 1994.
3. G. Asrar and R. Greenstone, eds., *1995 MTPE EOS Reference Handbook*, NASA Goddard Space Flight Center, Greenbelt, MD, 1995.
4. Schott, *Remote Sensing: The Image Chain Approach*, Oxford University Press, New York, 1997.
5. R.J.R. De Figueiredo, "Optimal Linear and Nonlinear Feature Extraction from Several Gaussian Pattern Class," *Proc. of Second Joint Int'l Conf. On Pattern Recognition*, Copenhagen, Denmark, August 1974.
6. S.A. Starks, R.J.R. de Figueiredo, and D.L. Van Rooy, "An Algorithm for Optimal Single Linear Feature Extraction from Several Gaussian Pattern Classes," *Int'l. J. of Computer and Info. Sciences*, 1977, Vol. 6, No. 1, pp. 41–54.

MAXIMUM ENTROPY APPROACH TO OPTIMAL SENSOR PLACEMENT FOR AEROSPACE NON-DESTRUCTIVE TESTING

R. OSEGUEDA, C. FERREGUT, M.J. GEORGE
J.M. GUTIERREZ, AND V. KREINOVICH
Future Aerospace Science and Technology Program (FAST)
Center for Structural Integrity of Aerospace Systems
University of Texas at El Paso, El Paso, TX 79968, USA†

Abstract. The ideal design of an airplane should include *built-in sensors* that are pre-blended in the perfect aerodynamic shape. Each built-in sensor is expensive to blend in and requires continuous maintenance and data processing, so we would like to use *as few sensors as possible.*

The *ideal* formulation of the corresponding optimization problem is, e.g., to minimize the average detection error for fault locations. However, there are *two obstacles* to this ideal formulation:

- First, this ideal formulation requires that we *know the probabilities* of different fault locations and the probabilities of different aircraft exploitation regimes. In reality, especially for a new aircraft, *we do not have those statistics* (and for the aging aircraft, the statistics gathered from its earlier usage may not be applicable to its current state). Therefore, instead of a well-defined optimization problem, we face a problem of not so well defined problem of optimization under uncertainty.
- Second, even if we know the probabilities, the corresponding optimization problem is very *computation-consuming* and difficult to solve.

In this paper, we overcome the first obstacle by using *maximum entropy* approach (MaxEnt) to select the corresponding probability distributions.

To overcome the second obstacle, we use the *symmetry* approach. Namely, the basic surface shapes are symmetric (with respect to some geometric transformations such as rotations or shifts). The MaxEnt approach results in distributions that are *invariant* with respect to these symmetries, and therefore, the resulting *optimality criterion* (be it the minimum of detection error, or the minimum of fault location error, etc.) is also *invariant* with respect to these same symmetries. It turns out that for an arbitrary optimality criterion that satisfies the natural symmetry conditions (crudely speaking, that the relative quality of two sensor

†E-mails: osegueda@utep.edu, ferregut@utep.edu, mjgeorge@utep.edu, jgutierr@utep.edu, and vladik@cs.utep.edu

placements should not change if we simply shift or rotate two placements), the line formed by the optimally placed sensors (optimally with respect to this criterion) can be described as an orbit of the corresponding Lie transformation groups. As a result, ***we describe the optimal sensor placements.***

A similar problem of optimal sensor placement is also discussed for ***space structures.***

Key words: Non-destructive testing, maximum entropy, aerospace structures, aging aircraft, futuristic aircraft, symmetry groups, geometry

1. Introduction

Checking structural integrity of aerospace systems is very important. Structural integrity is extremely important for airplanes, because in flight, the airframe is subjected to such stressful conditions that even a relatively small crack can be disastrous.

This problem becomes more and more important as the aircraft fleet ages.

At present, most airplanes do not have built-in sensors for structural integrity, and even those that have do not have a sufficient number of them, so additional sensors must be placed to test the structural integrity of an airframe.

It is important to test structural integrity in-flight. Each integrity violation (crack etc.) starts with a small disturbance that is only detectable in stressful in-flight conditions. Therefore, to detect these violations as early as possible, we should complement on-earth testing by in-flight measurements. Hence, we need ***sensors*** for in-flight tests.

The problem of sensor placement. Where should we place sensors for in-flight tests?

Most *existing* airplanes do not have built-in sensors for testing structural integrity (or at least do not have a sufficient number of these sensors), so, to test these airplanes, we must place these sensors ***outside*** the airframe. Sensors attached outside the airframe interfere with the airplane's well-designed aerodynamics; therefore, we should use as few sensors as possible.

This limitation leads to the following problem:

GIVEN: the number of sensors that we can locate on a certain surface of an airframe,

FIND: the optimal placements of these sensors, i.e., locations that allow us to detect the locations of the faults with the best possible accuracy.

For ***future*** aircraft, we have a similar problem of sensor placement. The ideal design of a future airplane should include built-in sensors that are pre-blended in the perfect aerodynamic shape. Each built-in sensor is expensive to blend in and requires continuous maintenance and data processing, so again, we would like to use as few sensors as possible.

The problem of optimal sensor placement is difficult, because it requires optimization under uncertainty. In both cases, the *ideal* formulation of the corresponding optimization problem is to minimize, e.g., the average detection error for fault locations.

However, this ideal formulation requires that we *know* the *probabilities* of different fault locations and the probabilities of different aircraft exploitation regimes. In reality, we do not know these probabilities:

- for a *new* aircraft, we do not have this statistics; and
- for the *aging* aircraft, the statistics gathered from its earlier usage may not be applicable to its current state.

Therefore, instead of a well-defined optimization problem, we face a not so well defined problem of *optimization under uncertainty.*

Since the problem is not well defined, we cannot simply use standard numerical optimization techniques.

To solve the optimal sensor placement problem, we will use symmetry. In this paper, we overcome the first obstacle – of not knowing the probabilities – by using *maximum entropy* approach (MaxEnt) to select the corresponding probability distributions.

To overcome the second obstacle – of computational complexity – we use the *symmetry* approach. Namely, the basic surface shapes are symmetric (with respect to some geometric transformations such as rotations or shifts). The MaxEnt approach results in distributions that are *invariant* with MaxEnt approach results in distributions that are *invariant* with respect to these symmetries, and therefore, the resulting *optimality criterion* (be it the minimum of detection error, or the minimum of fault location error, etc.) is also *invariant* with respect to these same symmetries. It turns out that for an arbitrary optimality criterion that satisfies the natural symmetry conditions (crudely speaking, that the relative quality of two sensor placements should not change if we simply shift or rotate two placements), the line formed by the optimally placed sensors (optimally with respect to this criterion) can be described as an orbit of the corresponding Lie transformation groups. As a result, *we describe the optimal sensor placements.*

The efficiency of symmetry ideas is in line with the general efficiency of these ideas in computer science in general (see, e.g., [5]) and in its maximum entropy problems in particular (see, e.g., [4]).

First stage: Geometric techniques. The problem of choosing an optimal sensor placement is formulated in *geometric* terms: we need to select *points* (sensor placements) on a *surface* of the given structure.

To solve this problem, we use the experience of solving similar geometric problems of optimization under uncertainty in image processing and image extrapolation [1–3]. Namely, astronomic image processing faces the problem of selecting the best family of images for use in extrapolation. It turns out that for every optimality criterion that satisfies the natural symmetry conditions (crudely speaking, that the relative quality of two image reconstructions should not change if we simply shift or rotate two images), the extrapolation shapes that are optimal with respect to this criterion can be described as orbits of the corresponding Lie transformation groups, which leads to exactly the shapes used in astronomy (such as spirals, planes, spheres, etc).

In this paper, we show that, since the basic surface shapes are symmetric, a similar symmetry-based approach can be applied to the problem of optimal sensor placement. For the simplest surfaces, this general approach describes several geometric patterns that every sensor placement, which is optimal with respect to reasonable (symmetric) optimality criterion, must follow.

Some of our results were announced in [6].

Second stage. After we have selected several possible sensor locations, we then use detailed numerical simulations:

- first, to *confirm* that these placement patterns indeed lead to better fault location, and
- second, to *select* a pattern that leads to the best results for each particular problem.

The results: in brief. As a result of this analysis, we get several possible optimal sensor placements.

A similar problem of optimal sensor placement is also important for *space* structures.

2. Preliminary analysis: we need the optimal sensor placement

2.1. IT IS DESIRABLE TO FIND THE OPTIMAL SENSOR PLACEMENTS

The quality of non-destructive testing essentially depends on the placement of the sensors: e.g., if all the sensors are concentrated in one area, and few are located in the remainder of the structure, then possible cracks and other faults in the under-covered area may go dangerously unnoticed. Therefore, it is important to choose a good sensor placement.

Currently, the choice of sensor placements is mainly made either *ad hoc*, or, at best, by testing a few possible placements and choosing the one that performs the best on a few benchmark examples. There are two drawbacks in this approach:

- In this approach, only a few possible placements are analyzed, so it is quite possible that we miss really good placements.
- Even when the placement that is good "on average" is indeed present among the tested placements, the very fact that we only test these placements on a few examples leads to the possibility that we will choose different placements, that work well for the tested examples, but that are, on average, much worse than the rejected placement.

In other words, often, the normal engineering *good enough* approach does not work for our problem.

It is, therefore, desirable to find the *optimal* (*best*) sensor placements.

2.2. "OPTIMAL" IN WHAT SENSE?

Since we do not know the exact optimality criterion, we will try to describe sensor placements that are the best relative to all possible reasonable optimality criteria. If we knew the exact probabilities of different exploitation regimes and of different faults, then we could formulate the exact optimality criterion and look for the sensor placement that is optimal relative to this criterion. In reality, however, as we have already mentioned in the Introduction, we do not know the exact optimality criterion.

Since we do not know *the* exact criterion, the natural idea is to do the following:

- consider ***all possible*** optimality criteria that are reasonable for this problem;
- describe all the placements that are optimal relative to each of these criteria;
- and finally, depending on the exact situation, choose the best placement among the "possibly best" ones.

In this manner, we still face the problem of choosing between several possible placements (and we may still make a wrong choice), but we are, at least, guaranteed that we do not initially miss the best placement.

This general program sounds ambitious and computationally intractable, but it is actually doable. Even when we know the optimality criterion, finding the optimal sensor placement is extremely computationally difficult and time-consuming. According to the above program, we intend to describe sensor placements that are optimal relative to *all* possible reasonable optimality criteria. Since there are many such criteria, it may seem, at first glance, that we need to repeat the (already time-consuming) computations so many times that the resulting required computation time will make this problem computationally intractable.

Fortunately, we will see that this problem is quite doable: namely, it is possible to describe all possibly optimal placements without actually solving all possible optimization problems, but using *geometric* arguments instead.

Before we start describing and using these arguments, we must describe in precise terms what we mean by a "reasonable" optimality criterion.

Optimality criteria can be arbitrarily complicated. Traditionally, the quality of different alternatives is described by a *numerical* optimality criterion, in which the quality of each alternative a from the set A of all possible alternatives is characterized by a real number $J(a)$, and we choose the alternative a for which this value $J(a)$ is the smallest possible (i.e., $J(a) \to \min$). For example, for the problem of placing the given number n of sensors, A is the set of all possible placements of these sensors, and $J(a)$ is, e.g., equal to the mean square average detection error of fault location based on the data from these sensors.

Such numerical criteria are useful and often sufficient, but in many cases, we end up with *several* alternatives with the same smallest possible value of the average error $J(a)$. In this case, it makes sense to select, among them, an alternative a for which, e.g., the *worst-case* error $J'(a)$ is the smallest possible. This very natural idea leads to a non-numerical optimality criterion, according to which *two* different functions $J(a)$ and $J'(a)$ are given, and an alternative a is considered to be *better* than an alternative b if either $J(a) < J(b)$, or $J(a) = J(b)$ and $J'(a) < J'(b)$.

This more complicated criterion can also result in several "best" alternatives, in which case we will be able, simultaneously, to optimize a third characteristic, etc. As a result, we can have arbitrarily complicated non-numerical optimality criteria. Since we want to describe placements that are optimal with respect to *all* possible reasonable criteria, we have to consider all these criteria.

How can we describe them?

A general description of an optimality criterion. In general, when we say that an optimality criterion is given, this means that for every two alternatives a and b from the set A of all alternatives, one of the following four possibilities holds: either a is better than b according to this criterion (we will denote it by $a < b$), or b is better than a ($b < a$), or a and b are of the same quality (we will denote it by $a \sim b$), or, according to the given criterion, the alternatives a and b are incompatible (we will denote this case by $a \parallel b$).

So, we can describe the optimality criterion as a pair of relations $\langle <, \sim \rangle$.

These two relations must satisfy natural consistency conditions, e.g., if a is better than b and b is better than c, then a should be better than c, etc. A pair that satisfies these natural consistency conditions is called a *pre-ordering* relation. In these terms, an optimality criterion is a *pre-ordering relation* on the set of all alternatives.

There is also one additional requirement that we have used before, when we talked about the necessity for complicated optimality criteria: that there should be *exactly one* optimal alternative. Indeed, if there are several alternatives that are equally good according to some criterion, it means that we still need to choose between them; thus, the current optimality criterion is not final. We are interested in *final* criteria, i.e., in pre-ordering relations in which there exists exactly one best alternative.

Now, that we have a general definition of an optimality criterion, we must describe all sensor placements that are optimal relative to these criteria. For this description, as we have mentioned, we will use the geometric techniques.

3. Geometric techniques

3.1. GEOMETRIC TRANSFORMATIONS: A SEEMING COMPLICATION

The idea of using symmetries first appeared not as a method for *solving* the problem, but rather as an additional unexpected complication that made its solution even harder. Namely, we started with simplified toy examples, and tried to use an optimization method to find the optimal placements for these toy problems. Since we were solving an extremely simplified problem, we expected that the optimization algorithm would soon give us a single optimal sensor placement. Instead, for each problem, different applications of the numerical algorithm, applications that started with different randomly chosen initial sensor placements, resulted in drastically different optimal sensor placements.

When we plotted these seemingly different solutions, we saw a simple explanation for this non-uniqueness: these "different" solutions turned out to be approximately one and the same solution, but differently rotated and/or shifted. How can we explain this behavior?

3.2. SYMMETRIES OF AEROSPACE SHAPES EXPLAIN THE OBSERVED

COMPLICATION

There is a simple geometric explanation for the above-described behavior. This explanation is based on the fact that most surfaces that form an airframe can be described, within a good accuracy, in simple geometric terms.

Basic geometric shapes of aerospace structures and their symmetries. Let us first describe the geometric shapes of basic aerospace structures:

- The airplane cabin can be described as a *cylinder*.
- The surface of the wings can be approximately described as a *plane* (same, for the tail).
- Finally, the plane's "nose" can be approximately described as either a part of the *sphere* (to be more precise, a half-sphere), or as a piece of a *cone*.

Each of these geometric shapes has certain geometric *symmetries*, i.e., geometric transformations that leave this shape invariant:

- a *cylinder* is invariant with respect to shifts along its axis and rotations around this axis;
- a *plane* is invariant with respect to shifts in the plane, rotations in this plane, and dilations (similarities);
- a *sphere* is invariant with respect to arbitrary rotations around its center;
- finally, a *cone* is invariant with respect to rotations around its axis and dilations centered at its vertex.

For perfectly symmetric shapes, optimal placement is non-unique. Let us first consider the idealized situation in which the shape is *precisely* symmetric (e.g., a perfect sphere, that is invariant with respect to arbitrary rotation T around its center). Let $P = \{p_1, \ldots, p_n\}$ be a sensor placement for which the optimality criterion (e.g., the average fault location error) is the smallest possible, and let T be one of the symmetries. Since the shape is invariant with respect to this symmetry, locations $T(p_1), \ldots, T(p_n)$ also belong to this same shape. Since natural optimality criteria are also invariant with respect to these geometric symmetries, the quality of the rotated placement $T(P) = \{T(p_1), \ldots, T(p_n)\}$ is equal to the quality of the original placement and therefore, the rotated placement is also optimal.

Thus, if P is an optimal placement, then for every symmetry T of the geometric shape, the placement $T(P)$ is also optimal. This explains non-uniqueness of optimal sensor placement for *perfectly symmetric* shapes.

For approximately symmetric shapes, optimal placement is also non-unique. Since optimal placement is non-unique for *perfectly symmetric* shapes, it is natural to expect that a similar complication occurs for the shapes that are *close* to the perfectly symmetric shapes (e.g., for a slightly deformed sphere).

Optimal sensor placement is non-unique even when for geometric shapes that are only locally symmetric. As we have mentioned earlier in this section, the actual shapes of aerospace structures are indeed close to perfectly symmetric ones, but they are only *locally* close to the perfectly symmetric shapes:

- For example, the shape of a cabin is close to a cylinder. An (infinite) cylinder is invariant with respect to rotations *and* shifts; however, the shape of a cabin is only a *piece* of this infinite cylinder.
- Similarly, a *wing* is only a piece of a plane, a *nose* is only a piece of a sphere or of a cone, etc.

In other words, the actual shapes are not themselves symmetric, they are only locally close to the symmetric shapes. However, for sensors testing structural integrity, local is all we need: the very need for numerous sensors comes from the fact that the effects of each newly appearing small structural fault are so small that they can be only detected by a sufficiently close sensor. So, the interaction of a fault and of a nearby sensor on, e.g., a small piece of a spherical surface depends only on the *local* properties of this surface and practically does not depend on whether this surface is the whole sphere or a piece of it.

Thus, if a surface locally coincides with the symmetric one, the local quality of each sensor placement on this surface coincides with the local quality of their placement on the perfectly symmetric shape, and therefore, the *optimal* placement on the actual surface locally coincides with the optimal sensor placement on the ideal symmetric surface.

Since the optimal placement on an ideal surface is non-unique, the placement on its piece is also non-unique.

3.3. DUE TO NON-UNIQUENESS, WE HAVE NOT A SINGLE OPTIMAL SENSOR PLACEMENT, BUT A FAMILY OF DIFFERENT OPTIMAL SENSOR PLACEMENTS

We have shown, both experimentally and theoretically, that, due to symmetry of the basic shapes of airframes, optimal sensor placements are non-unique: for every optimal placement P and for every symmetry T, the placement $T(P)$ is also optimal. Therefore, we cannot find a *unique* optimal sensor placement. Instead, we must look for a *family* of optimal sensor placements (that correspond to different symmetries T).

3.4. SO FAR, SYMMETRIES ONLY MADE OUR PROBLEM MORE COMPLICATED, BUT SYMMETRIES CAN ALSO HELP

So far, symmetries only made the optimal sensor placement problem more complicated. However, in general, symmetries are known to help in solving numerical problems.

For example, if we know that a solution $f(x, y, z)$ of a partial differential equation is invariant with respect to arbitrary rotations around 0, this means that the value of the desired function $f(x, y, z)$ depend only on a single parameter: distance $r = \sqrt{x^2 + y^2 + z^2}$ from 0. Therefore, instead of a *partial* differential equation that describes a function of three variables, we have a much easier-to-solve *regular* differential equation that describes an unknown function $f(r)$ of only one variable.

We will show that a similar simplification happens for the sensor placement problem.

3.5. TWO-STEP APPROACH

We start with a 2-dimensional geometric shape, and we want to find a finite number of points on this shape, i.e., in geometric terms, a 0-dimensional shape. Since moving directly from a 2D to 0D sets is complicated, we will do this transition in two, hopefully easier, steps – from 2D to 1D and then from 1D to 0D:

- On the first step, we will find a 1D curve or curves along which the optimal sensor placement will occur.
- Then, on the second step, we will find the optimal sensor placements on the chosen curves.

Let us first describe the first step.

3.6. FIRST STEP: FINDING THE OPTIMAL CURVE OR CURVES ON WHICH SENSORS WILL BE PLACED

Let us start with re-formulating our problem in precise mathematical terms.

What is given. We are given a geometric surface α that has several symmetries.

We can easily describe the set G of all these symmetries, i.e., in precise terms, the set of all geometric transformations (rotations, shifts, and dilations) that leave this surface invariant. Thus, we can assume that this set G is given as well.

An important comment about transformation groups. The set of transformations is very important because it is a particular case of a concept that plays a central role in modern theoretical physics: the concept of a *transformation group*.

Namely, it is easy to see that if transformations g and g' belong to this set G (i.e., leave the surface α invariant), then their composition $g \circ g'$ and the inverse transformation g^{-1} also leave the same surface invariant. A set of transformations that satisfies this property is called a *transformation group*.

The objective of the first step: from informal description to precise formulation. The goal of the first step is to find either a single curve or a family of curves that are optimal in some reasonable sense.

To describe this goal formally, we must explain which families of curves we will consider and what we mean by "optimal". Let is start with *families*.

In general, a *curve* can be described as a mapping that traces this curve, i.e., in more formal terms, a mapping $\vec{r}$ that maps real numbers t into points $\vec{r}(t)$ in 3D space. Correspondingly, a *family* of curves can be described as a family of such mappings characterized by one or several parameters $C_1, \ldots, C_p$, i.e., in more precise terms, as a mapping that maps tuples $(C_1, \ldots, C_p, t)$ of real numbers into points $\vec{r}(C_1, \ldots, C_p, t)$ of a 3D space. If we fix some values of p parameters $C_1, \ldots, C_p$, we get a curve from this family. (For example, the formula $\vec{r}(t) = (t, C_1 \cdot t + C_2)$ describes the family of all straight lines in a plane expect for the lines that are parallel to the y axis.)

How many parameters do we need in a sensor placement problem? In the simplest possible case of 1-parametric family ($p = 1$), the set of all points from all curves from this family already spans a 2D surface. Thus, we do not need more than one parameter to describe the lines of optimal sensor placements. So, in the following text, we will consider either single curves, or 1-parametric families of curves.

Now that we formalized the notion of a family, we must describe what "optimal" means. Here, the set of alternatives is the set of all curves (or of all 1-parametric families of curves) on the surface α. As before, the optimality criterion is a pre-ordering relation on this set for which there exists exactly one optimal curve (or family of curves).

We also want the optimality criterion to be *natural*, which means, in particular, that the relative quality of different placement curves should not change if we apply any transformation g that leaves the original surface α invariant. In precise terms, we require that the pre-ordering relation $\langle <, \sim \rangle$ that describes our optimality criterion satisfy the following two conditions:

- if $a > b$ and $g \in G$, then $g(a) > g(b)$;
- if $a \sim b$ and $g \in G$, then $g(a) \sim g(b)$.

A pre-ordering relation that satisfies these two conditions is called ***invariant*** with respect to the transformation group G.

So, we get the following precise formulation of the problem that correspond to the first step:

GIVEN: a surface α that is invariant with respect to a group G of geometric transformations.

WE KNOW: that on the set of all 1-parametric families of curves on a surface α, a pre-ordering relation is given that is invariant with respect to the transformation group G, and for which exactly one family is optimal.

FIND: the optimal 1-parametric family.

General solution to this problem. The problem, as formulated above, is a particular case of a ***general problem*** of finding optimal families of sets as formulated in the papers [1–3]. In these papers, we have actually solved this general problem by describing the ***general solution*** to it.

To formulate this general solution, we need to introduce two notions: of a ***subgroup*** and of an ***orbit***.

- A ***subgroup*** G_0 of a transformation group G is a subset $G_0 \subseteq G_0$ that is itself a transformation group.

  For example, the set of all rotations around the x-axis is a subgroup of the group of all rotations.

- To describe an ***orbit*** of a transformation group G, we must fix a point $\vec{r}$. If we apply all transformation from G to this point $\vec{r}$, then the resulting set $\{g(\vec{r}) \mid g \in G\}$ is exactly the orbit.

  For example, for the group G of all rotations around the x-axis, depending on the choice of the point $\vec{r}$, we get either a point (if $\vec{r}$ is on this axis), or a circle circling around the axis.

In these terms, the above-mentioned solution is as follows: *Every set from the optimal family consists of one or several orbits of subgroups of the original transformation group.*

Thus, to apply this general solution to our problem, we must, for all the geometric shapes enumerated above:

- find all subgroups G_0 of the corresponding transformation groups G; and then
- find all orbits of these subgroups.

This is a (somewhat tedious but) doable task. The results are as follows:

The results of Step 1: Optimal curves for sensor placement.

- For a *cylinder*, possibly optimal curves (i.e., orbits of subgroups) are:
  - *straight lines* parallel to the cylinder's axis;
  - *circles* orthogonal to the cylinder's axis; and
  - *cylindric spirals.*

    These spirals can be easily described in cylindric coordinates (z, ρ, φ), in which z is a coordinate along the cylinder's axis, ρ is a distance from this axis, and φ is an angle from some fixed direction orthogonal to the z-axis. In these coordinates, a spiral takes the form $\rho = \text{const}$ and $\varphi = k \cdot z$, for some constant k.

  Cylindric spirals are *generic* orbits; straight lines and circles can be viewed as their *degenerate* cases.
- For a *plane*, possibly optimal curves (i.e., orbits of subgroups) are: *straight lines*, *circles*, and *logarithmic spirals*, i.e., curves describe by the equation $\rho = C \cdot \exp(k \cdot \varphi)$ in polar coordinates. Here, logarithmic spiral is a generic shape.
- For a *sphere*, possibly optimal curves (i.e., orbits of subgroups) are *circles*.
- For a *cone*, possibly optimal curves (i.e., orbits of subgroups) are:
  - *straight line rays* going from the vertex of the cone;
  - *circles* that are orthogonal to the cone's axis; and
  - *conic spirals.*

    In cylindrical coordinates (z, ρ, φ), in which the cone is described by the equation $\rho = C \cdot z$, a conic spiral is described by the formula $\varphi = k \cdot z$ for some constant k.

  Conic spirals are the generic type of orbits.

Therefore, depending on the shape, sensors should be placed along one or several of these curves.

Important comments.

1. If the optimal sensor placement is not along a single curve, but along *several* curves, then the same ideas of transformation groups can be used to choose appropriate families (as orbits of *discrete* subgroups). Let us give a few examples:

- If we have several straight lines on the cylinder, these straight lines must be *equidistant* in the sense that the angular distant between every two neighboring lines is the same.
- If we have several circles around the cylinder, then these circles should be equidistant.

– If we have several straight lines on a plane, then these straight lines should be:

- either parallel and equidistant,
- or parallel at distances that form a geometric progression, or
- pass through the same point and form equidistant angles.

– If we have several circles on a plane, then these circles must be:

- either parallel, equal, and equidistant,
- or concentric, with their radii forming a geometric progression, etc.

2. In *space structures*, we face yet another shape: a paraboloid ($y = z = c \cdot x^2$). This structure is invariant with respect to rotations and *re-scalings* $x' = \lambda x$, $y' = \lambda^2 y$, $z' = \lambda^2 z$. For this group, we can also describe the resulting orbits as spirals ($\rho = C \cdot \exp(k \cdot \varphi)$ in cylindric coordinates).

3.7. SECOND STEP: FINDING THE ACTUAL SENSOR PLACEMENTS (MAIN IDEA)

The problem that corresponds to the second step, i.e., the problem of selecting a 0D subset from a 1D curve can be formulated and solve in a similar manner as the problem that we solved at the first step.

– We started with a surface α with a transformation group G.
– On the first step, optimal curves for sensor placements from orbits of subgroups G_0 of this group G.
– Similarly, on the second step, optimal sensor placements form orbits of subgroups G_1 of the corresponding groups G_0.

From the mathematical viewpoint, the main difference between these two steps is that on the second step, we start already with a 1D transformation group G_0 and thus, its subgroups G_1 are *discrete*. Thus, we face the problem of describing all orbits of discrete subgroups of the above groups.

Due to lack of space, we are not able to enumerate all possible orbits of this type here, but we will briefly enumerate the ones that correspond to generic curves: on a cylinder, we get equidistant points on a cylindric curve; on a plane and on a cone, we get points on the corresponding spiral whose distances from the center of this spiral form a geometric progression.

In all these families, there are still a few parameters whose choice depends on what exactly our goal is. The specific values of these parameters are determine by computer simulations.

4. Computer simulations: in brief

Symmetry approach enables us to select several possible sensor placements. To choose one of these placements, we test different sensor placements on different fault locations using the detailed (accurate) model of an aerospace structure.

Acknowledgment

This work was supported by the Future Aerospace Science and Technology Program (FAST) Center for Structural Integrity of Aerospace Systems, effort sponsored by the Air Force Office of Scientific Research, Air Force Materiel Command, USAF, under grant number F49620-95-1-0518. It was also supported in part by NASA under cooperative agreement NCCW-0089 and by NSF under grants No. DUE-9750858 and EEC-9322370.

The authors are thankful to all the participants of the 1997 International Maximum Entropy Workshop for valuable discussions.

References

1. A. Finkelstein, O. Kosheleva, and V. Kreinovich, "Astrogeometry, error estimation, and other applications of set-valued analysis", *ACM SIGNUM Newsletter*, 1996, Vol. 31, No. 4, pp. 3–25.
2. A. Finkelstein, O. Kosheleva, and V. Kreinovich, "Astrogeometry: towards mathematical foundations", *International Journal of Theoretical Physics*, 1997, Vol. 36, No. 4, pp. 1009–1020.
3. A. Finkelstein, O. Kosheleva, and V. Kreinovich, "Astrogeometry: geometry explains shapes of celestial bodies", *Geombinatorics*, 1997, Vol. 6, No. 4, pp. 125–139.
4. V. Kreinovich, H. T. Nguyen, and E. A. Walker, "Maximum entropy (MaxEnt) method in expert systems and intelligent control: new possibilities and limitations", In: Ken M. Hanson and Richard N. Silver, Eds., *Maximum Entropy and Bayesian Methods*, Kluwer Academic Publishers, Dordrecht, 1990, pp. 93–100.
5. H. T. Nguyen and V. Kreinovich, *Applications of continuous mathematics to computer science*, Kluwer, Dordrecht, 1997 (to appear).
6. R. Osegueda *et al.*, "Non-Equilibrium Thermodynamics Explains Semiotic Shapes: Applications to Astronomy and to Non-Destructive Testing of Aerospace Systems", *Proceedings of the International Conference on Intelligent Systems and Semiotics (ISAS'97)*, National Institute of Standards and Technology Publ., Gaithersburg, MD, 1997.

MAXIMUM ENTROPY UNDER UNCERTAINTY

HENRYK GZYL
Escuela de Física y Matemáticas Facultad de Ciencias
Universidad Central de Venezuela
A.P. 52120, Caracas 1050-A, Venezuela
e-mail:hgzyl@reacciun.ve

Abstract. A method for finding maximum entropy distributions when the data is noisy is provided. It does not rely on making a priori assumptions on the nature of the noise contaminating the measurements.

1. The Method

This note is devoted to a variation in the theme of obtaining a maxentropic solution to the problem of finding a probability distribution $p = \{p_1, \cdots, p_n\}$ such that

$$\sum_{j=1}^{n} A_{ij} p_j + e_i = a_i + e_i = b_i \qquad i = 1, \cdots, m. \tag{1}$$

the rows of the matrix $\{A_{ij} : i = 1, \cdots, m, j = 1, \cdots, n\}$ are to be thought of as the m-components of a vector valued function $A, A_i : \{1, \cdots, n\} \to \mathbf{R}$ with $A_i(j) = A_{ij}$. The $A_i's$ are the observables of the sytem. The meaning of (1) is the following: the $b_i's$ are the components of a data vector b, which results when $a_i = \sum A_{ij} p_j$ are measured and the measurements are contaminated with the additive noise e_i. We will not make assumptions on the nature of the noise, but we shall assume that the experimental square root mean deviations σ_i are known. If we took a glance at many of the papers in the 12 volumes listed under [1] – [12], we will find that our problem is solved by considering some a priori probability $Q = \{q_1, \cdots, q_n\}$ and then maximizing

$$\alpha(P, Q) = S(P, Q) - \alpha \sum_{i=1}^{m} \frac{1}{\sigma_i^2} \left(\sum_{i=1}^{n} A_{ij} P_j - b_i \right)^2 \tag{2}$$

where α is some regularization parameter and

$$S(P, Q) = -\sum_{i=1}^{n} p_j \log(p_j / q_j) \tag{3}$$

is called the relative entropy (of P .w.r.t. Q), the Kulback-Leibler information number, I-divergence and some other names. In the essay "Maxent, mathematics and information" in the last of them, Csiszar critizises the usual bayesian justification for using (2) as starting point for finding P, stating that even Bayesians find the argument questionable [13]. Here we propose a variant of (1) and (2) for solving such problems, and present applications elsewhere. The method we present can be grounded in firm and elementary justification.

Instead of proposing to look for probability distributions P satisfying (1), we demand that they satisfy

$$\sum_{i=1}^{m} \frac{1}{\sigma_i^2} \left(\sum_{j=1}^{n} A_{ij} P_j - b_i \right)^2 \leq T^2 \tag{4}$$

or if you prefer, that $E_p(A) \in B_\sigma(b,T)$. Here $E_p(A)$ is the vector in $\mathbf{R^m}$ with components $E_p(A_i)$ and $B_\sigma(b,T) \doteq \left\{ y \in \mathbf{R^m} \mid \|\mathbf{y}-\mathbf{b}\|_\sigma^2 \leq \mathbf{T^2} \right\}$ with $\|x\|_\sigma^2 = (x, \sigma^{-2}x)$, where σ^{-1} is the diagonal matrix with entries σ_i^{-1}.

The $E_p(A_i)$ are just $\sum_{j=1}^{n} A_{ij}P_j$, what else?. We shall denote by A_j^* the vector in $\mathbf{R^m}$ made up of the elements in the j-th column of A.

To find a P satisfying (4) we shall look for a P^* in

$$\mathbf{P}(a,T) = \left\{ P \mid \sum p_j = 1, \;\; E_p(A) \in B_\sigma(b,T) \right\} \tag{5}$$

such that

$$\sup\{S(P,Q) \mid P \in \mathbf{P}(a,T)\} = S(P^*,Q). \tag{6}$$

That is P^* is a probability distribution, compatible with the observed data having the highest entropy. Note that $\mathbf{P}(a,T) = \bigcup_\alpha \mathbf{P}(\alpha)$ where $\alpha \in B_\sigma(O,T)$ and

$$\mathbf{P}(\alpha) = \left\{ P \mid \sum p_j = 1, \;\; E_P(A) = \alpha \right\} \tag{7}$$

and recall that, see [14], or any of [1] – [12],

$$\sup\{S(P,Q) \mid P \in \mathbf{P}(\alpha)\} = \inf\{\log Z(\lambda) + (\lambda,\alpha) \mid \lambda \in \mathbf{R^m}\} \tag{8}$$

where $Z(\lambda) = \sum_{j=1}^{n} e^{-\langle \lambda, A_j^* \rangle} q_j$. If $\lambda(\alpha)$ denotes the point al which the minimum on the right hand side of 8 is reached, then

$$P_j(\lambda(\alpha)) = \frac{e^{-\langle \lambda(\alpha), A_j^* \rangle}}{Z(\lambda(\alpha))} q_j \tag{9}$$

is the standard maxentropic probability distribution satisfying $E_{P(\lambda(\alpha))}(A) = \alpha$.

Note that $P(a,T)$ is a convex closed set and that $S(P,Q)$ is concave (in P), thus has a maximum on $P(a,T)$ if the set is not empty. To find λ note that, if

$P \in P(a,T)$ and $P(\lambda)$ is given by

$$P_i(\lambda) = \frac{e^{-\langle \lambda, A_j^* \rangle}}{Z(\lambda)} q_j \tag{10}$$

with $Z(\lambda)$ as above, then from $S(P, P(\lambda)) \leq 0$ we easily obtain

$$S(P,Q) \leq S(\lambda) := \log Z(\lambda) + \langle \lambda, \alpha \rangle \tag{11}$$

for $\alpha = E_P(A)$. Thus

$$\begin{aligned} \sup\{S(P,Q) \mid P \in \mathbf{P}(a,T)\} &= \sup\{\sup\{S(P,Q) \mid P(\alpha)\} \mid \alpha \in B_\sigma(b,T)\} \\ \sup_\alpha \left\{\inf_\lambda \{S(\lambda,\alpha)\} \mid \lambda \in \mathbf{R}^m\right\} &= \inf_\lambda \left\{\sup_\alpha \{\log Z(\lambda) + \langle \lambda, \alpha \rangle\}\right\}, \end{aligned}$$

where we used (8) at the second step and standard min-max at the last.

Note that $\alpha \in B_\sigma(b,T)$ is equivalent to $\alpha = b + y, y \in B_\sigma(O,T)$. Thus $(\lambda, \alpha) = (\lambda, b) + (\lambda, y) = (\lambda, b) + T\left(\sigma\lambda, \frac{\sigma^{-1}y}{T}\right)$.

Since $y \in B_\sigma(O,T)$ is equivalent to $\frac{\sigma^{-1}y}{T} \in B(0,1)$ the standard closed unit ball in the euclidean metric. Thus

$$\begin{aligned} &\sup\{S(P,Q) \mid p \in \mathbf{P}(a,T)\} \\ = &\inf_\lambda \{\log Z(\lambda) + (\lambda, b) + T \sup\{(\sigma\lambda, \gamma) \mid \gamma \in B(0,1)\}\} \\ = &\inf_\lambda \{\log Z(\lambda) + (\lambda, b) + T\|\sigma\lambda\|\} = \inf S_T(\lambda). \end{aligned} \tag{12}$$

where the last line drops out from thc fact that $\sup\{(\sigma\lambda, \gamma)\, \gamma \in B(0,1)\}$ is eached at $\frac{\sigma\lambda}{\|\sigma\lambda\|}$. To summanize

Theorem 1 *If* $\mathbf{P}(a,T)$ *is not empty and* $\inf\{S_T(\lambda)\, \lambda \in \mathbf{R}^m\}$ *is reached at* $\lambda^* \in \mathbf{R}^m$, *then*

$$\sup\{S(P,Q) \mid P \in \mathbf{P}(a,T)\} = S_T(\lambda^*) \tag{13}$$

where $S_T(\lambda) = \log Z(\lambda) + (\lambda, b) + T\|\sigma\lambda\|$ *and the maxentropic solution to* (4) *is provided by* $P(\lambda^*)$.

Proof: Read again if need be.

Comment: It is on the left hand side of (13) that our variant to maximize (2) lies. The rest is the procedure to identify the $P(\lambda^*)$.

We must dutifuly aknowledge [14], whose ideas in the context of maximum entropy in the mean, we follow closely. See the paper by Bercher et al. in [11].

References

1. C.R. Smith and W.T. Grandy ,Jr. (eds) "Bayesian Methods in Inverse Problems". Kluwer Pubs, Dordrecht, 1985
2. C.R.Smith and G.J. Erickson (eds.) "Maximum Entropy and Bajesian Spectral Analysis and Estimation Problems." Kluwer Acad. Pubs, Dordrecht, 1987
3. G.J. Erickson and C.R. Smith (eds.) "Maximum Entropy and Bayesian Methods is Science and Enginecring". Vol. I.Foundations. Kluwer Acad. Pubs. Dordrecht, 1988
4. G.J. Erickson and C.R. Smith (eds) "Maximum Entropy and Bayesian Methods in Science and Engineering". Vol. II, Applications, Kluwer Acad.Pubs, Dordrecht, 1988
5. J.Skilling (eds.) "Maximum Entropy Dordrecht". Vol. II, Applications.Kluwer Acad. Pubs, Dordrecht, 1988
6. P.F. Fougere (eds) "Maximun Entropy and Bayesiam Methods". Kluwer Acad. Pubs., Dordrecht, 1990
7. W.T. Grandy, Jr. and L.H. Schick (eds) "Maximum Entropy and Bayesian Methods, Kluwer Acad. Pubs., Dordrecht, 1991"
8. C.R. Smith, G.J.Erickson and P.O. Neudorfer (eds.) "Maximum Entropy and Bayesian Methods", Kluwer Acad. Pubs., Dordrecht 1992.
9. A.Mohammad-Djafri and G. Demoment (eds). "Maximum Entropy and Bayesiam Methods", Kluwer Acad. Pubs., Dordrecht, 1993
10. G.R. Heidbreder (ed.) "Maximum Entropy and Bayesiam Methods". Kluwer Acad. Pubs., Dordrecht, 1996
11. J.Skilling and S.Sibisi (eds.) "Maximum Entropy and Bayesian Methods," Kluwer Acad. Pubs, Dordrecht, 1996
12. K.M. Hanson and R.N. Silver (eds.) "Maximum Entropy and Bayesiam Methods," Kluwer Acad. Pubs. Dordrech, 1997.
13. I.Csiszar, in [12] pp.35-50.
14. F.Gamboa, "Mimmizacion de information de Kullback et maximization de l'entropie sous une contrainte quadratique" CRAS, Paris t.306, Seriel, pp. 425-427, 1988.

Fundamental Theories of Physics

Series Editor: Alwyn van der Merwe, *University of Denver, USA*

1. M. Sachs: *General Relativity and Matter.* A Spinor Field Theory from Fermis to Light-Years. With a Foreword by C. Kilmister. 1982 ISBN 90-277-1381-2
2. G.H. Duffey: *A Development of Quantum Mechanics.* Based on Symmetry Considerations. 1985 ISBN 90-277-1587-4
3. S. Diner, D. Fargue, G. Lochak and F. Selleri (eds.): *The Wave-Particle Dualism.* A Tribute to Louis de Broglie on his 90th Birthday. 1984 ISBN 90-277-1664-1
4. E. Prugovečki: *Stochastic Quantum Mechanics and Quantum Spacetime.* A Consistent Unification of Relativity and Quantum Theory based on Stochastic Spaces. 1984; 2nd printing 1986 ISBN 90-277-1617-X
5. D. Hestenes and G. Sobczyk: *Clifford Algebra to Geometric Calculus.* A Unified Language for Mathematics and Physics. 1984 ISBN 90-277-1673-0; Pb (1987) 90-277-2561-6
6. P. Exner: *Open Quantum Systems and Feynman Integrals.* 1985 ISBN 90-277-1678-1
7. L. Mayants: *The Enigma of Probability and Physics.* 1984 ISBN 90-277-1674-9
8. E. Tocaci: *Relativistic Mechanics, Time and Inertia.* Translated from Romanian. Edited and with a Foreword by C.W. Kilmister. 1985 ISBN 90-277-1769-9
9. B. Bertotti, F. de Felice and A. Pascolini (eds.): *General Relativity and Gravitation.* Proceedings of the 10th International Conference (Padova, Italy, 1983). 1984 ISBN 90-277-1819-9
10. G. Tarozzi and A. van der Merwe (eds.): *Open Questions in Quantum Physics.* 1985 ISBN 90-277-1853-9
11. J.V. Narlikar and T. Padmanabhan: *Gravity, Gauge Theories and Quantum Cosmology.* 1986 ISBN 90-277-1948-9
12. G.S. Asanov: *Finsler Geometry, Relativity and Gauge Theories.* 1985 ISBN 90-277-1960-8
13. K. Namsrai: *Nonlocal Quantum Field Theory and Stochastic Quantum Mechanics.* 1986 ISBN 90-277-2001-0
14. C. Ray Smith and W.T. Grandy, Jr. (eds.): *Maximum-Entropy and Bayesian Methods in Inverse Problems.* Proceedings of the 1st and 2nd International Workshop (Laramie, Wyoming, USA). 1985 ISBN 90-277-2074-6
15. D. Hestenes: *New Foundations for Classical Mechanics.* 1986 ISBN 90-277-2090-8; Pb (1987) 90-277-2526-8
16. S.J. Prokhovnik: *Light in Einstein's Universe.* The Role of Energy in Cosmology and Relativity. 1985 ISBN 90-277-2093-2
17. Y.S. Kim and M.E. Noz: *Theory and Applications of the Poincaré Group.* 1986 ISBN 90-277-2141-6
18. M. Sachs: *Quantum Mechanics from General Relativity.* An Approximation for a Theory of Inertia. 1986 ISBN 90-277-2247-1
19. W.T. Grandy, Jr.: *Foundations of Statistical Mechanics.* Vol. I: *Equilibrium Theory.* 1987 ISBN 90-277-2489-X
20. H.-H von Borzeszkowski and H.-J. Treder: *The Meaning of Quantum Gravity.* 1988 ISBN 90-277-2518-7
21. C. Ray Smith and G.J. Erickson (eds.): *Maximum-Entropy and Bayesian Spectral Analysis and Estimation Problems.* Proceedings of the 3rd International Workshop (Laramie, Wyoming, USA, 1983). 1987 ISBN 90-277-2579-9
22. A.O. Barut and A. van der Merwe (eds.): *Selected Scientific Papers of Alfred Landé.* [*1888-1975*]. 1988 ISBN 90-277-2594-2

Fundamental Theories of Physics

23. W.T. Grandy, Jr.: *Foundations of Statistical Mechanics.*
Vol. II: *Nonequilibrium Phenomena.* 1988 ISBN 90-277-2649-3
24. E.I. Bitsakis and C.A. Nicolaides (eds.): *The Concept of Probability.* Proceedings of the Delphi Conference (Delphi, Greece, 1987). 1989 ISBN 90-277-2679-5
25. A. van der Merwe, F. Selleri and G. Tarozzi (eds.): *Microphysical Reality and Quantum Formalism, Vol. 1.* Proceedings of the International Conference (Urbino, Italy, 1985). 1988 ISBN 90-277-2683-3
26. A. van der Merwe, F. Selleri and G. Tarozzi (eds.): *Microphysical Reality and Quantum Formalism, Vol. 2.* Proceedings of the International Conference (Urbino, Italy, 1985). 1988 ISBN 90-277-2684-1
27. I.D. Novikov and V.P. Frolov: *Physics of Black Holes.* 1989 ISBN 90-277-2685-X
28. G. Tarozzi and A. van der Merwe (eds.): *The Nature of Quantum Paradoxes.* Italian Studies in the Foundations and Philosophy of Modern Physics. 1988 ISBN 90-277-2703-1
29. B.R. Iyer, N. Mukunda and C.V. Vishveshwara (eds.): *Gravitation, Gauge Theories and the Early Universe.* 1989 ISBN 90-277-2710-4
30. H. Mark and L. Wood (eds.): *Energy in Physics, War and Peace.* A Festschrift celebrating Edward Teller's 80th Birthday. 1988 ISBN 90-277-2775-9
31. G.J. Erickson and C.R. Smith (eds.): *Maximum-Entropy and Bayesian Methods in Science and Engineering.*
Vol. I: *Foundations.* 1988 ISBN 90-277-2793-7
32. G.J. Erickson and C.R. Smith (eds.): *Maximum-Entropy and Bayesian Methods in Science and Engineering.*
Vol. II: *Applications.* 1988 ISBN 90-277-2794-5
33. M.E. Noz and Y.S. Kim (eds.): *Special Relativity and Quantum Theory.* A Collection of Papers on the Poincaré Group. 1988 ISBN 90-277-2799-6
34. I.Yu. Kobzarev and Yu.I. Manin: *Elementary Particles. Mathematics, Physics and Philosophy.* 1989 ISBN 0-7923-0098-X
35. F. Selleri: *Quantum Paradoxes and Physical Reality.* 1990 ISBN 0-7923-0253-2
36. J. Skilling (ed.): *Maximum-Entropy and Bayesian Methods.* Proceedings of the 8th International Workshop (Cambridge, UK, 1988). 1989 ISBN 0-7923-0224-9
37. M. Kafatos (ed.): *Bell's Theorem, Quantum Theory and Conceptions of the Universe.* 1989 ISBN 0-7923-0496-9
38. Yu.A. Izyumov and V.N. Syromyatnikov: *Phase Transitions and Crystal Symmetry.* 1990 ISBN 0-7923-0542-6
39. P.F. Fougère (ed.): *Maximum-Entropy and Bayesian Methods.* Proceedings of the 9th International Workshop (Dartmouth, Massachusetts, USA, 1989). 1990 ISBN 0-7923-0928-6
40. L. de Broglie: *Heisenberg's Uncertainties and the Probabilistic Interpretation of Wave Mechanics.* With Critical Notes of the Author. 1990 ISBN 0-7923-0929-4
41. W.T. Grandy, Jr.: *Relativistic Quantum Mechanics of Leptons and Fields.* 1991 ISBN 0-7923-1049-7
42. Yu.L. Klimontovich: *Turbulent Motion and the Structure of Chaos.* A New Approach to the Statistical Theory of Open Systems. 1991 ISBN 0-7923-1114-0
43. W.T. Grandy, Jr. and L.H. Schick (eds.): *Maximum-Entropy and Bayesian Methods.* Proceedings of the 10th International Workshop (Laramie, Wyoming, USA, 1990). 1991 ISBN 0-7923-1140-X

Fundamental Theories of Physics

44. P.Pták and S. Pulmannová: *Orthomodular Structures as Quantum Logics.* Intrinsic Properties, State Space and Probabilistic Topics. 1991 ISBN 0-7923-1207-4
45. D. Hestenes and A. Weingartshofer (eds.): *The Electron.* New Theory and Experiment. 1991 ISBN 0-7923-1356-9
46. P.P.J.M. Schram: *Kinetic Theory of Gases and Plasmas.* 1991 ISBN 0-7923-1392-5
47. A. Micali, R. Boudet and J. Helmstetter (eds.): *Clifford Algebras and their Applications in Mathematical Physics.* 1992 ISBN 0-7923-1623-1
48. E. Prugovečki: *Quantum Geometry.* A Framework for Quantum General Relativity. 1992 ISBN 0-7923-1640-1
49. M.H. Mac Gregor: *The Enigmatic Electron.* 1992 ISBN 0-7923-1982-6
50. C.R. Smith, G.J. Erickson and P.O. Neudorfer (eds.): *Maximum Entropy and Bayesian Methods.* Proceedings of the 11th International Workshop (Seattle, 1991). 1993 ISBN 0-7923-2031-X
51. D.J. Hoekzema: *The Quantum Labyrinth.* 1993 ISBN 0-7923-2066-2
52. Z. Oziewicz, B. Jancewicz and A. Borowiec (eds.): *Spinors, Twistors, Clifford Algebras and Quantum Deformations.* Proceedings of the Second Max Born Symposium (Wrocław, Poland, 1992). 1993 ISBN 0-7923-2251-7
53. A. Mohammad-Djafari and G. Demoment (eds.): *Maximum Entropy and Bayesian Methods.* Proceedings of the 12th International Workshop (Paris, France, 1992). 1993 ISBN 0-7923-2280-0
54. M. Riesz: *Clifford Numbers and Spinors* with Riesz' Private Lectures to E. Folke Bolinder and a Historical Review by Pertti Lounesto. E.F. Bolinder and P. Lounesto (eds.). 1993 ISBN 0-7923-2299-1
55. F. Brackx, R. Delanghe and H. Serras (eds.): *Clifford Algebras and their Applications in Mathematical Physics.* Proceedings of the Third Conference (Deinze, 1993) 1993 ISBN 0-7923-2347-5
56. J.R. Fanchi: *Parametrized Relativistic Quantum Theory.* 1993 ISBN 0-7923-2376-9
57. A. Peres: *Quantum Theory: Concepts and Methods.* 1993 ISBN 0-7923-2549-4
58. P.L. Antonelli, R.S. Ingarden and M. Matsumoto: *The Theory of Sprays and Finsler Spaces with Applications in Physics and Biology.* 1993 ISBN 0-7923-2577-X
59. R. Miron and M. Anastasiei: *The Geometry of Lagrange Spaces: Theory and Applications.* 1994 ISBN 0-7923-2591-5
60. G. Adomian: *Solving Frontier Problems of Physics: The Decomposition Method.* 1994 ISBN 0-7923-2644-X
61 B.S. Kerner and V.V. Osipov: *Autosolitons.* A New Approach to Problems of Self-Organization and Turbulence. 1994 ISBN 0-7923-2816-7
62. G.R. Heidbreder (ed.): *Maximum Entropy and Bayesian Methods.* Proceedings of the 13th International Workshop (Santa Barbara, USA, 1993) 1996 ISBN 0-7923-2851-5
63. J. Peřina, Z. Hradil and B. Jurčo: *Quantum Optics and Fundamentals of Physics.* 1994 ISBN 0-7923-3000-5
64. M. Evans and J.-P. Vigier: *The Enigmatic Photon.* Volume 1: The Field $\boldsymbol{B}^{(3)}$. 1994 ISBN 0-7923-3049-8
65. C.K. Raju: *Time: Towards a Constistent Theory.* 1994 ISBN 0-7923-3103-6
66. A.K.T. Assis: *Weber's Electrodynamics.* 1994 ISBN 0-7923-3137-0
67. Yu. L. Klimontovich: *Statistical Theory of Open Systems.* Volume 1: A Unified Approach to Kinetic Description of Processes in Active Systems. 1995 ISBN 0-7923-3199-0; Pb: ISBN 0-7923-3242-3

Fundamental Theories of Physics

68. M. Evans and J.-P. Vigier: *The Enigmatic Photon.* Volume 2: Non-Abelian Electrodynamics. 1995 ISBN 0-7923-3288-1
69. G. Esposito: *Complex General Relativity.* 1995 ISBN 0-7923-3340-3
70. J. Skilling and S. Sibisi (eds.): *Maximum Entropy and Bayesian Methods.* Proceedings of the Fourteenth International Workshop on Maximum Entropy and Bayesian Methods. 1996 ISBN 0-7923-3452-3
71. C. Garola and A. Rossi (eds.): *The Foundations of Quantum Mechanics – Historical Analysis and Open Questions.* 1995 ISBN 0-7923-3480-9
72. A. Peres: *Quantum Theory: Concepts and Methods.* 1995 (see for hardback edition, Vol. 57) ISBN Pb 0-7923-3632-1
73. M. Ferrero and A. van der Merwe (eds.): *Fundamental Problems in Quantum Physics.* 1995 ISBN 0-7923-3670-4
74. F.E. Schroeck, Jr.: *Quantum Mechanics on Phase Space.* 1996 ISBN 0-7923-3794-8
75. L. de la Peña and A.M. Cetto: *The Quantum Dice.* An Introduction to Stochastic Electrodynamics. 1996 ISBN 0-7923-3818-9
76. P.L. Antonelli and R. Miron (eds.): *Lagrange and Finsler Geometry.* Applications to Physics and Biology. 1996 ISBN 0-7923-3873-1
77. M.W. Evans, J.-P. Vigier, S. Roy and S. Jeffers: *The Enigmatic Photon.* Volume 3: Theory and Practice of the $\boldsymbol{B}^{(3)}$ Field. 1996 ISBN 0-7923-4044-2
78. W.G.V. Rosser: *Interpretation of Classical Electromagnetism.* 1996 ISBN 0-7923-4187-2
79. K.M. Hanson and R.N. Silver (eds.): *Maximum Entropy and Bayesian Methods.* 1996 ISBN 0-7923-4311-5
80. S. Jeffers, S. Roy, J.-P. Vigier and G. Hunter (eds.): *The Present Status of the Quantum Theory of Light.* Proceedings of a Symposium in Honour of Jean-Pierre Vigier. 1997 ISBN 0-7923-4337-9
81. M. Ferrero and A. van der Merwe (eds.): *New Developments on Fundamental Problems in Quantum Physics.* 1997 ISBN 0-7923-4374-3
82. R. Miron: *The Geometry of Higher-Order Lagrange Spaces.* Applications to Mechanics and Physics. 1997 ISBN 0-7923-4393-X
83. T. Hakioğlu and A.S. Shumovsky (eds.): *Quantum Optics and the Spectroscopy of Solids.* Concepts and Advances. 1997 ISBN 0-7923-4414-6
84. A. Sitenko and V. Tartakovskii: *Theory of Nucleus.* Nuclear Structure and Nuclear Interaction. 1997 ISBN 0-7923-4423-5
85. G. Esposito, A.Yu. Kamenshchik and G. Pollifrone: *Euclidean Quantum Gravity on Manifolds with Boundary.* 1997 ISBN 0-7923-4472-3
86. R.S. Ingarden, A. Kossakowski and M. Ohya: *Information Dynamics and Open Systems.* Classical and Quantum Approach. 1997 ISBN 0-7923-4473-1
87. K. Nakamura: *Quantum versus Chaos.* Questions Emerging from Mesoscopic Cosmos. 1997 ISBN 0-7923-4557-6
88. B.R. Iyer and C.V. Vishveshwara (eds.): *Geometry, Fields and Cosmology.* Techniques and Applications. 1997 ISBN 0-7923-4725-0
89. G.A. Martynov: *Classical Statistical Mechanics.* 1997 ISBN 0-7923-4774-9
90. M.W. Evans, J.-P. Vigier, S. Roy and G. Hunter (eds.): *The Enigmatic Photon.* Volume 4: New Directions. 1998 ISBN 0-7923-4826-5
91. M. Rédei: *Quantum Logic in Algebraic Approach.* 1998 ISBN 0-7923-4903-2
92. S. Roy: *Statistical Geometry and Applications to Microphysics and Cosmology.* 1998 ISBN 0-7923-4907-5

MIX
Papier aus verantwortungsvollen Quellen
Paper from responsible sources
FSC® C105338

If you have any concerns about our products,
you can contact us on
ProductSafety@springernature.com

In case Publisher is established outside the EU,
the EU authorized representative is:
Springer Nature Customer Service Center GmbH
Europaplatz 3, 69115 Heidelberg, Germany

Printed by Libri Plureos GmbH
in Hamburg, Germany